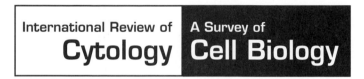

International Review of
Cytology A Survey of **Cell Biology**

VOLUME 135

International Review of Cytology

A Survey of Cell Biology

Edited by

Kwang W. Jeon
Department of Zoology
The University of Tennessee
Knoxville, Tennessee

Martin Friedlander
Jules Stein Eye Institute
UCLA School of Medicine
Los Angeles, California

VOLUME 135

Academic Press, Inc.
Harcourt Brace Jovanovich, Publishers
San Diego   New York   Boston   London   Sydney   Tokyo   Toronto

Academic Press, Inc.
1250 Sixth Avenue, San Diego, California 92101-4311

United Kingdom Edition published by
Academic Press Limited
24–28 Oval Road, London NW1 7DX

Library of Congress Catalog Number: 52-5203

International Standard Book Number: 0-12-364535-2

PRINTED IN THE UNITED STATES OF AMERICA
92 93 94 95 96 97   EB   9 8 7 6 5 4 3 2 1

CONTENTS

Cerebellar Lectins

Jean-Pierre Zanetta, Sabine Kuchler, Sylvain Lehmann, Ali Badache,
Susanna Maschke, Philippe Marschal, Pascale Dufourcq,
and Guy Vincendon

The Role of Jasmonic Acid and Related Compounds in the Regulation of Plant Development

Yasunori Koda

Restriction Fragment Length Polymorphism Analysis of Plant Genomes and Its Application to Plant Breeding

C. Gebhardt and F. Salamini

Regulatory Functions of Soluble Auxin-Binding Proteins

Shingo Sakai

Subcellular and Molecular Mechanisms of Bile Secretion

Susan Jo Burwen, Douglas L. Schmucker, and Albert L. Jones

CONTRIBUTORS

Numbers in parentheses indicate the pages on which the authors' contributions begin.

Ali Badache (123), *Laboratoire de Neurobiologie Moléculaire des Interactions Cellulaires, Centre de Neurochimie du CNRS, 67000 Strasbourg, France*

Susan Jo Burwen (269), *Cell Biology and Aging Section, Veterans Administration Medical Center, San Francisco, California 94121 and the Departments of Medicine and Anatomy and the Liver Center, University of California, San Francisco, San Francisco, California 94143*

Pascale Dufourcq (123), *Laboratoire de Neurobiologie Moléculaire des Interactions Cellulaires, Centre de Neurochimie du CNRS, 67000 Strasbourg, France*

C. Gebhardt (201), *Max-Planck-Institut für Züchtungsforschung, D-5000 Köln 30, Germany*

Silvia Hein (39), *Instituto de Histología y Patología, Universidad Austral de Chile, Valdivia, Chile*

Albert L. Jones (269), *Cell Biology and Aging Section, Veterans Administration Medical Center, San Francisco, California 94121 and the Departments of Medicine and Anatomy and the Liver Center, University of California, San Francisco, San Francisco, California 94143*

Yasunori Koda (155), *Department of Botany, Faculty of Agriculture, Hokkaido University, Sapporo 060, Japan*

Sabine Kuchler (123), *Laboratoire de Neurobiologie Moléculaire des Interactions Cellulaires, Centre de Neurochimie du CNRS, 67000 Strasbourg, France*

Sylvain Lehmann (123), *Laboratoire de Neurobiologie Moléculaire des Interactions Cellulaires, Centre de Neurochimie du CNRS, 67000 Strasbourg, France*

Philippe Marschal (123), *Laboratoire de Neurobiologie Moléculaire des Interactions Cellulaires, Centre de Neurochimie du CNRS, 67000 Strasbourg, France*

Susanna Maschke (123), *Laboratoire de Neurobiologie Moléculaire des Interactions Cellulaires, Centre de Neurochimie du CNRS, 67000 Strasbourg, France*

Dorothy M. Morré (1), *Department of Foods and Nutrition, Purdue University, West Lafayette, Indiana 47907*

Andreas Oksche (39), *Department of Anatomy and Cytobiology, Justus Liebig University of Giessen, Giessen, Germany*

Esteban M. Rodríguez (39), *Instituto de Histología y Patología, Universidad Austral de Chile, Valdivia, Chile*

Shingo Sakai (239), *Institute of Biological Sciences, University of Tsukuba, Tsukuba, Ibaraki 305, Japan*

F. Salamini (201), *Max-Planck-Institut für Züchtungsforschung, D-5000 Köln 30, Germany*

Douglas L. Schmucker (269), *Cell Biology and Aging Section, Veterans Administration Medical Center, San Francisco, California 94121 and the Departments of Medicine and Anatomy and the Liver Center, University of California, San Francisco, San Francisco, California 94143*

Guy Vincendon (123), *Laboratoire de Neurobiologie Moléculaire des Interactions Cellulaires, Centre de Neurochimie du CNRS, 67000 Strasbourg, France*

Carlos R. Yulis (39), *Instituto de Histología y Patología, Universidad Austral de Chile, Valdivia, Chile*

Jean-Pierre Zanetta (123), *Laboratoire de Neurobiologie Moléculaire des Interactions Cellulaires, Centre de Neurochimie du CNRS, 67000 Strasbourg, France*

Intracellular Actions of Vitamin A

Dorothy M. Morré
Department of Foods and Nutrition, Purdue University,
West Lafayette, Indiana 47907

I. Introduction

Although there have been extensive studies of the biological and biochemical responses to vitamin A (retinoids) in a number of systems (Sporn *et al.*, 1984; Sherman, 1986; Jetten, 1985), the mechanism of retinoid action is not fully understood. Because the effects of retinoids are tissue specific, it is likely that more than one independent mechanism may be involved in eliciting responses to retinoids. No single function can explain adequately their widespread effects (Pawson *et al.*, 1982). Possibly, some of the retinoid-induced alterations in cellular behavior involve changes at more than one level (e.g., both genomic and membrane).

Attempts to clarify the underlying mechanisms of action of the retinoids have not been successful due to inconsistencies of their effects. For example, some studies show that retinoids are effective in controlling the proliferation of certain types of malignant cells while in other instances they are ineffective or they actually stimulate growth (Lotan, 1980; Boutwell and Verma, 1981). While vitamin A (particularly, retinoic acid) normally is an inducer of differentiation (Lotan and Nicolson, 1977; Roberts and Sporn, 1984; Amatruda and Koeffler, 1986), other reports indicate that retinoids can inhibit differentiation of mesodermal elements (i.e., in chondrogenesis and osteogenesis) (Lewis *et al.*, 1978; Takigawa *et al.*, 1980; DiSimone and Reddi, 1981) and block terminal differentiation of murine preadipocytes into mature adipocytes in a dose-dependent reversible manner (Murray and Russell, 1980). Additionally, retinoids are required for normal reproduction and embryonic development but excessive amounts can result in teratogenic effects (Shenefelt, 1972; Geelen, 1979).

II. Mechanistic Actions of Vitamin A

A. Genomic Expression

A number of reports have been published to support the theory that retinoids can influence genomic expression by activating or repressing specific genes. Retinoids appear to work via high-affinity receptor proteins, as indicated by their high structural specificity and the low concentrations at which they work. Recently, several forms of a retinoic acid-binding receptor have been reported (Giguere *et al.*, 1987; Petkovich *et al.*, 1987; Benbrook *et al.*, 1988; Brand *et al.*, 1988; Krust *et al.*, 1989; Zelent *et al.*, 1989) which belong to a superfamily of nuclear receptors that are gene ligand-transcription factors. These receptors may act similarly to or interact with steroid and thyroid hormones by repressing or activating gene expression upon binding retinoic acid (Evans, 1988). For example, expression of thyroid hormone (T_3) and retinoic acid receptors in HeLa and CV1 cells at half-maximal concentrations of 0.3 and 0.1 μM resulted in ligand-dependent inhibition of epidermal growth factor receptor and *c-erb B-2/neu* promoter activity (Hudson *et al.*, 1990). Ligand-activated T_3 and retinoic acid receptors, as demonstrated by deletion mapping, acted via a 36 base pair 5' fragment of the epidermal growth factor (EGF) receptor gene *in vivo* and in the presence of nuclear extract bound to this same region with high affinity *in vitro*. Glass *et al.* (1989) showed that the human T_3 receptor interacted with the human retinoic acid receptor to form a heterodimer that resulted in increased binding of the retinoic acid receptor to a subset of T_3 response elements. They suggested that more elaborate control of transcription can occur due to formation of the heterodimers.

B. Embryonic Development

High doses of retinoids, including retinol, retinyl esters, and retinoic acid, induce modifications of several morphogenetic processes, including duplication of structures in the developing limbs of urodeles (Maden, 1983; Thoms and Stocum, 1984) and chickens (Summerbell, 1983) and the formation of feathers on the feet of chicken embryos (Dhouailly and Hardy, 1978; Dhouailly *et al.*, 1980). Additional modifications seen *in vitro* with excess retinoids included formation of mucous glands from vibrissa follicles in the upper lip skin of embryonic mice (Hardy and Bellows, 1978) and from the developing cheek pouch of the newborn hamster (Mock and Main, 1978). Normally, these mucous glands are not found at these locations but, once initiated by excess retinoids, continue to develop *in vitro* in an orderly manner, even in the absence of excess retinoid (Hardy and Bellows, 1978; Covant and Hardy, 1988).

In experiments designed to determine whether the retinoid acted through the epithelium or the mesenchymal stroma, Covant and Hardy (1990) used explants of cheek pouch which were grown for 7 days in either standard medium or medium supplemented with 1.8×10^{-5} M retinyl acetate. Then the explants were separated into epithelium and mesenchyme by trypsinization and cultured for an additional 1–2 weeks in standard medium. No mucous glands were formed in epithelium or a recombinant of epithelium and stroma exposed to retinyl acetate, nor were glands formed in stroma that had not been exposed to retinyl acetate. This indicated that direct exposure of the stroma to high levels of retinoids may be the essential factor for gland initiation. Earlier, Yuspa and Harris (1974) showed that keratinocytes in monolayer culture respond directly to excess retinyl acetate with suppression of keratinization and increased mucous gland production.

Direct evidence for a role of retinoids in development was found in high-performance liquid chromatography (HPLC) analysis of developing chick limb buds (Thaller and Eichele, 1987). When limb buds were dissected into two, retinol was found in equal amounts in the two parts, whereas retinoic acid was enriched in the posterior part. This suggested that there was a gradient of retinoic acid across the anterior–posterior axis of the limb bud, with the highest concentration on the posterior side, and led various investigators to conclude that retinoic acid may be a natural morphogen involved in vertebrate development (Brockes, 1989; Eichele, 1989). Retinoic acid appears to respecify pattern formation during limb bud regeneration. For example, application of retinoic acid locally to the anterior side of the developing limb bud of stage 20–21 chick embryo resulted in mirror image duplication in the anterior–posterior axis. Six-digit double posterior limbs develop instead of the normal three-digit buds (Tickle et al., 1982, 1985; Summerbell, 1983). This effect mimics that found when a zone of polarizing activity is grafted into the anterior edge and creates an anteroposterior mirror image duplication of the digits. Thaller and Eichele (1987) showed that retinoic acid is produced locally in the limb field from retinol via retinal. Another morphogenetically active compound that is generated in situ from retinol through a 3,4-didehydroretinoic acid has been isolated and identified as 3,4-didehydroretinoic acid (Thaller and Eichele, 1990). When added to fractionated extracts of whole chick embryos, both 3,4-didehydroretinoic acid and retinoic acid were equipotent in evoking digit duplications.

When Maden et al. (1989) examined other regions of the chick embryo they found that retinoid-binding proteins were localized in the developing nervous system. Maden et al. (1989) detected cellular retinoic acid-binding protein (CRABP) in the developing chick limb that formed a gradient opposite in direction to that formed by retinoic acid. Interestingly, Niazi and Saxena (1978) showed that extra limb segments were produced in the regenerating limbs of toads if they were treated with retinyl palmitate. The same specific effects of retinyl palmitate on the proximodistal axis were observed in newts and axolotls in that complete limbs could be regenerated from amputations through the hand (Maden, 1982, 1983; Thoms and Stocum, 1984). Also, pattern duplication of the anterior posterior axis was demonstrated.

C. Differentiation

A number of investigators (Jetten, 1987; Jetten and Shirley, 1986; Rearick *et al.,* 1987; Eckert and Green, 1984; Fuchs and Green, 1980) have concluded that retinoids inhibit squamous cell differentiation in tracheobronchial and epidermal epithelial cells based on reduction of transglutaminase type I, sulfatransferase, and keratin expression (Edmondson *et al.,* 1990). Kim *et al.* (1985) and Jetten *et al.* (1986) found that retinoids enhanced the production of mucin-glycoproteins in tracheobronchial epithelial cells. Jetten (1980, 1985) showed that retinoic acid increased the synthesis of EGF receptor in a variety of systems. Chopra *et al.* (1989) examined the trophic effects of vitamin A on cell multiplication and morphology of differentiating epithelial cell cultures from human tracheobronchial epithelium grown in serum-free medium. Vitamin A at 10^{-6} and 10^{-7} M inhibited cell replication and enhanced the secretion of [^{3}H]glucosamine-labeled glycoconjugates. The production of plasma membrane vesicles was increased and the cells acquired a highly secretory pattern of ultrastructure.

A good *in vitro* system that has been utilized to study regulation of growth and differentiation by vitamin A is the cultured keratinocyte, which exhibits morphological and biochemical changes in response to retinoids (Sporn *et al.,* 1973; Yuspa and Harris, 1974; Yuspa *et al.,* 1981, 1983; Brown *et al.,* 1985; Creek *et al.,* 1989; Batova *et al.,* 1990; Siegenthaler *et al.,* 1990). Anti-RBP antibodies used in immunofluorescence studies indicate that retinol-binding protein (RBP) interacts with keratinocytes in the epidermis (Forsum *et al.,* 1977) and is concentrated around epidermal cells at about 30–40% of the plasma concentration (Torma and Vahlquist, 1983). Biochemically, the addition of retinoic acid to keratinocytes inhibits calcium and phorbol ester-induced terminal differentiation (Yuspa *et al.,* 1981, 1983), cornified envelope formation (Yaar *et al.,* 1981; Yuspa *et al.,* 1982; Green and Watt, 1982), and phorbol ester-induced ornithine decarboxylase activity (Verma *et al.,* 1978, 1979). Retinoids added to cultured keratinocytes also induce tissue transglutaminase (Yuspa *et al.*, 1981; Lichti *et al.*, 1985), regulate the expression of various keratins (Fuchs and Green, 1981; Eckert and Green, 1984; Gilfix and Eckert, 1985), increase the number of cell surface EGF receptors (Jetten, 1980; Strickland *et al.*, 1984), and stimulate the incorporation of labeled sugars into glycoconjugates (De Luca and Yuspa, 1974; Adamo *et al.*, 1979a; Ledger *et al.*, 1984). Additionally, growth inhibition of human keratinocytes by retinoic acid may be in part due to the induction of the synthesis and secretion of transforming growth factor β (TGF-β) (Batova *et al.*, 1990). Treatment of cultured human keratinocytes and keratinocytes immortalized by transfection with human papillomavirus type 16 DNA with retinoic acid stimulated latent TGF-β in a dose- and time-dependent manner. Secretion of TGF-β was induced with concentrations of retinoic acid as low as 10^{-10} M, with a maximal induction at 10^{-7} M retinoic acid. Although both cell types expressed specific high-affinity receptors and TGF-β and exhibited inhibition of [^{3}H]thymidine uptake in response to TGF-β treatment, retinoic acid treatment did not affect the number of TGF-β receptors or their affinity for TGF-β.

The induction of differentiation is accompanied by several molecular changes, including an increase in the expression of NAD glycohydrolase (Hemmi and Breitman, 1982) and transglutaminase type II activity (Davies *et al.*, 1985) and a reduction in the levels of the protooncogene c-*myc* (Bentley and Groudine, 1986; Westin *et al.*, 1982). Some of the new retinobenzoic acids (in particular, Am 80, Am 580, Ch 55, and Fv 80) have been evaluated in terms of the induction of differentiation of HL-60 cells. As with retinoic acid, the enhanced expression of c-*myc* of the cells was strongly suppressed by these retinobenzoic acids before the appearance of morphological and biochemical indications of differentiation (Hashimoto *et al.*, 1987). Ch 55 suppressed the protooncogene c-*mos*, accompanied by early marker changes associated with cell differentiation (Ogiso *et al.*, 1987). Specific binding of the new retinobenzoic acids to CRABP (Jetten *et al.*, 1987; Kawamura and Hashimoto, 1980) showed that some of the Am series of retinoic acid analogs exhibited reasonable binding to CRABP. However, the most active compound (Ch 55) did not bind to CRABP from rat testis (Jetten *et al.*, 1987), rabbit tracheal epithelial cells (Jetten *et al.*, 1987), or from bovine adrenal glands (Kawamura and Hashimoto, 1980).

Elevation of the level of CRABP in a population of differentiating cultured human keratinocytes compared to keratinocytes maintained in a nondifferentiated state led Siegenthaler *et al.* (1988) to speculate that the state of differentiation of the cells may affect conversion of retinol into retinoic acid. Subsequently, Siegenthaler *et al.* (1990) found that the conversion of retinoic acid from retinol occurred in the differentiating keratinocytes but not in the nondifferentiating population. In the differentiating cells, the retinoic acid was bound to endogenous CRABP, suggesting that in these cells CRABP may be an important intermediate between enzymatic generation of retinoic acid and its binding to nuclear receptors. However, when retinal was used as a substrate, retinoic acid was formed in both cell types but at a rate approximately 3.5 times higher in the differentiating cells than in the undifferentiated cells. In both cell populations, retinal was reduced to retinol at a similar rate in the presence of NADH. Analyses of the enzymes catalyzing the conversion of retinol into retinoic acid in normal cultured keratinocytes showed that the enzymatic system was independent of alcohol dehydrogenase and suggested that the conversion of retinol to retinoic acid may be rate-limiting. The use of inhibitors such as ethanol did not significantly inhibit the conversion of retinol to retinoic acid.

Mori (1922) reported epithelial tissue changes as a result of vitamin A deficiency. Later, Wolbach and Howe (1925) used vitamin A-deficient rodents to demonstrate that vitamin A was necessary for control of proliferation and differentiation in a variety of epithelial tissues. They showed that differentiation of stem cells into mature epithelial cells did not occur in the vitamin A-deficient animals. Keratin accumulated and cellular proliferation was excessive. This led to a reasonable suggestion that retinoids may be important in the prevention of cancer. The initial report associating vitamin A and cancer was made in 1926 by Fujimaki, who found that rats fed a vitamin A-deficient diet developed carcinomas of the stomach. Since then, numerous other investigators have reported that vitamin A, particularly retinoic acid, acts to prevent, inhibit, or ameliorate tumorigenesis at a variety of organ sites (for a review, see Sporn et al., 1976). Due to its size and complexity, this area of research can be discussed only briefly in this review.

D. Skeletal Abnormalities

A variety of bone abnormalities, including rapid bone resorption (Bélanger and Clark, 1967; Moore and Sharman, 1978), increased bone mineral turnover (Mellanby, 1947; Wolbach, 1947; Clark and Smith, 1964), and destruction of bone matrix (Clark and Bassett, 1962), have been found in growing rats given excess vitamin A. Chronic ingestion of a twofold excess of vitamin A (as stabilized retinyl acetate) by aged rats over their adult life span resulted in a 15% increase in vertebral trabecular bone density. In man, acute vitamin A toxicity causes hypercalcemia (Frame et al., 1974; Ragavan et al., 1982; Farrington et al., 1981). Some synthetic retinoids in intermediate doses have been associated with skeletal hyperostosis and calcification of tendons and ligaments (Pittsley and Yoder, 1983; DiGiovanna et al., 1986).

III. Transport and Uptake

The only retinoid-binding protein that has been found in plasma is RBP; however, many specific retinoid-binding proteins have been isolated from interstitial fluid and both intracellular retinol- and retinoic acid-binding proteins have been identified (Blomhoff *et al.*, 1990) (Table I). The retinoid-binding proteins can be categorized into the two major classes of serum and cellular binding proteins.

A. Serum Binding Protein

Following absorption in the intestines, retinol is incorporated into chylomicrons as retinyl esters and transported to the hepatic parenchymal cells via lymph and blood (Goodman and Blaner, 1984). Subsequently, the chylomicron remnant retinyl esters taken up by parenchymal cells in rats are hydrolyzed, probably at the plasma membrane or early endosome, by a retinyl ester hydroxylase (Harrison and Gad, 1989) before retinol is transferred within 2 to 4 hr to perisinusoidal stellate cells for storage as retinyl esters (Blomhoff *et al.*, 1985, 1988, 1990) (Fig. 1). When antibodies against RBP were included in the liver perfusate, the transfer of retinol to stellate cells was completely blocked, suggesting that RBP mediates paracrine transfer of retinol from parenchymal to stellate cells (Blomhof *et al.*, 1985)

Morphological studies show that when large doses of vitamin A are consumed, the stellate cells in liver as well as in other tissues (i.e., intestine, kidneys, heart, large blood vessels, ovaries, and testes) take up and store retinyl esters (Wake, 1980). Prior to mobilization, the stored retinyl esters must be hydrolyzed. Then retinol is transported via a specific carrier protein, RBP, that is synthesized primarily in the liver and secreted into the plasma as a retinol–RBP complex (holo-RBP) (Smith and Goodman, 1979). Apo-RBP is eliminated by glomerular filtration in the kidney (Kanai *et al.*, 1968; Sporn *et al.*, 1984) after dissociation from the 55-kDa protein transthyretin (TTR). Normally, the holo-RBP circulates reversibly bound in the plasma as a 1:1 molar complex with TTR, a tetramer which also binds thyroid hormones. Newcomer *et al.* (1984) predicted from the three-dimensional structure determined by X-ray crystallography that RBP should contain a hydrophobic pocket able to bind one molecule of retinol with the β-ionone ring located innermost and the isoprene tail stretched out almost to the surface of the protein. The binding avidity of retinol to the RBP–TTR complex is higher than to RBP alone (Fex and Johannesson, 1987). Noy and Xu (1990), for example, found that binding of RBP to TTR increased its avidity to retinol about 2-fold. The half-life of holo-RBP

TABLE I
Retinoid-Binding Proteins and Receptors[a]

Protein[b]	Approximate mass (kDa)	Main ligand	Suggested function
RBP	21	Retinol	Blood plasma
IRBP	140	Retinol, retinal	transport
			Intercellular transport in visual cycle
Four proteins secreted from pig uterus	22	Retinol	Transport to the fetus
Two luminal proteins in rat epididymis	20	Retinoic acid	Intracellular transport
CRBP(I)	16	Retinol	Donor for LRAT[c] reaction, intracellular transport
CRBP(II)	15	Retinol	Donor for LRAT reaction
CRBP(III) from fish eye	15	Retinol	?
CRABP(I)	16	Retinoic acid	Intracellular transport, regulate free retinoic acid concentration
CRABP(II) from neonatal rat	15	Retinoic acid	Intracellular transport, regulate free retinoic acid concentration
CRABP(II) from embryonal chick	16	Retinoic acid	Intracellular transport, regulate free retinoic acid concentration
CRALBP	36	Retinal	Enzymatic reactions in the visual cycle
RARα	50	Retinoic acid	Ligand-dependent transcription factor
RARβ	50	Retinoic acid	Ligand-dependent transcription factor
RARγ	50	Retinoic acid	Ligand-dependent transcription factor
RXRα	50	Retinoic acid	Ligand-dependent transcription factor

[a] With permission from Blomhof *et al.* (1990). Copyright 1990 by the AAAS.

[b] RBP, Retinol-binding protein; IRBP, intracellular retinol-binding protein; CRBP (I and II), cellular retinol-binding proteins I and II; CRABP (I and II), cellular retinoic acid-binding proteins I and II; CRALBP, cellular retinaldehyde-binding protein; RARα, RARβ, RARγ, retinoic acid receptors α, β, and γ.

[c] LRAT, Lecithin–retinol acyltransferase.

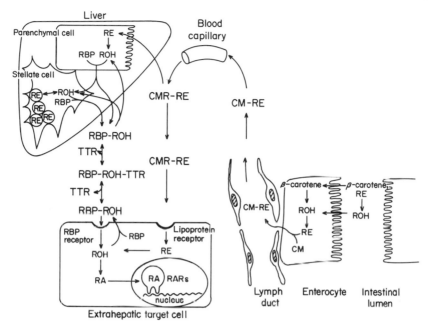

FIG. 1 Major pathways for retinoid transport in the body. Dietary retinyl esters (REs) are hydrolyzed to retinol (ROH) in the intestinal lumen before absorption by enterocytes, and carotenoids are absorbed and then partially converted to retinol in the enterocytes. In the enterocytes, retinol reacts with fatty acids to form esters before incorporation into chylomicrons (CMs). Chylomicrons then reach the general circulation by way of the intestinal lymph, and chylomicron remnants (CMRs) are formed in blood capillaries. Chylomicron remnants, which contain almost all the absorbed retinol, are mainly cleared by the liver parenchymal cells and, to some extent, also by cells in other organs. In liver parenchymal cells, retinyl esters are rapidly hydrolyzed to retinol, which then binds to retinol-binding protein (RBP). Retinol–RBP is secreted and transported to hepatic stellate cells. Stellate cells may then secrete retinol–RBP directly into plasma. Most retinol–RBP in plasma is reversibly complexed with transthyretin (TTR). The uncomplexed retinol-RBP is presumably taken up in a variety of cells by cell surface receptors specific for RBP. Most of the retinol taken up will then recycle to plasma, either on the "old" RBP or bound to a newly synthesized RBP. RA, Retinoic acid; RAR, retinoic acid receptor. (Redrawn with permission from Blomhoff et al., 1990. Copyright 1990 by the AAAS.)

associated with TTR in serum under normal physiological conditions is approximately 11–16 hr, whereas in the apo- form it is removed much more rapidly. Harrison et al. (1980) and Rask et al. (1983) have studied the subcellular localization of RBP in rat liver. Its concentration was high in endoplasmic reticulum (microsomes) and increased in the endoplasmic reticulum with vitamin A deficiency (Rask et al., 1983). Additionally, Harrison et al. (1980) and Rask et al. (1983) reported findings consistent with synthesis and secretion of RBP by rat liver via the classical secretory pathway involving Golgi apparatus, endoplasmic reticulum, and, from colchicine inhibition studies (Smith *et al.*, 1980), via microtubule-guided secretory vesicles.

The RBP transport system is involved in regulating the retinol supply to tissue and likely protects tissues from the surface-active properties of retinol. Normally, the mobilization of vitamin A from liver and its availability to peripheral tissues are under tight regulation by factors that control the rate of production and secretion of RBP (Goodman, 1974; Smith and Goodman, 1979). Retinol, stored as retinyl ester, is released as retinol by the liver and circulates as a retinol–RBP–TTR complex (Goodman, 1980). With chronic excess vitamin A intake, the capacity of the RBP transport system is eventually exceeded and the retinyl esters spill into the blood and circulate unbound to the RBP transport system but associated with serum lipoproteins (Mallia *et al.*, 1975; Smith and Goodman, 1976). The unbound retinol has the potential to overstimulate the retinol effect as well as penetrate the lipoprotein bilayers, disrupt membrane integrity, and precipitate the release of lysosomal hydroxylases (i.e., cathepsin), which then degrade the extracellular matrix of the tissue (Gorgacz *et al.*, 1975).

B. Cellular Binding Proteins

By analogy with steroid hormones and with other lipophilic milecules that traverse hydrophilic cellular regions, transport through the cytoplasm is apparently facilitated by specific cellular binding proteins. Retinoids are normally found associated with specific intracellular binding proteins (Chytil and Ong, 1984, 1987; Pfeffer *et al.*, 1986). Two examples have been described and characterized for retinol, namely, cellular retinol-binding protein (CRBP) (Chytil and Ong, 1984, 1987) and cellular retinol-binding protein II (CRBP II) (Ong *et al.*, 1984; Chytil and Ong, 1987). Although CRBP has been found in all rat tissues that have been examined (Kato *et al.*, 1985; Blaner *et al.*, 1986), its precise cellular functions remain unknown. Goodman (1986) has suggested that likely it serves as a protein that transports retinol from one locus to another within the cell. Studies by Takase *et al.* (1979) and Liau *et al.* (1981, 1985) have shown that CRBP can facilitate the transfer of retinol to the nucleus. Ong *et al.* (1988) suggested that CRBP may be involved in esterification of retinol, particularly in the liver. In rats, the tissue levels are so highly regulated and maintained that administration of high levels (10 times normal) of retinol did not result in changes in CRBP levels in any tissues examined (Kato *et al.*, 1985). Results from studies using a sensitive RNase protection assay to determine absolute CRBP mRNA levels and a specific radioimmunoassay for

CRBP showed that the tissue levels of CRBP mRNA and of CRBP protein were highly correlated and suggested that tissue levels of CRBP are regulated mainly through factors that regulate the tissue levels of CRBP mRNA (Rajan et al., 1990).

Unlike CRBP, the tissue distribution of CRBP II is more restrictive. CRBP II is a 134-amino acid intracellular protein that was first isolated from rat (Ong et al., 1984) and later from humans (Page and Ong, 1987). Transcription of the CRBP II gene in adult animals is exclusively in the polarized absorptive epithelial cells located on small intestinal villi (Crow and Ong, 1985), whereas CRBP II likely is involved in intestinal metabolism of recently absorbed or newly formed retinol (Ong et al., 1987). Initial detection of the CRBP II mRNA in the rat small intestine is at the time of development of the absorptive epithelium (Li et al., 1986). The gene is highly conserved in rats, mice, and humans (Demmer et al., 1987). A greater degree of esterification of retinol occurred when it was presented as a CRBP II–retinol complex to rat small intestinal microsomes in vitro than when it was delivered to either CRBP or serum RBP, both of which are not synthesized in the intestine (Ong et al., 1987). Additionally, CRBP II may be needed to direct newly absorbed retinol to the endoplasmic reticulum where a lecithin–retinol acyltransferase (LRAT) resides to esterify the retinol (MacDonald and Ong, 1988a,b). Levin et al. (1988) suggested that CRBP II might be uniquely suited for interaction with carotenoid-derived retinal in the enterocyte since it, but not CRBP, could bind all-trans retinol. CRBP II has been detected in rat embryonal liver and lung. In chick embryonic duodenal organ culture, results of pulse–chase experiments suggest that $1,25-(OH)_2$ vitamin D_3 (calcitriol) markedly decreases the synthesis but not the degradation rate of CRBP II (Finlay et al., 1990). Nishiwaki et al. (1990) have isolated a third retinol-binding protein (CRBP III) from fish eyes.

Another binding protein, cellular retinoic acid-binding protein (CRABP), that specifically interacts with retinoic acid but not with retinol or retinal is immunologically different from CRBP and CRBP II, although it has a similar molecular weight and a related amino acid sequence (Eriksson et al., 1981). CRABP is detectable in many tissues late in embryogenesis but is much more limited in its distribution in normal adult tissues (Ong and Chytil, 1976; Ong et al., 1987; Sani and Corbett, 1977). Levels of CRABP rise 3- to 4-fold in regenerating axolotl limb and peak at the time when retinoic acid is most effective in causing pattern duplications (Maden et al., 1989). Sani (1979) reported that CRABP is associated with the plasma membrane, but other workers (Schindler et al., 1981) have not been able to show the presence of CRABP on the plasma membrane.

Additionally, two other proteins that bind retinoids have been found in the eye. The cellular retinaldehyde-binding protein (CRALBP) has a molecular weight of 33,000 and is found in the Müller cells of the retina and in the retinal pigment epithelia, but not in the rod outer segments (Stubbs *et al.*, 1979). Both 11-*cis*-retinal and 11-*cis*-retinol are bound with high stereospecificity by CRALBP and can be enzymatically interconverted when bound to the protein (Saari and Bredberg, 1987). A specific retinal-binding protein restricted to the cells of the retina and pineal gland has been described and appears to be structurally unrelated to other retinoid-binding proteins (Futterman *et al.*, 1977; Saari *et al.*, 1982). Presumably, it is involved in reactions of the visual process.

C. Cell Systems Lacking Binding Protein

Although a role has been proposed for cellular retinoic acid- and retinol-binding proteins in binding via specific receptors to specific sites in the chromatin and thereby altering gene expression (Chytil and Ong, 1979, 1983), several cell systems that respond to retinoids lack the binding proteins (Douer and Koeffler, 1982; Libby and Bertram, 1982). For example, human myeloid leukemia (HL-60) cells do not contain detectable levels of either cellular retinol (CRBP)- or cellular retinoic acid (CRABP)-binding proteins, yet these cells respond to retinoic acid and some of the other retinoids. The Ch series of benzoic acid analogs of retinoic acid also do not bind to CRABP or CRBP but are very effective inducers of differentiation in HL-60 cells (Kagechika *et al.*, 1984; Jetten *et al.*, 1987). These observations indicate that CRBP and CRABP are not involved in the mechanism of action of retinoids in all cells. Nervi *et al.* (1989) demonstrated that nuclei isolated from HL-60 cells contain a receptor that exhibits high binding affinity for retinoic acid as well as for the retinoids of the Ch series. They show that HL-60 cells contain predominantly transcripts encoded by the retinoic acid receptor α (RARα) gene.

D. Cellular Uptake

Various mechanisms for cellular uptake of retinol have been proposed; however, the mechanism by which retinol is transferred from RBP is not known. One possibility is that retinol transfer from RBP to the target cell may occur via specific receptors for RBP that exist in the plasma membranes (Heller, 1975; Rask and Peterson, 1976; Chen and Heller, 1977; McGuire *et al.*, 1981; Ottonello and Maraini, 1981; Pfeffer *et al.*, 1986; Sivaprasadarao and Findlay, 1988). Although a number of investigators have claimed that cells of the pigment epithelium of the retina (Heller, 1975; Maraini and Gozzoli, 1975; Chen and Heller, 1977), the intestinal mucosa (Rask and Peterson, 1976), and the testis (Bhat and Cama, 1979) contain a receptor for the retinol–RBP complex, evidence for the presence of such a receptor has been insufficient to prove the existence of a receptor (Daniels *et al.*, 1985; Noy *et al.*, 1986; Cooper *et al.*, 1987). Results from *in vitro* studies with model systems that do not contain any receptors show that retinol spontaneously dissociates from the retinol–RBP complex and suggest that retinol may partition into the plasma membrane without utilizing a cell surface receptor (Fex and Johannesson, 1987; Noy and Xu, 1990). Other reports indicate that retinol will move rapidly between lipid bilayers and membranes (Noy, 1988; Fex and Johannesson, 1988). In model systems consisting of small unilamellar vesicles, the rate of association with bilayers depends strongly on the composition of the fatty acyl chains of the lipids. Bilayers comprising mixed-chain fatty acids had a higher affinity for retinol than did bilayers with symmetric chains (Noy and Xu, 1990). Unlike phospholipids (McLean and Phillips, 1981) but similar to unesterified cholesterol (McLean and Phillips, 1981; Fugler *et al.*, 1985) and long-chain fatty acids (Storch and Kleinfeld, 1986), retinol can transfer rapidly and spontaneously between liposomal membranes as well as between membranes and erythrocytes (Fex and Johannesson, 1988; Rando and Bangerter, 1982; Stillwell and Bryant, 1983; Stillwell *et al.*, 1982).

Other possibilities are that the holo-RBP (retinol plus apo-RBP) may be internalized by receptor-mediated endocytosis (Blomhoff *et al.*, 1985, 1990) (Fig. 1) or that retinol, a hydrophobic molecule, may partition between RBP in the serum and the cell membrane, with no involvement of membrane receptors (Lan *et al.*, 1984; Hamilton and Cristola, 1986; Hodam *et al.*, 1990). Some evidence suggests that cellular uptake of retinol without apo-RBP may occur. To study retinol delivery, Hodam *et al.* (1990) compared the uptake of [^{3}H]retinol fed to human keratinocytes either by adding [^{3}H]retinol directly to the medium or bound to purified human RBP. Incubated with either free [^{3}H]retinol or [^{3}H]retinol–RBP complex, the cells accumulated [^{3}H]retinol in a time-dependent manner but with very different kinetics. Uptake of the free [^{3}H]retinol was rapid (20–30% in 1 hr), linear with time for 1 hr, maximal by 3 hr, and no further accumulation occurred over the next 2 hr. In contrast, uptake of [^{3}H]retinol from the complex was slow (0.5% in 1 hr) but linear with time for at least 24 hr. Excess free unlabeled retinol did not inhibit uptake of retinol from either source but a mixture of unlabeled apo/holo-RBP did inhibit uptake of [^{3}H]retinol from RBP. The authors hypothesized that free retinol was rapidly taken up at the cell surface due to the lipophilic nature of the vitamin and that the human keratinocyte cells did not have specific RBP or retinol receptors on their surfaces.

Creek *et al.* (1989) compared the uptake, metabolism, and cellular fate of [^{3}H]retinol that was delivered to primary mouse keratinocytes either bound to purified rat RBP or free in solution. The keratinocytes accumulated 15- to 20-fold more [³H]retinol when delivered free in solution than when bound to RBP. The delivered retinol was rapidly released (within 30 min) from the cells. The major metabolite formed from [³H]retinol irregardless of mode of retinol delivery was retinyl palmitate. These results agreed with previous studies that showed that in human epidermis about 70% of the vitamin occurs as fatty acyl esters (Vahlquist et al., 1985).

Sivaprasadarao and Findlay (1988) used initial-velocity studies of [³H]retinol uptake from the [³H]retinol–RBP complex in human placental brush-border membrane vesicles to show that uptake was rapid and time- and temperature-dependent and could be reversed by the addition of native or apo-RBP, but not by serum albumin. After 5 min, most of the bound RBP was displaced from the putative receptors, thus indicating that the bound RBP has a very high dissociation rate constant.

Several investigators (Ganguly et al., 1980; Chen and Heller, 1977) have contended that free retinoids would be unable to penetrate cells. Conversely, results from in vitro studies indicate that retinoids not complexed with RBP can penetrate cells passively and relatively rapidly (Wiggert et al., 1977; Saari et al., 1980; Sivaprasadarao and Findlay, 1988). Uptake of free [³H]retinol was temperature-independent, partially reversible, and showed no requirement for a specific protein for reversibility (Sivaprasadarao and Findlay, 1988). Treatment of the membrane vesicles with p-chloromercuribenzenesulfonate (PCMBS) (an inhibitor of ¹²⁵I-labeled RBP binding) also inhibited the uptake of retinol from RBP but uptake of free retinol was unaffected. Binding of RBP was abolished if PCMBS was added after the attainment of steady-state equilibrium, but the retinol already taken up from RBP was not affected. These findings suggested that binding of RBP to its specific receptor was necessary for subsequent delivery of retinol to the membrane. Moreover, the rate of uptake of [³H]retinol from RBP was decreased by TTR without substantially altering the steady-state uptake levels, thus suggesting that membranes take up retinol from uncomplexed RBP.

[³H]Retinol provided as a complex with RBP to cultured Sertoli cells accumulated in a time- and temperature-dependent manner (Shingleton et al., 1989). The rate of uptake of retinol was biphasic. At 32°C, the accumulation was linear for approximately 1 hr, after which accumulation continued at a linear but decreased rate for 23 hr. The change in rate of accumulation occurred when the amount of retinol was approximately equal to the cellular content of CRBP. Unlabeled retinol–RBP in excess competed with delivery of the labeled retinol, indicating that RBP delivery of retinol was a saturable and competable process. Free [³H]retinol, on the other hand, associated with Sertoli cells in a noncompetable manner. [³H]Retinol accumulation was not affected by either energy inhibitors or lysosomal poisons, indicating that endocytosis of the retinol–RBP complex was not the mechanism by which RBP delivered retinol to Sertoli cells.

Another member of the protein superfamily is β-lactoglobulin (BLG) (Godovac-Zimmermann et al., 1985; Sawyer et al., 1985). Like serum RBP, BLG also binds retinol in a molar ratio of 1 : 1 (Futterman and Heller, 1972; Fugate and Song, 1980). Evidence for this being a physiological function of BLG comes from observations that there is a receptor for the BLG–retinol complex in neonatal calf intestine (Papiz et al., 1986). Additional evidence that BLG can enhance intestinal uptake of vitamin A comes from in vitro studies with everted gut sacs prepared from suckling rats and incubated in Krebs Ringer buffer containing either free [^{3}H]retinol or BLG–[^{3}H]retinol to which a trace amount of [^{14}C]polyethylene glycol was added as a nonabsorbable marker (Said et al., 1989). Uptake of retinol from BLG–retinol was saturable, energy-dependent, as evidenced by lack of effect of metabolic inhibitors, and partially temperature-dependent. Serum RBP also enhanced uptake. Both BLG–retinol and RBP–retinol inhibited uptake of retinol from BLG–[^{3}H]retinol in a concentration-dependent manner. These findings suggested the possibility of a receptor on the brush border membrane that was able to recognize BLG.

IV. Antioxidant Properties

Retinoids have been implicated as scavengers of superoxide or free radicals (Lichiti and Lucy, 1969) and thereby prevent peroxidation of membrane lipids. Additionally, retinol has been reported to participate in vitro in radical reactions (Lichiti and Lucy, 1969; Pokrovsky et al., 1974), and retinyl palmitate has been shown to act as an antioxidant in vivo (Sanjecv et al., 1981). This antioxidant activity may contribute to the apparent protection provided by vitamin A against chemical carcinogenesis in animal models. Vitamin A deficiency seems, however, to exert no profound effects on enzymes of lung or liver (superoxide dismutase, catalase, glutathione peroxidase, NADPH-dependent lipid peroxidation) of the rat involved in in vivo protection against oxygen toxicity (Dogra et al., 1983).

V. Effects on Membranes

A. Effect on Lysosomes

Vitamin A toxicity in vitro possibly related to lysosome labilization was first indicated from the classical studies of Fell and Mellanby (1952) with chick fetuses and limb bud rudiments of chick embyos. Other examples

16   DOROTHY M. MORRÉ

included accumulation of lipids (Moore, 1967), general effects on membranes such as erythrocyte expansion (Dingle and Lucy, 1962) and swelling of mitochondria (Lucy *et al.*, 1963; Moore, 1967; Keiser *et al.*, 1964), and ruptured lysosomes (Lucy and Dingle, 1964) in response to excess vitamin A. Also, lysosomal labilization associated with an increase in free activity of acid phosphatase in hyper- and hypovitaminosis (Dingle, 1961; Dingle *et al.*, 1961; Dingle and Lucy, 1962; de Duve *et al.*, 1962; Glauert *et al.*, 1963; Basset and Packer, 1965) has been noted *in vitro,* but *in vivo* demonstrations have been limited only to a few reports (Dingle *et al.,* 1966) where crude mitochondrial–lysosomal pellets were assayed.

When retinol was added to highly purified preparations of secondary lysosomes from rat liver, labilization occurred over the range of 1 to 100 μM (Morré, 1988) (Fig. 2). No significant stabilization was observed at

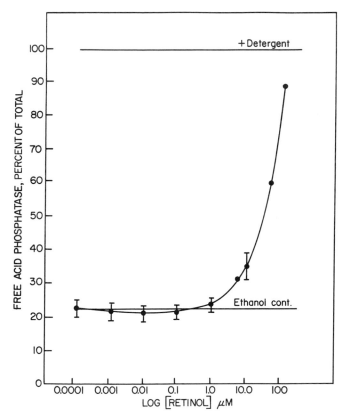

FIG. 2   Free acid phosphatase as a function of retinol concentration. Purified lysosome fractions were incubated for 15 min with varying concentrations of retinol. (Reproduced from Morré, 1988, with permission from Springer-Verlag.)

any concentration over the range of 0.1 nM to 1 μM with or without preincubation with the vitamin. Likewise, addition of retinyl palmitate, the most prevalent physiological form of vitamin A in membranes, was largely without effect on lysosome stability. Similar findings were reported from earlier *in vitro* experiments (de Duve *et al.*, 1962). Thus, it now seems unlikely that lysosome labilization is a major cellular response even to vitamin A excess.

B. Effect on Insulin Release

The addition of vitamin A to intact pancreatic islets resulted in a number of ultrastructural alterations. Chertow *et al.* (1979) found that vitamin A at 10^{-4} M retinol decreased the volume fraction of secretory granules to 80% of control, increased the volume fraction of material within the rough endoplasmic reticulum to 147% of control, increased the outer mitochondrial compartment to 116% with a reciprocal decrease of the inner compartment volume fractions, and increased the number per unit area of phagolysosomes to 190% of control. The alterations in outer mitochondrial compartment volume are consistent with earlier findings in tissue and cell models *in vivo* (Dingle and Lucy, 1962) and indicate a direct effect of vitamin A on mitochondrial function.

Chertow *et al.* (1979) found that vitamin A (retinol) had diphasic effects on biphasic insulin release. At 10^{-4} M, retinol inhibited second-phase 9.7 and 13.9 mM glucose-induced insulin release to 50%, 10 m M glyceraldehyde-induced release to 64%, and 20 mM leucine-induced release to 66% of control. Kinetic analysis on glucose-induced release of insulin indicated a change in the V_{max} but not in the K_m on exposure to vitamin A, which suggested that a potentiator of insulin release was inhibited by vitamin A. This could have resulted from impairment of transport of proinsulin to the Golgi apparatus. Alternatively, there could have been a decrease in packaging of insulin into secretory granules within the Golgi apparatus which would account for the decrease in the volume fraction of secretory granules without a concomitant change in the number of secretory granules. Additionally, impaired transfer of hormone from the rough endoplasmic reticulum to the Golgi apparatus would have resulted in the observed increase in the volume fraction of the rough endoplasmic reticulum due to an accumulation of material within the rough endoplasmic reticulum. At 4 mM, calcium did significantly antagonize the inhibitory effect of vitamin A. More recently, Kato *et al.* (1985) have utilized immunohistochemistry and radioimmunoassay to demonstrate relatively high levels of CRBP, CRABP, TTR, and RBP in rat islets.

Additionally, Chertow *et al.* (1989) used rat insulinoma (RINm5F) cells, a β cell line, to show that the addition of either retinol or retinoic acid to the incubation medium increased KCl-induced insulin release, which supported the earlier findings that vitamin A has a role in insulin release and further supported earlier findings (Chertow *et al.*, 1987) that retinoic acid can substitute for retinol in this role.

C. Involvement of Golgi Apparatus

1. Morphological Alterations

Vitamin A has been reported to be concentrated in the hepatocyte Golgi apparatus (Nyquist *et al.*, 1971), and many of the response features of cells and tissues to vitamin A involved cellular activities associated with Golgi apparatus (Wolf, 1980; D. M. Morré *et al.*, 1981, 1988; Martin and Morré, 1988). Both vitamin A deficiency and excess influence Golgi apparatus architecture (D. M. Morré *et al.*, 1981), based on quantitation of parameters from measurements of electron micrographs using morphometric methods. Golgi apparatus of liver consisted of three to four stacked cisternae plus associated secretory vesicles and small vesicular profiles. In vitamin A-deficient animals, liver Golgi apparatus was characterized by small stacks with highly fenestrated saccules. Total membrane surface occupied by Golgi apparatus was also less.

In livers of rats fed diets containing high amounts of vitamin A, Golgi apparatus membrane surface was increased and more variable, ranging from nearly normal in appearance to much larger than normal (D. M. Morré *et al.*, 1981). Saccules were larger and many lacked the numerous fenestrations normally present. Unusual cisternal configurations also were encountered. Except for slightly swollen mitochondria with vitamin A excess, the Golgi apparatus response was the major ultrastructural alteration in liver found to respond consistently to changes in vitamin status. Previously, a reduction in the Golgi apparatus was shown in epithelial cells of vitamin A-deficient calves (Hayes *et al.*, 1970). In agreement with our findings, Prutkin (1975) previously had demonstrated that vitamin A applications to neoplastic rabbit epithelium caused an increase in Golgi apparatus membranes. Likewise, Chertow *et al.* (1975) showed in semiquantitative studies with bovine parathyroid pieces that retinol caused Golgi apparatus dilatations.

2. Biochemical Changes

Biochemical verification of a response of the Golgi apparatus to vitamin A was from measurement of galactosyltransferase and uridine 5'-

diphosphate phosphatase in isolated preparations of Golgi apparatus from livers of rats (Martin and Morré, 1988; Table II). Large doses of vitamin A were administered either by gavage with vitamin A dissolved in olive oil or by feeding diets containing excess amounts of vitamin A. Galactosyltransferase activity in Golgi apparatus was significantly decreased under conditions of both chronic and acute administration of excess vitamin A and negatively correlated with total vitamin A content as well as specific forms of vitamin A, as determined by HPLC analyses. In contrast to galactosyltransferase, activity of uridine 5'-diphosphate phosphatase was increased significantly in the experimental groups.

In other experiments, retinol was provided to rats by gavage (500 mg/kg body weight) and 16–18 hr later 0.4 mCi [^{35}S]methionine was administered, followed 15 min later by a chase of excess unlabeled methionine (Morré *et al.*, 1988). Following the pulse, the rats were sacrificed at various times during the chase and endoplasmic reticulum, Golgi apparatus, plasma membrane, and lysosome fractions were prepared (Croze and Morré, 1984; Evers *et al.*, 1984) and analyzed for radioactivity. An effect of retinol was found that was consistent with an altered flux of labeled proteins through the secretory pathway (Morré *et al.*, 1988) (Fig. 3). Both endoplasmic reticulum and Golgi apparatus in livers of both retinol-treated and control rats were nearly equally labeled during the 7.5-min pulse, and specific radioactivity of the proteins continued to rise during the subsequent chase. At about 40 min postchase, the specific radioactivity of endoplasmic reticulum reached a plateau and that of the Golgi apparatus began to decline. The specific labeling of endoplasmic reticulum was less with retinol-treated rats than that for controls at 30 and 60 min postchase,

TABLE II

Specific Activities of Galactosyltransferase and Uridine 5'-diphosphate Phosphatase in Golgi Apparatus Isolated from Liver of Rats[a]

Enzyme	Control	Excess
Galactosyltransferase	216 ± 14	127 ± 32[b]
Uridine 5'-diphosphate phosphatase	4 ± 1	7 ± 2

[a] Rats were gavaged with either olive oil (control) or olive oil containing 40 mg of retinol (excess) for each of 3 consecutive days. Values are means ± SEM of fractions from six rats per treatment group. Units of specific activity are nanomoles of acceptor-dependent disappearance of UDP-galactose per hour per milligram of protein for galactosyltransferase and micromoles of inorganic phosphate formed per hour per milligram of protein for uridine 5'-diphosphate phosphatase.

[b] Significantly less than control ($P < 0.05$).

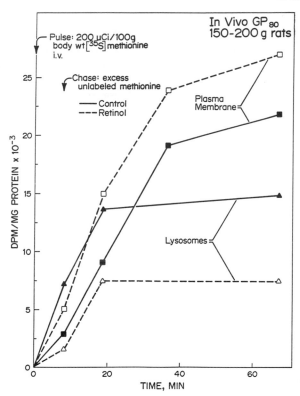

FIG. 3   Short-time labeling and turnover kinetics of [^{35}S]methionine-labeled proteins of rat liver cell fractions. Olive oil was administered without (solid line) or with (dashed line) retinol (500 mg/kg body weight) 16 hr prior to sacrifice. The pulse of 0.4 mCi [^{35}S]methione administered via the tail vein was followed after 7.5 min by a chase of excess unlabeled methionine. The experiment was repeated three times. For each experiment, livers were pooled from three rats to give a total of nine rats/point. Values are average specific activities from three experiments ± standard deviations among experiments.

whereas, with Golgi apparatus, labeling was less at all time points. There was a small but significant increase in specific radioactivity of plasma membranes from livers of rats receiving retinol. In contrast to plasma membrane, lysosome labeling was significantly less at all time points with retinol preincubation than in the control without retinol preincubation.

The differences in labeling of plasma membrane and lysosomes also were found with analysis by immunoprecipitation for content of a membrane-located glycoprotein, GP$_{80}$. Specific labeling of GP$_{80}$ in plasma membranes was about 25% higher in the retinol-treated rats compared to controls, whereas in lysosomes there was an even greater difference (2-

fold lower) but in the opposite direction (Morré et al., 1988) (Fig. 3). Results of this study indicate that with excess vitamin A there was a switching of membrane trafficking away from lysosome formation, with much less of an effect on secretory vesicle formation and delivery to the plasma membrane. While the lysosomes were labeled less in this study, overall delivery to the plasma membrane was increased as expected if secretory activities were enhanced. Impaired trafficking from the endoplasmic reticulum through the Golgi apparatus could result in incomplete fucosylation in the presence of excess vitamin A, as observed by Büchsel and Reutter (1982) (Table III), as well as an accumulation of underglycosylated ganglioside precursor and deficiency of fully glycosylated gangliosides (D. M. Morré et al., 1981).

Membrane flux kinetics with slices (7.5-min [³H]leucine pulse followed by removal of radioactivity and transfer to excess unlabeled leucine) agreed with those observed in intact animals (Morré et al., 1985). The improved chase conditions showed more clearly the accelerated transit through the Golgi apparatus. The half-time for transit was reduced from about 15 min for control slices to about 8 min for retinol-treated slices. Line slope analyses showed both accelerated loss of label from endoplasmic reticulum and accelerated accumulation of label in the plasma membrane as a result of a 30-min preincubation of the slices with 25 µg of retinol/ml. In the same series of experiments, lysosomes were labeled during the pulse period with nearly linear kinetics in the liver slices followed by rapid turnover. Turnover of labeled lysosomal proteins was unaffected by retinol preincubation but the initial rate of labeling was

TABLE III

Incorporation of L-Fucose[a]

Fraction	dpm/mg protein/hr	
	Retinol-treated	Untreated controls
Serum	6786 ± 363[b]	4507 ± 619
Whole liver	1135 ± 188	1193 ± 35
Cytosol	1170 ± 174	1074 ± 12
Plasma membrane	2329 ± 409	5861 ± 727

[a] Wistar rats were given all-*trans* retinol (1.5×10^6 IU/kg body weight/day) p.o. for 3 days. The control animals received the same volume of wheat germ oil daily. On day 4, L-[¹⁴C] fucose was injected i.p. at a dose of 100 µCi/kg body weight. One hour later, the animals were killed, and the radioactivity was determined in the acid-precipitable fractions. (From Büchsel and Reutter, 1982.)

[b] Mean ± SD of four determinations.

reduced by about 30%, in contrast to plasma membrane labeling which was increased. These results parallel those with the intact animals, were consistent with an accelerated flux through the Golgi apparatus toward the plasma membrane, and showed reduced labeling of lysosomes as a result of retinol treatment. Indicated also was a reduced opportunity for processing and a diversion of some proteins normally destined for lysosomes to the cell surface.

In a later study, Morré and Morré (1987) found alterations in morphology of the liver in rats receiving excess vitamin A. Consistently, there was an increase in the number of transition vesicles in the *cis*-Golgi apparatus zone. The transfer of membrane materials from the endoplasmic reticulum to the Golgi apparatus is thought to occur via these small 50- to 60-nm transition vesicles (Ziegel and Dalton, 1962; Morré *et al.*, 1971). The vesicles bleb from a specialized part-rough, part-smooth region of the endoplasmic reticulum (transitional endoplasmic reticulum) and fuse to form new *cis*-face Golgi apparatus cisternae or merge with existing Golgi apparatus membrane.

3. Cell-Free System

The recent development of cell-free systems between endoplasmic reticulum and Golgi apparatus has permitted us to move from the static images of the exocytic pathway provided from electron microscopy to an *in vitro* system where the reconstitution of membrane transfer occurs in a cell-free environment. Fractions of part-rough, part-smooth elements of the endoplasmic reticulum (transition elements) form small vesicles *in vitro* when incubated in the presence of cytosol and ATP and ATP-regenerating system (Morré *et al.*, 1986). These small vesicles resembled morphologically the transition vesicles seen *in vivo* in electron micrographs. By metabolically labeling the donor membranes with [^{3}H]leucine in tissue slices prior to their isolation, radioactive membrane proteins were transferred to nonradiolabeled Golgi apparatus immobilized on nitrocellulose. The system offers the advantage in that it does not require viral infection, which allows the use of the same membrane donor/acceptor combinations to study the transport of a variety of different radiolabeled molecules that may move between the endoplasmic reticulum and Golgi apparatus. Experiments were done to determine whether the donor transition elements isolated from liver (Morré *et al.*, 1986) might be responsive to retinol, in terms of altered transition vesicle formation and/or transfer to Golgi apparatus acceptor. Nowack *et al.* (1990) showed that, at an optimum concentration of 1 μg/ml, the rate and amount of transfer were approximately doubled to a plateau of about 60 min incubation in the cell-free system. In the complete system (ATP, ATP-regenerating system, and cytosol) plus retinol, there were approximately twice the numbers of 50- to 70-nm vesicular profiles as there were without retinol (Table IV). Equivalent findings were obtained when the retinol response was monitored both directly by electron microscopy and by isolation of transition vesicles by preparative free-flow electrophoresis. The transition vesicle-enriched shoulder of the electrophoretic separation as determined from the A_{280} was increased with the retinol-incubated preparations (Fig. 4).

In order to define a possible role for retinol in stimulation of transition vesicle formation, the experimental system was bisected into vesicle formation and vesicle fusion steps. Vesicle formation, but not fusion, was found to be retinol responsive both by electron microscopy and from quantitation of numbers of vesicles produced after separation by preparative free-flow electrophoresis from the bulk of the starting material. Addition of GTP to the transfer system overcame the retinol effect, suggesting that there was a GTP-requiring step in the vesicle formation process.

To approach the molecular basis of retinol effects on cell-free formation of transition vesicles, experiments were designed to examine the effects of retinol on GTP hydrolysis by the transitional endoplasmic reticulum of vesicle formation. Retinol inhibited GTP hydrolysis (Zhao et al., 1990). The inhibition was noncompetitive and specific for GTP as substrate. The K_i for retinol was 0.3 mM (8.6 μg/ml). The apparent K_m for GTP was approximately 0.3 mM. Enrichment of solubilized membrane fractions by chromatography on a DE-52 column showed an even greater degree of inhibition of GTP hydrolysis by retinol than that observed with the starting endoplasmic reticulum membranes.

GTP-binding proteins likely are involved in most, if not all, vesicle formation–fusion events in eukaryotic secretory pathways (Bourne,

TABLE IV

Effect of Retinol on the Number of 50 to 70 nm Vesicular Profiles within the Space between the Transitional Endoplasmic Reticulum and the cis-Golgi Apparatus (Transition Vesicles)[a]

Condition	Retinol	Transition vesicle profiles per 100 total membrane profiles
Complete	None	2.3 ± 0.7
	1 μg/ml	4.8 ± 1.2
−ATP	None	1.5 ± 0.3
	1 μg/ml	1.9 ± 0.6

[a] Results are based on analyses of three micrographs at a primary magnification of 14,000 photographed across the entire width of the pellet for each of three different membrane preparations ± SD.

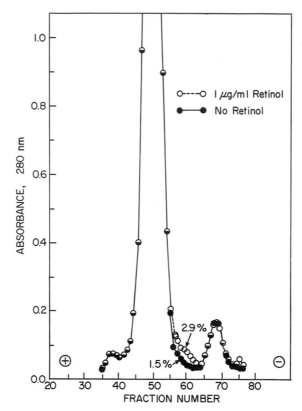

FIG. 4 Free-flow electrophoresis profile of transitional endoplasmic reticulum membranes incubated with and without added retinol. The transition elements were incubated for 10 min at 37°C in the presence of nucleoside triphosphate, the ATP-regenerating system, and the >10-kDa cytosolic fraction. The primary peak contained unfractionated endoplasmic reticulum membranes. The shoulder regions were enriched in the transition vesicles as determined by electron microscopy and amount of protein (values are percentage of total protein). (Reproduced from Nowack *et al.*, 1990, with permission of Wissenschaftliche Verlagsgesellschaft mbH, Stuttgart, Germany.)

1988). G proteins may be involved in one or more portions of the secretory pathway by facilitation of vesicle formation with stably bound GTP at the site of the transitional endoplasmic reticulum, migration of the vesicles to and attachment to the *cis*-Golgi apparatus, and interaction with a docking protein to trigger GTP hydrolysis and vesicle fusion. Retinol may facilitate cell-free formation of transition vesicles through inhibition of GTP hydrolysis, especially under conditions where GTP might be expected to be rate-limiting, such as at low cytosol concentrations without added GTP (Nowack *et al.*, 1990).

4. Video-Enhanced Optical Microscopy

A direct response of the Golgi apparatus to vitamin A has been confirmed as well by video-enhanced optical microscopy. D. M. Morré *et al.* (submitted) added retinol to bovine mammary gland epithelial cells (BME) that were plated on glass coverslips with medium contained in 400-μl Plexiglas wells cemented to the coverslips to facilitate video microscopy (Spring and Franke, 1981). The BME cells contain a well-developed Golgi apparatus identified in the light microscope images from its characteristic position adjacent to the nucleus. Retinol (1 μl) dissolved in ethanol was added to the culture well and mixed carefully. Control preparations received only ethanol (1 μl) added in an identical manner.

When retinol (1 μg/ml, final concentration) was added to the cells, the Golgi apparatus exhibited a markedly accelerated pattern of movements. Within 1 min after retinol addition, rapid undulations of cellular components with an emergence and disappearance of apparent vesicular components were detected at the plane of focus and were maximal after about 3 min. Then the activity pattern returned to normal but could again be initiated with a second addition of retinol; however, no response was observed at 0.05 μg/ml. Ethanol alone was without effect. The appearance of the Golgi apparatus in the electron microscope was not changed by the retinol treatment either before or after glutaraldehyde fixation, although accumulations of secretory vesicles and other membranes were observed at the *trans* face when activity was stimulated by retinol addition.

To further aid in identification of the Golgi apparatus region by light microscopy, the sodium-selective ionophore monensin was added to the cells in the wells, followed by glutaraldehyde fixation. The same cellular regions that responded by increased activity with retinol responded by formation of numerous swollen vacuoles in the presence of monensin.

D. Plasma Membrane

1. Glycosylation Pattern

A deficiency of vitamin A results in reduction in biosynthesis of both $\alpha_2\mu$ globulin (Haars and Pitot, 1979) and α_1-macroglobulin (Kiorpes *et al.*, 1976) and elongation of (mannose)$_9$(N-acetylglucosamine)$_2$-dolichylpyrophosphate, as well as an accumulation of (mannose)$_5$(N-acetylglucosamine)$_2$-dolichylpyrophosphate (Rosso *et al.*, 1981). Additionally, decreased incorporation of [^{3}H]glucosamine into mucin glycoproteins of rat goblet cells (De Luca *et al.*, 1970, 1971) and mannose into liver mannolipids and glycoproteins (De Luca *et al.*, 1975) has been shown with deficiency of vitamin A. Correction of these defects by vitamin A repletion lead De Luca *et al.* (1975) to hypothesize that vitamin A may function *in vivo* as a lipid intermediate. However, it was not possible to demonstrate the *in vivo* formation of retinyl-phosphate-mannose (De Luca *et al.*, 1982; Creek *et al.*, 1987) or the *in vitro* mannosylation of any endogenous retinyl-phosphate (Creek *et al.*, 1986). Additional experiments showed that *in vitro*-synthesized retinyl-phosphate-mannose was unable to function as a mannose donor during elongation of dolichylpyrophosphate oligosaccharides in the membranes of the Thy-E negative mouse lymphoma cells (De Luca *et al.*, 1987). These mutant cells are unable to synthesize dolichylphosphate-mannose (Chapman *et al.*, 1980). De Luca *et al.* (1987) have shown that mannose incorporation into GDP-mannose, Dol-P-mannose, and lipid-linked oligosaccharides and glycopeptides is decreased in vitamin A deficiency but returns to control levels following retinoid administration. In livers of hamsters, the reduction in uptake of label from [2-^{3}H]mannose into the precursor GDP-mannose correlated with impaired biosynthesis of the precursor GDP-mannose (Rimoldi *et al.*, 1990).

A systematic comparison between the *in vivo* incorporation patterns for [2-^{3}H]mannose, [4,5-^{3}H]galactose, and [5^{3}H]glucose into sugar phosphates and nucleotides showed that increased hexose transport in early vitamin A depletion is specific for mannose (Shankar *et al.*, 1990). Progression of deficiency resulted in a smaller increase in [2-^{3}H]mannose and a significant decrease in [^{3}H]mannose-phosphate and GDP-[^{3}H]mannose, suggesting a decreased mannose kinase activity. After 6 and 8 weeks of deficiency, synthesis of glucosyl-phosphate was reduced 50–90%, which suggested an impaired glucokinase activity. Other investigators have shown that glycosylation reactions are influenced by excess vitamin A (Olson, 1983). Büchsel and Reutter (1982) observed a 67% decrease in the incorporation rate of labeled L-fucose together with an increase in turnover of protein-bound fucose in livers of rats fed 1.5×10^6 IU of all-*trans* retinol/kg body weight.

These findings, along with those of polyacrylamide gel electrophoresis analyses of fucose-labeled polypeptides of the plasma membrane, suggested that excess doses of vitamin A resulted in a rapid turnover of incompletely fucosylated glycoproteins of the plasma membrane. Additionally, certain fully glycosylated members of the ganglioside series become deficient, with a corresponding accumulation of underglycosylated precursor gangliosides in livers of rats fed high doses of vitamin A (D. J. Morré *et al.*, 1981). Also, increases have been observed in: (1) *in vivo* mannosylation (Hassell *et al.*, 1978), (2) incorporation of mannose into procollagen (Dion *et al.*, 1981; Jetten and De Luca, 1985), (3) presence of complex versus high-mannose chains on the surface glycoproteins of cultured spontaneously transformed moust fibroblasts (Balb/c 3T12 cells) (Sasak *et al.*, 1980), (4) sialylation of melanoma cells (Lotan, 1980), and (5) the activity of uridine diphosphate galactose:β-D-galactosyl-α-1,3-

galactosyltransferase in both Golgi apparatus fractions from rat liver (Martin and Morré, 1988) and F9 cells (Cummings and Mattox, 1988). In contrast, lactosaminoglycan types of glycopeptides decreased (Muramatsu and Muramatsu, 1983).

2. Cellular Adhesion

Retinoids cause increased cellular adhesion to plastic substrate, but this ability does not always correlate with either their capacity to maintain epithelial differentiation or inhibit carcinogen-induced neoplastic transformation (Bertram, 1980). Nor do all cell types respond similarly to the retinoids. For example, the addition of retinyl acetate to a C3H/10T 1/2 clone 8 mouse fibroblast line increased adhesion, maintenance of epithelial differentiation, and inhibition of neoplastic transformation, whereas the analog, retinylidene dimedone, inhibited transformation but caused a significant increase in adhesion. Conversely, retinoids such as all-*trans* retinoic acid, 13-*cis* retinoic acid, and the C_{17} carboxylic analog of retinoic acid that were essentially inactive in the transformation assay also exhibited little cellular adhesion. In another cell line, BALB/3T12 cells, both Adamo et al. (1979b) and Jetten et al. (1979) found that concentrations of retinoic acid similar to those used by Bertram (1980) increased cellular adhesion.

3. Redox Activities

Studies with hormones, growth factors, and anticancer drugs suggest a relationship between redox activities of the plasma membrane and growth (Crane et al., 1990). Mammalian cells possess a transplasma membrane redox system capable of transfer of electrons from cytoplasmic reducing agents, such as NADH, to external impermeable oxidants (Crane et al., 1985). However, this system also may transfer electrons to a natural electron acceptor, oxygen, via a terminal NADH oxidase that is subject to regulation by hormones and growth factors (D. J. Morré et al., 1991). For retinoic acid inhibition of growth, the plasma membrane has long been suggested by numerous investigators as a possible site of action (Lipkin et al., 1978; Manino et al., 1978; Yen et al., 1984; Amatruda and Koeffler, 1986). In view of the response of the plasma membrane redox system to inhibition by a number of growth inhibitory drugs (Sun and Crane, 1984, 1985; Sun et al., 1983, 1984, 1985), Sun et al. (1987) reported that retinoic acid also inhibited the transmembrane reductase of the plasma membrane redox system of HL-60 cells with either diferric transferrin or ferricyanide as an impermeant electron acceptor.

Golub et al. (1988) showed evidence for a cascade of events, beginning at the plasma membrane, that were induced by retinoic acid in HL-60 cells. Retinoic acid caused an inhibition of electron transport, with a subsequent decrease in the rate at which NAD^+ was generated. The result was a down-regulation of NAD^+-dependent inosine monophosphate dehy-drogenase with a consequent fall in GTP. The fact that retinoic acid can induce differentiation even when immobilized outside of the cell (Yen *et al.*, 1984) suggests that retinoic acid reacts with a receptor molecule.

With uninduced HL-60 cells, neither retinoic acid at 0.1 n*M* nor cal-citriol at 1 p*M* inhibited reduction of NADH-ferricyanide reductase (B. Joshi *et al.*, unpublished) (Fig. 5). However, when combined, the activity was inhibited by about 50% in uninduced cells and by more than 70% in cells induced with dimethylsulfoxide (DMSO) (Fig. 5). Reduction of ferri-cyanide by HeLa cells was inhibited also by retinoic acid (Sun *et al.*, 1984, 1987). Ferricyanide acts as an external electron acceptor for transmem-brane electron transport.

Conversely, electron transport to oxygen, as measured by effects on NADH oxidase, was stimulated by combined retinoic acid and calcitriol at concentrations below which neither retinoic acid nor calcitriol alone was effective. A markedly synergistic stimulation was observed (Fig. 6), espe-cially in the cells which had not been induced to differentiate with DMSO.

4. Cell Alkalinization

Retinoic acid, which induces the differentiation of HL-60 cells to granulo-cytes (Breitman *et al.*, 1980), produces cell alkalinization from pH 7.03 to pH 7.37 after exposure of HL-60 cells to 1 μM retinoic acid for 5 days (Ladoux *et al.*, 1987). The half-maximal effect of retinoic acid is found at 10 n*M*. Alkalinization by retinoic acid develops slowly and precedes the differentiation of the cells, whereas, conversely, alkalinization occurs after differentiation of the cells when induced by DMSO. An increase in intracellular pH has been implicated consistently as a mitogenic signal (Busa and Nuccitelli, 1984) and is a consistent response to stimulation of plasma membrane redox enzymes in other systems (Crane *et al.*, 1990).

VI. Summary

Because the effects of vitamin A vary with tissue type and often with the form of vitamin A itself, a complete understanding of the mechanism(s) of action still has not been attained. The action of vitamin A may be at the level of genomic expression, at the membrane level, or both. Intercellular and intracellular transport of vitamin A are facilitated by specific binding

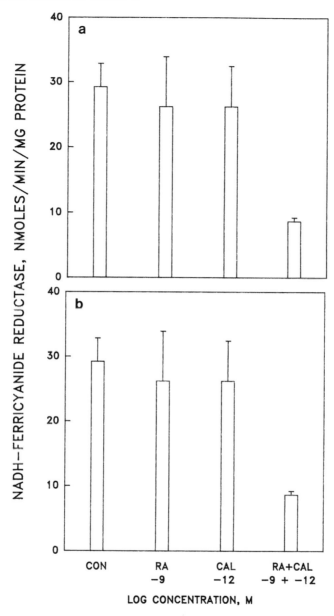

FIG. 5 Synergistic effect of low levels of retinoic acid (R. A.) and calcitriol (CAL) on activity of NADH-ferricyanide reductase in plasma membranes of (a) uninduced and (b) induced HL-60 cells. (B. Joshi, M. S. Thesis, 1991, Purdue University, W. Lafayette, IN).

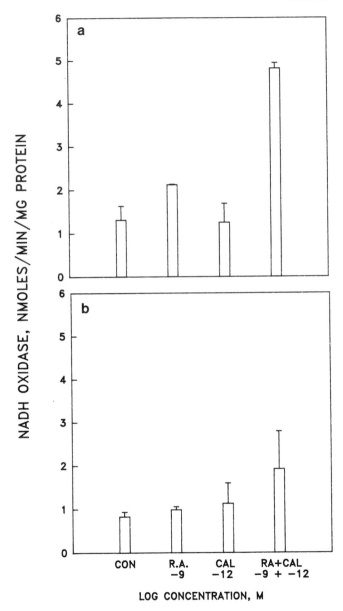

FIG. 6   Synergistic effect of low levels of retinoic acid (R. A.) and calcitriol (CAL) on activity of NADH oxidase in plasma membranes of (a) uninduced and (b) induced HL-60 cells (B. Joshi, M. S. Thesis, 1991, Purdue University, W. Lafayette, IN).

proteins but probably not in the cellular uptake of vitamin A. Subcellularly, vitamin A may exert a direct effect on transit through the Golgi apparatus, as observed from both biochemical and morphological studies. In my laboratory, recent work using cell-free systems has shown that retinol stimulates transition vesicle formation from endoplasmic reticulum in a GTP-requiring step.

Acknowledgments

Supported in part by grants from the National Institutes of Health (GM44675) and Phi Beta Psi Sorority and by a project of the Purdue Agricultural Experiment Station, 8546-561264, Paper No. 13201 of the Purdue University Agricultural Experiment Station, West Lafeyette, IN 47907.

References

Adamo, S., De Luca, L. M., Silverman-Jones, C., and Yuspa, S. H. (1979a). *J. Biol. Chem.* **254**, 3279–3287.
Adamo, S., De Luca, L. M., Akalovsky, I., and Bhat, P. V. (1979b). *JNCI, J. Natl. Cancer Inst.* **62**, 1473–1478.
Amatruda, T. T., III, and Koeffler, H. P. (1986). *In* "Retinoids and Cell Differentiation" (M. I. Sherman, ed.), p. 79. CRC Press, Boca Raton, Florida.
Basset, B. E., and Packer, L. (1965). *J. Cell Biol.* **27**, 448–450.
Batova, A., Pirisi, L., and Creek, K. E. (1990). *Proc. Am. Assoc. Cancer Res.* **31**, A1990.
Bélanger, L. F., and Clark, I. (1967). *Anat. Rec.* **158**, 443–452.
Benbrook, D., Lernhardt, E., and Pfahl, M. (1988). *Nature (London)* **333**, 669–671.
Bentley, D. L., and Groudine, M. (1986). *Nature (London)* **321**, 702–706.
Bertram, J. S. (1980). *Cancer Res.* **40**, 3141–3146.
Bhat, M. K., and Cama, H. R. (1979). *Biochim. Biophys. Acta* **587**, 275–281.
Blaner, W. S., Das, K., Mertz, J. R., Das, S. R., and Goodman, D. S. (1986). *J. Lipid Res.* **27**, 1084–1088.
Blomhoff, R., Rasmussen, M., Nilsson, A., Norum, K. R., Berg, T., Blaner, W. S., Kato, M., Mertz, J. R., Goodman, D. S., Eriksson, U., and Peterson, P. A. (1985). *J. Biol. Chem.* **260**, 13560–13565.
Blomhoff, R., Norum, K. R., and Berg, T. (1988). *J. Biol. Chem.* **260**, 13571–13575.
Blomhoff, R., Green, M. H., Berg, T., and Norum, K. R. (1990). *Science* **250**, 399–404.
Bourne, H. R. (1988). *Cell (Cambridge, Mass.)* **53**, 669–671.
Boutwell, R. K., and Verma, A. K. (1981). *Ann. N. Y. Acad. Sci.* **359**, 275–280.
Brand, N., Petkovich, M., Krust, A., Chambon, P., de Thé, H., Marchio, A., Tiollais, P., and Dejean, A. (1988). *Nature (London)* **332**, 850–853.
Breitman, T. R., Selonick, S. E., and Collins, S. J. (1980). *Proc. Natl. Acad. Sci. U.S.A.* **77**, 2936–2940.
Brockes, J. P. (1989). *Neuron* **2**, 1285–1294.
Brown, R., Gray, R. H., and Bernstein, I. A. (1985). *Differentiation (Berlin)* **28**, 268–278.
Büchsel, R., and Reutter, W. (1982). *Cancer Res.* **42**, 2450–2456.
Busa, W. B., and Nuccitelli, R. (1984). *Am. J. Physiol.* **246**, 409–438.

Chapman, A., Fujimoto, K., and Kornfeld, S. (1980). *J. Biol. Chem.* **255**, 4441–4446.
Chen, C. C., and Heller, J. (1977). *J. Biol. Chem.* **252**, 5212–5221.
Chertow, B. S., Buschmann, R. J., and Henderson, W. J. (1975). *Lab. Invest.* **32**, 190–200.
Chertow, B. S., Buschmann, R. J., and Kaplan, R. L. (1979). *Diabetes* **28**, 754–761.
Chertow, B. S., Blaner, W. S., Baranetsky, N. G., Sivitz, W. I., and Cordle, M. B. (1987). *Proc. Natl. Acad. Sci. U.S.A.* **82**, 2488–2492.
Chertow, B. S., Moore, M. R., Blaner, W. S., Wilford, M. R., and Cordle, M. B. (1989). *Diabetes* **38**, 1544–1548.
Chopra, D. P., Klinger, M. M., and Sullivan, J. K. (1989). *J. Cell Sci.* **93**, 133–142.
Chytil, F., and Ong, D. E. (1979). *Fed. Proc., Fed. Am. Soc. Exp. Biol.* **38**, 2510–2514.
Chytil, F., and Ong, D. E. (1983). *Adv. Nutr. Res.* **5**, 13–29.
Chytil, F., and Ong, D. E. (1984). *In* "The Retinoids" (M. B. Sporn, A. B. Roberts, and D. S. Goodman, eds.), Vol. 2, pp. 89–123. Academic Press, Orlando, Florida.
Chytil, F., and Ong, D. E. (1987). *Annu. Rev. Nutr.* **7**, 321–335.
Clark, I., and Bassett, C. A. L. (1962). *J. Exp. Med.* **115**, 147–156.
Clark, I., and Smith, H. R. (1964). *J. Biol. Chem.* **239**, 1266–1271.
Cooper, R. B., Noy, N., and Zakim, D. (1987). *Biochemistry* **26**, 5890–5896.
Covant, H. A., and Hardy, M. H. (1988). *J. Exp. Zool.* **246**, 139–149.
Covant, H. A., and Hardy, M. H. (1990). *J. Exp. Zool.* **253**, 271–279.
Crane, F. L., Sun, I., Clark, M. G., and Grebing, G. (1985). *Biochim. Biophys. Acta* **811**, 233–264.
Crane, F. L., Sun, I. L., and Löw, H. (1990). *In* "Oxidoreduction at the Plasma Membrane: Relation to Growth and Transport" (F. L. Crane, D. J. Morré, and H. Löw, eds.), Vol. 1, pp. 101–125. CRC Press, Boca Raton, Florida.
Creek, K. E., Rimoldi, D., Clifford, A. J., Silverman-Jones, C. S., and De Luca, L. M. (1986). *J. Biol. Chem.* **261**, 3490–3500.
Creek, K. E., Shankar, S., and De Luca, L. M. (1987). *Arch. Biochem. Biophys.* **254**, 482–490.
Creek, K. E., Silverman-Jones, C. S., and De Luca, L. M. (1989). *J. Invest. Dermatol.* **92**, 283–289.
Crow, J. A., and Ong, D. E. (1985). *Proc. Natl. Acad. Sci. U.S.A.* **82**, 4707–4711.
Croze, E. M., and Morré, D. J. (1984). *J. Cell. Physiol.* **114**, 46–57.
Cummings, R. D., and Mattox, S. A. (1988). *J. Biol. Chem.* **263**, 511–519.
Daniels, C., Noy, N., and Zakim, D. (1985). *Biochemistry* **24**, 3286–3292.
Davies, P. J. A., Murtaugh, M. P., Moore, W. T., Johnson, G. S., and Lucas, D. (1985). *J. Biol. Chem.* **260**, 5166–5174.
de Duve, C., Wattiaux, R., and Wibo, M. (1962). *Biochem. Pharmacol.* **9**, 97–116.
De Luca, L. M., and Yuspa, S. H. (1974). *Exp. Cell Res.* **86**, 106–110.
De Luca, L. M., Schumacher, M., and Wolf, G. (1970). *J. Biol. Chem.* **17**, 4551–4558.
De Luca, L. M., Schumacher, M., and Nelson, D. P. (1971). *J. Biol. Chem.* **246**, 5762–5765.
De Luca, L. M., Silverman-Jones, C. S., and Barr, R. M. (1975). *Biochim. Biophys. Acta* **409**, 342–359.
De Luca, L. M., Brugh, M. S., Silverman-Jones, C. S., and Shidoji, Y. (1982). *Biochem. J.* **208**, 159–170.
De Luca, L. M., Silverman-Jones, C. S., Rimoldi, D., Creek, K. E., and Warren, C. D. (1987). *Chem. Scr.* **27**, 193–198.
Demmer, L. A., Birkenmeier, E. H., Sweetser, D. A., Levin, M. S., Zollman, S., Sparkes, R. S., Mohandas, T., Lusis, A. J., and Gordon, J. I. (1987). *J. Biol. Chem.* **262**, 2458–2467.
Dhouailly, D., and Hardy, M. H. (1978). *Wilhelm Roux Arch. Dev. Biol.* **185**, 195–200.
Dhouailly, D., Hardy, M. H., and Sengel, P. (1980). *J. Morphol.* **58**, 63–78.

DiGiovanna, J. J., Helfgott, R. K., Gerber, L. H., and Peck, G. L. (1986). *N. Engl. J. Med.* **315**, 1177–1182.

Dingle, J. T. (1961). *Biochem. J.* **79**, 509–512.

Dingle, J. T., and Lucy, J. A. (1962). *Biochem. J.* **84**, 611–621.

Dingle, J. T., Lucy, J. A., and Fell, H. B. (1961). *Biochem. J.* **79**, 497–500.

Dingle, J. T., Sharman, I. M., and Moore, T. (1966). *Biochem. J.* **98**, 476–484.

Dion, L. D., De Luca, L. M., and Colburn, N. H. (1981). *Carcinogenesis (London)* **2**, 951–958.

DiSimone, D. P., and Reddi, A. H. (1981). *J. Cell Biol.* **91**, 147a.

Dogra, S. C., Khanduja, K. L., Gupta, M. P., and Sharma, R. R. (1983). *Acta Vitaminol. Enzymol.* **5**, 47–52.

Douer, D., and Koeffler, H. P. (1982). *J. Clin. Invest.* **69**, 277–283.

Eckert, R. L., and Green, H. (1984). *Proc. Natl. Acad. Sci. U.S.A.* **81**, 4321–4325.

Edmondson, S. W., Wu, R., and Mossman, B. T. (1990). *J. Cell. Physiol.* **142**, 21–30.

Eichele, G. (1989). *Trends Genet.* **5**, 246–251.

Eriksson, U., Sundelin, J., Rask, L., and Peterson, P. A. (1981). *FEBS Lett.* **135**, 70–72.

Evans, R. M. (1988). *Science* **240**, 889–895.

Evers, D. C., Anderson, J. N., and Morré, D. J. (1984). *Eur. J. Cell Biol.* **35**, 81–89.

Farrington, K., Miller, P., Varghese, Z., Baillod, R. A., and Moorehead, J. F. (1981). *Br. Med. J.* **282**, 1999–2002.

Fell, H. B., and Mellanby, E. (1952). *J. Physiol. (London)* **116**, 320–349.

Fex, G., and Johannesson, G. (1987). *Biochim. Biophys. Acta* **901**, 255–264.

Fex, G., and Johannesson, G. (1988). *Biochim. Biophys. Acta* **944**, 249–255.

Finlay, J. A., Strom, M., Ong, D. E., and DeLuca, H. F. (1990). *Biochemistry* **29**, 4914–4921.

Forsum, U., Rask, L., Tjernlund, U. M., and Peterson, P. A. (1977). *Arch. Dermatol. Res.* **258**, 85–88.

Frame, B., Jackson, C., Reynolds, W., and Umphrey, J. (1974). *Ann. Intern. Med.* **80**, 44–48.

Fuchs, E., and Green, H. (1980). *Cell (Cambridge, Mass.)* **19**, 1033–1042.

Fuchs, E., and Green, H. (1981). *Cell (Cambridge, Mass.)* **25**, 617–625.

Fugate, R. D., and Song, P. S. (1980). *Biochim. Biophys Acta* **625**, 28–42.

Fugler, L., Clejan, S., and Bittman, R. (1985). *J. Biol. Chem.* **260**, 4098–4102.

Fujimaki, Y. J. (1926). *Cancer Res.* **10**, 469–477.

Futterman, S., and Heller, J. (1972). *J. Biol. Chem.* **247**, 5168–5172.

Futterman, S., Saari, J. C., and Blair, S. J. (1977). *J. Biol. Chem.* **252**, 3267–3271.

Ganguly, J., Rao, M. R. S., Murthy, S. K., and Sarada, K. (1980). *Vitam. Horm. (N.Y.)* **38**, 1–54.

Geelen, J. A. G. (1979). *CRC Crit. Rev. Toxicol.* **6**, 351–375.

Giguere, V., Ong, E. S., Segui, P., and Evans, R. M. (1987). *Nature (London)* **330**, 624–629.

Gilfix, B. M., and Eckert, R. L. (1985). *J. Biol. Chem.* **260**, 14026–14029.

Glass, C. K., Lipkin, S. M., Devary, O. V., and Rosenfeld, M. G. (1989). *Cell (Cambridge, Mass.)* **59**, 697–708.

Glauert, A. M., Daniel, M., Lucy, J. A., and Dingle, J. T. (1963). *J. Cell Biol.* **17**, 111–121.

Godovac-Zimmermann, J., Conti, A., Liberatori, J., and Braunitzer, G. (1985). *Biol. Chem. Hoppe-Seyler* **366**, 431–434.

Golub, E. S., Pagan, T., Sun, I., and Crane, F. L. (1988). *In* "Plasma Membrane Oxidoreductases in Control of Animal and Plant Growth" (F. L. Crane, D. J. Morré, and H. Löw, eds.), pp. 313–321. Plenum, New York.

Goodman, D. S. (1974). *Vitam. Horm. (N.Y.)* **32**, 167–180.

Goodman, D. S. (1980). *Ann. N. Y. Acad. Sci.* **359**, 69–78.

Goodman, D. S. (1986). *Harvey Lect.* **81**, 111–132.

Goodman, D. S., and Blaner, W. S. (1984). *In* "The Retinoids" (M. B. Sporn, A. B. Roberts, and D. S. Goodman, eds.), Vol. 2, pp. 1–39. Academic Press, Orlanda, Florida.

Gorgacz, E. J., Nielsen, S. W., Frier, H. I., Eaton, H. D., and Rousseau, J. E., Jr. (1975). *Am. J. Vet. Res.* **36**, 171–180.

Green, H., and Watt, F. M. (1982). *Mol. Cell. Biol.* **2**, 1114–1117.

Haars, L. J., and Pitot, H. C. (1979). *J. Biol. Chem.* **254**, 9401–9407.

Hamilton, J. A., and Cristola, D. P. (1986). *Proc. Natl. Acad. Sci. U.S.A.* **83**, 82–86.

Hardy, M. H., and Bellows, C. G. (1978). *J. Invest. Dermatol.* **71**, 236–241.

Harrison, E. H., and Gad, M. (1989). *J. Biol. Chem.* **264**, 17142–17147.

Harrison, E. H., Smith, J. E., and Goodman, D. S. (1980). *Biochim. Biophys. Acta* **628**, 489–497.

Hashimoto, Y., Kagechika, H., Kawachi, E., and Shudo, K. (1987). *Chem. Pharm. Bull.* **35**, 3190–3194.

Hassell, J. R., Silverman-Jones, C. S., and De Luca, L. M. (1978). *J. Biol. Chem.* **253**, 1627–1631.

Hayes, K. C., Combs, H. L., and Faherty, T. B. (1970). *Lab. Invest.* **22**, 81–89.

Heller, J. J. (1975). *J. Biol. Chem.* **250**, 3613–3619.

Hemmi, H., and Breitman, T. R. (1982). *Biochem. Biophys. Res. Commun.* **109**, 669–674.

Hodam, J. R., St. Hilaire, P., and Creek, K. E. (1990). *Proc. Am. Assoc. Cancer Res.* **31**, A1967.

Hudson, L. G., Santon, J. B., Glass, C. K., and Gill, G. N. (1990). *Proc. Am. Assoc. Cancer Res.* **31**, A2095.

Jetten, A. M. (1980). *Nature (London)* **284**, 626–629.

Jetten, A. M. (1985). *In* "Growth and Maturation Factors" (G. Guroff, ed.), Vol. 3, p. 221. Wiley (Interscience), New York.

Jetten, A. M. (1987). *Dermatologica* **175**, Suppl. 1, 37–44.

Jetten, A. M., and De Luca, L. M. (1985). *Carcinogenesis (London)* **6**, 337–342.

Jetten, A. M., and Shirley, J. E. (1986). *J. Biol. Chem.* **261**, 15097–15101.

Jetten, A. M., Jetten, M. E. R., Shapiro, S. S., and Poon, J. P. (1979). *Exp. Cell Res.* **119**, 189–199.

Jetten, A. M., Fitzgerald, D. J., and Nettesheim, P. (1986). *In* "Nutritional Diseases in Comparative Pathobiology" (G. Migaki, ed.), p. 33. Alan R. Liss, New York.

Jetten, A. M., Anderson, K., Deas, M. A., Kagechita, D. H., Lotan, R., Rearick, J. J., and Shudo, K. (1987). *Cancer Res.* **47**, 3523–3527.

Kagechika, H., Kawachi, E., Hasimoto, Y., and Shudo, K. (1984). *Chem. Pharm. Bull.* **32**, 4209–4212.

Kanai, M., Raz, A., and Goodman, D. S. (1968). *J. Clin. Invest.* **47**, 2025–2044.

Kato, M., Blaner, W. S., Mertz, J. R., Das, K., Kato, K., and Goodman, D. S. (1985). *J. Biol. Chem.* **260**, 4832–4838.

Kawamura, H., and Hashimoto, Y. (1980). *Gann* **71**, 501–506.

Keiser, H., Weissmann, G., and Bernheimer, A. W. (1964). *J. Cell Biol.* **22**, 101–113.

Kim, K. C., Rearick, J. I., Nettesheim, P., and Jetten, A. M. (1985). *J. Biol. Chem.* **260**, 4021–4027.

Kiorpes, T. C., Molica, S. J., and Wolf, G. (1976). *J. Nutr.* **106**, 1659–1667.

Krust, A., Kastner, P., Petkovich, M., Zelent, A., and Chambon, P. (1989). *Proc. Natl. Acat. Sci. U.S.A.* **86**, 5310–5314.

Ladoux A., Cragoe, E. J., Jr., Geny, B., Abita, J. P., and Frelin, C. (1987). *J. Biol. Chem.* **262**, 811–816.

Lan, N. C., Karin, M., Nguyen, T., Weiz, A., Birnbaum, M. J., Eberhardt, N. L., and Baxter, J. D. (1984). *J. Steroid Biochem.* **20**, 77–88.

Ledger, P. W., Reano, A., Bonnefoy, J. Y., and Trivolet, J. (1984). *Br. J. Dermatol.* **3**, 38–42.

Levin, M. S., Lock, B., Yang, N.-C. C., Li, E., and Gordon, J. I. (1988). *J. Biol. Chem.* **263**, 17715–17723.

Lewis, C. A., Pratt, R. M., Pennypacker, J. P., and Hassell, J. R. (1978). *Development (Cambridge, UK), Suppl.*, p. 31–47.

Li, E., Demmer, L. A., Sweetser, D. A., Ong, D. E., and Gordon, J. I. (1986). *Proc. Natl. Acad. Sci. U.S.A.* **83**, 5779–5783.

Liau, G., Ong, D. E., and Chytil, F. (1981). *J. Cell Biol.* **91**, 63–68.

Liau, G., Ong, D. E., and Chytil, F. (1985). *Arch. Biochem. Biophys.* **237**, 354–360.

Libby, P. R., and Bertram, J. S. (1982). *Carcinogenesis (London)* **3**, 481–484.

Lichiti, F. U., and Lucy, T. A. (1969). *Biochem. J.* **112**, 221–230.

Lichti, U., Ben, T., and Yuspa, S. H. (1985). *J. Biol. Chem.* **260**, 1422–1426.

Lipkin, G., Knescht, M. E., and Rosenberg, M. (1978). *Science* **38**, 635–643.

Lotan, R. (1980). *Biochim. Biophys. Acta* **605**, 33–91.

Lotan, R., and Nicolson, G. L. (1977). *J. Natl. Cancer Inst. (U.S.)* **59**, 1717–1722.

Lucy, J. A., and Dingle, J. T. (1964). *Nature (London)* **204**, 156–160.

Lucy, J. A., Luscombe, M., and Dingle, J. T. (1963). *Biochem. J.* **89**, 419–425.

MacDonald, P. N., and Ong, D. E. (1988a). *Biochem. Biophys. Res. Commun.* **156**, 157–163.

MacDonald, P. N., and Ong, D. E. (1988b). *J. Biol. Chem.* **263**, 12478–12482.

Maden, M. (1982). *Nature (London)* **295**, 672–675.

Maden, M. (1983). *J. Embryol. Exp. Morphol.* **77**, 273–295.

Maden, M., Ong, D. E., Summerbell, D., and Chytil, F. (1989). *Development (Cambridge, UK), Suppl.*, p. 109–119.

Mallia, A. K., Smith, J. E., and Goodman, D. S. (1975). *J. Lipid Res.* **16**, 180–188.

Manino, R. J., Ballmer, K., and Burger, M. M. (1978). *Science* **201**, 824–826.

Maraini, G., and Gozzoli, F. (1975). *Invest. Ophthamol.* **14**, 785–787.

Martin, B. R., and Morré, D. M. (1988). *J. Nutr.* **118**, 968–975.

McGuire, B. W., Orgebin-Crist, M. C., and Chytil, F. (1981). *Endocrinology (Baltimore)* **108**, 658–667.

McLean, L. R., and Phillips, M. C. (1981). *Biochemistry* **20**, 2893–2900.

Mellanby, D. (1947). *J. Physiol. (London)* **105**, 382–399.

Mock, D., and Main, J. H. P. (1978). *J. Dent. Res.* **58**, 635–637.

Moore, T. (1967). *In* "The Vitamins" (W. H. Sebrell and R. S. Harris, eds.), pp. 245–266. Academic Press, New York.

Moore, T., and Sharman, I. M. (1978). *Int. J. Vitam. Nutr. Res.* **49**, 14–20.

Mori, S. J. (1922). *JAMA, J. Am. Med. Assoc.* **79**, 197–200.

Morré, D. J., and Morré, D. M. (1987). *Cell Biol. Int. Rep.* **11**, 89–93.

Morré, D. J., Mollenhauer, H. H., and Bracker, C. E. (1971). *In* "Results and Problems in Cell Differentiation: Origin and Continuity of Cell Organelles" (J. Reinert and H. Ursprung, eds.), Vol. 2, pp. 82–126. Springer-Verlag, Berlin.

Morré, D. J., Creek, K. E., Morré, D. M., and Richardson, C. L. (1981). *Ann. N. Y. Acad. Sci.* **359**, 367–382.

Morré, D. J., Paulik, M., and Nowack, D. (1986). *Protoplasma* **132**, 110–113.

Morré, D. J., Crane, F. L., Eriksson, L. C., Löw, H., and Morré, D. M. (1991). *Biochim. Biophys. Acta* **1057**, 140–146.

Morré, D. M. (1988). *In* "Biomembranes. Basic and Medical Research" (G. Benga and J. M. Tager, eds.), pp. 81–97. Springer-Verlag, New York.

Morré, D. M., Morré, D. J., and Walter, M. (1981). *Eur. J. Cell Biol.* **25**, 28–35.

Morré, D. M., Morré, D. J., Reutter, W., and Mollenhauer, H. H. (1985). *J. Cell Biol.* **101**, 60a.

Morré, D. M., Morré, D. J., Bowen, S., Reutter, W. W., and Windel, K. (1988). *Eur. J. Cell Biol.* **46**, 306–315.

Morré, D. M., Spring, H., Trendlenburg, M., Montag, M., Mollenhauer, B. A., Mollenhauer, H. H., Morré, D. J. (1991). *J. Nutr.* (submitted for publication).

Muramatsu, H., and Muramatsu, T. (1983). *FEBS Lett.* **163,** 181–184.

Murray, T., and Russell, T. R. (1980). *J. Supramol. Struct.* **14,** 255–266.

Nervi, C., Grippo, J. F., Sherman, M. I., George, M. D., and Jetten, A. M. (1989). *Proc. Natl. Acad. Sci. U.S.A.* **86,** 5854–5858.

Newcomer, M. E., Jones, T. A., Aqvist, J., Sundelin, J., Eriksson, U., Rask, I., and Peterson, P. A. (1984). *EBMO J.* **3,** 1451–1454.

Niazi, I. A., and Saxena, S. (1978). *Folia Biol. (Krakow)* **26,** 3–8.

Nishiwaki, S., Kato, M., Okuno, M., Kanai, M., and Muto, Y. (1990). *Biochim. Biophys. Acta* **1037,** 192–199.

Nowack, D. D., Paulik, M., Morré, D. J., and Morré, D. M. (1990). *Biochim. Biophys. Acta* **1051,** 250–258.

Noy, N. (1988). *Biophys. J.* **53,** 7a.

Noy, N., and Xu, Z.-J. (1990). *Biochemistry* **29,** 3878–3883.

Noy, N., Donnelly, T. M., and Zakim, D. (1986). *Biochem. J.* **25,** 2013–2021.

Nyquist, S., Crane, F. L., and Morré, D. J. (1971). *Science* **173,** 939–941.

Ogiso, Y., Kitagawa, K., Nishino, H., Iwashima, A., and Shudo, K. (1987). *Exp. Cell Res.* **173,** 262–267.

Olson, J. A. (1983). *Semin. Oncol.* **10,** 290–293.

Ong, D. E., and Chytil, F. (1976). *Proc. Natl. Acad. Sci. U.S.A.* **73,** 3976–3978.

Ong, D. E., Kakkad, B., and MacDonald, P. N. (1984). *J. Biol. Chem.* **259,** 1476–1482.

Ong, D. E., Kakkad, B., and MacDonald, P. N. (1987). *J. Biol. Chem.* **262,** 2729–2736.

Ong, D. E., MacDonald, P. N., and Gubitosi, A. M. (1988). *J. Biol. Chem.* **263,** 5789–5796.

Ottonello, S., and Maraini, G. (1981). *Exp. Eye Res.* **32,** 69–75.

Page, D. L., and Ong, D. E. (1987). *J. Lipid Res.* **28,** 739–745.

Papiz, M. Z., Sawyer, L., Eliopulos, E. E., North, A. C. T., Findlay, J. B. C., Sivaprasadarao, R., Jones, T. A., Newcomer, M. E., and Kraulis, P. J. (1986). *Nature (London)* **324,** 383–385.

Pawson, B. A., Ehmann, C. W., Itri, L. M., and Sherman, M. I. (1982). *J. Med. Chem.* **25,** 1269–1277.

Petkovich, M., Brand, N. J., Krust, A., and Chambon, P. (1987). *Nature (London)* **330,** 444–450.

Pfeffer, B. A., Clark, V. M., Flannery, J. G., and Bok, D. (1986). *Invest. Ophthalmol. Visual Sci.* **27,** 1031–1040.

Pittsley, R. A., and Yoder, F. W. (1983). *N. Engl. J. Med.* **308,** 1012–1014.

Pokrovsky, A. A., Lasheneva, V. V., and Kon, I. Y. (1974). *Int. J. Nutr. Vitam. Res.* **44,** 477–486.

Prutkin, L. (1975). *Cancer Res.* **35,** 364–369.

Ragavan, V. V., Smith, J. E., and Belezikian, J. P. (1982). *Am. J. Med. Sci.* **283,** 161–164.

Rajan, N., Blaner, W. S., Soprano, D. R., Suhara, A., and Goodman, D. S. (1990). *J. Lipid Res.* **31,** 321–329.

Rando, R. R., and Bangerter, F. W. (1982). *Biochem. Biophys. Res. Commun.* **104,** 430–436.

Rask, L., and Peterson, P. A. (1976). *J. Biol. Chem.* **251,** 6360–6366.

Rask, L., Valtersson, C., Anundi, H., Kvist, S., Eriksson, U., Dallner, G., and Peterson, P. A. (1983). *Exp. Cell Res.* **143,** 91–102.

Rearick, J. I., Albro, P. W., and Jetten, A. M. (1987). *J. Biol. Chem.* **262,** 13069–13074.

Rimoldi, D., Creek, K. E., and De Luca, L. M. (1990). *Mol. Cell. Biochem.* **93,** 129–140.

Roberts, A. B., and Sporn, M. B. (1984). *In* "The Retinoids" (M. B. Sporn, A. L. B. Roberts, and D. S. Goodman, eds.), Vol. 2. Academic Press, Orlando, Florida.

Rosso, G. C., Bendrick, C. J., and Wolf, G. (1981). *J. Biol. Chem.* **256,** 8341–8347.

Saari, J. C., and Bredberg, D. L. (1987). *J. Biol. Chem.* **262,** 7618–7622.
Saari, J. C., Bredberg, L., and Futterman, S. (1980). *Invest. Ophthamol.* **19,** 1301–1308.
Saari, J. C., Bredberg, L., and Garwin, G. G. (1982). *J. Biol. Chem.* **257,** 13329–13333.
Said, H. M., Ong, D. E., and Shingleton, J. L. (1989). *Am. J. Clin. Nutr.* **49,** 690–694.
Sani, B. P. (1979). *Biochem. Biophys. Res. Commun.* **75,** 7–12.
Sani, B. P., and Corbett, T. H. (1977). *Cancer Res.* **37,** 209–213.
Sanjecv, P. K., Vasudevan, D. M., and Jayanthy-Bay, N. (1981). *Acta Vitaminol. Enzymol.* **4,** 214–218.
Sasak, W., De Luca, L. M., Dion, L. D., and Silverman-Jones, C. S. (1980). *Cancer Res.* **40,** 1944–1949.
Sawyer, L., Papiz, M. Z., North, A. C. T., and Eliopoulos, S. E. (1985). *Biochem. Soc. Trans.* **13,** 265–266.
Schindler, J., Matthaei, K. I., and Sherman, M. I. (1981). *Proc. Natl. Acad. Sci. U.S.A.* **78,** 1077–1080.
Shankar, S., Creek, K. E., and De Luca, L. M. (1990). *J. Nutr.* **120,** 361–374.
Shenefelt, R. E. (1972). *Teratology* **5,** 103–118.
Sherman, M. I. (1986). *In* "Retinods and Cell Differentiation" (M. I. Sherman, ed.), p. 162. CRC Press, Boca Raton, Florida.
Shingleton, J. L., Skinner, M. K., and Ong, D. E. (1989). *Biochemistry* **28,** 9641–9647.
Siegenthaler, G., Saurat, J.-H., and Ponec, M. (1988). *Exp. Cell Res.* **178,** 114–126.
Siegenthaler, G., Saurat, J.-H., and Ponec, M. (1990). *Biochem. J.* **268,** 371–378.
Sivaprasadarao, A., and Findlay, J. B. C. (1988). *Biochem. J.* **255,** 571–579.
Smith, J. E., and Goodman, D. S. (1976). *N. Engl. J. Med.* **294,** 805.
Smith, J. E., and Goodman, D. S. (1979). *Fed. Proc., Fed. Am. Soc. Exp. Biol.* **38,** 2504–2509.
Smith, J. E., Deen, D. D., Sklan, D., and Goodman, D. W. (1980). *J. Lipid Res.* **21,** 229–237.
Sporn, M. B., Dunlop, N. M., and Ysupa, S. H. (1973). *Science* **182,** 722–723.
Sporn, M. B., Dunlop, N. M., Newton, D. I., and Smith, J. M. (1976). *Fed. Proc., Fed. Am. Soc. Exp. Biol.* **35,** 1332–1338.
Sporn, M. B., Roberts, A. B., and Goodman, D. S., eds. (1984). "The Retinoids," Vols. 1 and 2. Academic Press, Orlando, Florida.
Sporn, M. B., Roberts, A. B., Roche, N. S., Kagechika, H., and Shudo, K. (1986). *J. Am. Acad. Dermatol.* **15,** 756.
Spring, H., and Franke, W. W. (1981). *Eur. J. Cell Biol.* **24,** 298–308.
Stillwell, W., and Byrant, L. (1983). *Biochim. Biophys. Acta* **731,** 483–486.
Stillwell, W., Ricketts, M., Hudson, H., and Nahmias, S. (1982). *Biochim. Biophys. Acta* **688,** 653–659.
Storch, J., and Kleinfeld, A. M. (1986). *Biochemistry* **25,** 1717–1726.
Strickland, J. E., Jetten, A. M., Kawamura, H., and Yuspa, S. H. (1984). *Carcinogenesis (London)* **5,** 735–740.
Stubbs, G. W., Saari, J. C., and Futterman, S. (1979). *J. Biol. Chem.* **254,** 8529–8533.
Summerbell, D. (1983). *J. Embryol. Exp. Morphol.* **78,** 269–289.
Sun, I. L., and Crane, F. L. (1984). *Biochem. Int.* **9,** 299–306.
Sun, I. L., and Crane, F. L. (1985). *Biochem. Pharmacol.* **34,** 617–623.
Sun, I. L., Crane, F. L., Chou, J. Y., Löw, H., and Grebing, G. (1983). *Biochem. Biophys. Res. Commun.* **116,** 210–216.
Sun, I. L., Crane, F. L., Löw, H., and Grebing, C. (1984). *J. Bioenerg. Biomembr.* **16,** 209–221.
Sun, I. L., Crane, F. L., Grebing, C., and Löw, H. (1985). *Exp. Cell Res.* **156,** 528–536.
Sun, I. L., Garcia-Canero, R., Liu, W., Toole-Simms, W., Crane, F. L., Morré, D. J., and Löw, H. (1987). *Biochem. Biophys. Res. Commun.* **145,** 467–473.

Takase, S., Ong, D. E., and Chytil, F. (1979). *Proc. Natl. Acad. Sci. U.S.A.* **76,** 2204–2208.
Takigawa, M., Ishida, H., Takano, T., and Suzuki, F. (1980). *Proc. Natl. Acad. Sci. U.S.A.* **77,** 1481–1485.
Thaller, C., and Eichele, G. (1987). *Nature (London)* **327,** 625–628.
Thaller, C., and Eichele, G. (1990). *Nature (London)* **345,** 815–819.
Thoms, S. D., and Stocum, D. L. (1984). *Dev. Biol.* **103,** 319–328.
Tickle, C., Alberts, B. M., Walpert, L., and Lee, J. (1982). *Nature (London)* **296,** 564–565.
Tickle, C., Lee, J., and Eichele, G. (1985). *Dev. Biol.* **109,** 72–95.
Torma, H., and Vahlquist, A. (1983). *Arch. Dermatol. Res.* **275,** 324–328.
Vahlquist, A., Torma, H., Rollman, O., and Berne, B. (1985). *In* "Retinoids: New Trends in Research and Therapy" (J. H. Sarat, ed.), pp. 159, 167. Karger, Basel.
Verma, A., Rice, H., Shapas, B., and Boutwell, R. K. (1978). *Cancer Res.* **38,** 793–801.
Verma, A., Shapas, B. G., Rice, H. M., and Boutwell, R. K. (1979). *Cancer Res.* **38,** 419–425.
Wake, K. (1980). *Int. Rev. Cytol.* **66,** 303–353.
Westin, E. H., Wong-Staal, F., Gelman, E. P., Dalla Favera, R., Papas, T., Lautenberg, J. A., Eva, A., Reddy, P. E., Tronick, S. R., Aaronson, S. A., and Gallo, R. C. (1982). *Proc. Natl. Acad. Sci. U.S.A.* **79,** 2490–2494.
Wiggert, B., Russell, P., Lewis, M., and Chader, G. (1977). *Biochem. Biophys. Res. Commun.* **79,** 218–225.
Wolbach, S. B. (1947). *J. Bone Jt. Surg.* **29,** 171–192.
Wolbach, S. B., and Howe, P. R. (1925). *J. Exp. Med.* **47,** 753–778.
Wolf, G. (1980). *In* "Nutrition and the Adult. Micronutrients" (R. B. Alfin-Slater and D. Krischevsky, eds.), p. 97. Plenum, New York.
Yaar, M., Stanley, J. R., and Katz, S. I. (1981). *J. Invest. Dermatol.* **76,** 363–366.
Yen, A., Reece, S. L., and Albright, K. L. (1984). *Exp. Cell Res.* **152,** 493–499.
Yuspa, S. H., and Harris, C. C. (1974). *Exp. Cell Res.* **86,** 95–105.
Yuspa, S. H., Lichti, U., Ben, T., and Hennings, H. (1981). *Ann. N. Y. Acad. Sci.* **359,** 260–274.
Yuspa, S. H., Ben, T., and Steinert, P. (1982). *J. Biol. Chem.* **257,** 9906–9908.
Yuspa, S. H., Ben, T., and Lichti, U. (1983). *Cancer Res.* **43,** 5707–5712.
Zelent, A., Krust, A., Petkovich, M., Kastner, P., and Chambon, P. (1989). *Nature (London)* **339,** 714–717.
Zhao, J., Morré, D. J., Paulik, M., Yim, J., and Morré, D. M. (1990). *Biochim. Biophys. Acta* **1055,** 230–233.
Ziegel, R. F., and Dalton, A. J. (1962). *J. Cell Biol.* **15,** 45–54.

Cell Biology of the Subcommissural Organ**

Esteban M. Rodríguez,* Andreas Oksche,† Silvia Hein,* and Carlos R. Yulis*

* Instituto de Histología y Patología, Universidad Austral de Chile, Valdivia, Chile

† Department of Anatomy and Cytobiology, Justus Liebig University of Giessen, Giessen, Germany

I. Introduction

The subcommissural organ (SCO) is a complex of nonneuronal secretory cells covering and penetrating the posterior commissure (Figs. 1, 4, 6–9). This complex protrudes toward the third ventricle, occupies the posterior portion of the diencephalic roof caudal to the pineal organ, and marks the entrance to the Sylvian aqueduct. In this area a clear-cut borderline between the diencephalon (epithalamus) and the mesencephalon (pretectal area) can hardly be drawn. Due to its topographical position between the ventricular cavity and a well-developed circumscribed vascular system, the SCO qualifies as one of the circumventricular organs (Hofer, 1959; Leonhardt, 1980). Its secretory cells are specialized ependymal and ependyma-derived (hypendymal) elements of neuroepithelial origin. In principle, these cells belong to the class of neuroglia, irrespective of certain modifications and specializations.

The bulk of the secretory products of the SCO, apparently a complex containing different glycoproteins, is released into the ventricular cerebrospinal fluid (CSF). However, this secretory material also gains access to the vascular network of the SCO region and to the CSF-containing subarachnoid space or its equivalents in lower vertebrates. In the ventricle, the secretion is condensed into a threadlike structure, Reissner's fiber (RF), that terminates in the ampulla caudalis of the central canal (Figs. 2, 3, 5, 19, 20). At this point, the material of RF is assumed to undergo chemical changes and finally to reach local blood capillaries or sinuslike enlarged vessels.

** Dedicated to Professor Berta Scharrer on the occasion of her 85th birthday.

39

In evolutionary terms, the SCO is an ancient and persistent structure of the vertebrate brain. In comparison to the diverse systems of peptidergic neurons, which have become stepwise deciphered (cf. Scharrer, 1978, 1990), the function of the SCO is still an enigma. Concerning this secretory parenchyma, the knowledge of its ultrastructure, including the intracellular compartments of synthesis and release of the secretory material, had for a long time outpaced the insight into the chemical nature of its secretory product. This has now been largely changed due to the introduction of new methods and technologies. However, in spite of a remarkable progress in the field of analysis, the systemic role played by the SCO in the brain and in the entire organism still remains open to discussion. A principal aim of this review is to establish structure–function relationships in order to provide a sound basis for further experimental work and to augment the interest in a mysterious secretory structure of the brain.

The existence of the peculiar ependymal structure, called the "subcommissural organ" (Dendy and Nicholls, 1910), has been known for longer than a century. In 1860, Reissner observed within the central canal of *Petromyzon fluviatilis* a peculiar fibrous structure (Fig. 2). In 1900, Studnička described uncommonly tall ependymal cells covering the posterior commissure (Fig. 1); he emphasized the highly specialized character of this ependyma. This observation was confirmed by Dendy (1902). Sargent (1904) postulated an interrelationship between SCO and RF, and regarded RF as a SCO-dependent structure.

Concerning the early period of SCO research, a comprehensive account was given by Bargmann (1943), in close connection with an analysis of the adjacent pineal organ. With reference to this classical review and the historical remarks integrated into Sections VI and XIII, this introductory historical outline focuses only on a few crucial developments.

After decades of extended descriptive work using conventional histological methods, a strong impetus for SCO research arose from a discovery of Stutinsky (1950), who showed that the secretion of the SCO can be selectively stained with Gomori's chrome–alum hematoxylin. For detailed references on the demonstration of the secretory material of the SCO, see Leonhardt (1980). For an account on enzyme-histochemical findings, see Ziegels (1976).

An experimental approach to the SCO was introduced by Olsson (1958a,b; cf. Leonhardt, 1980). Olsson succeeded in transecting RF within the spinal cord. He showed that the proximal fragment of RF displays growth in a rostrocaudal direction due to a continuous shift of secretory material released by the SCO; at the same time, the material located in the distal stump was consumed and disappeared (Section VI.)

In 1954, Oksche started a series of studies on the influence on the SCO of seasonal and environmental factors and several pharmacological agents

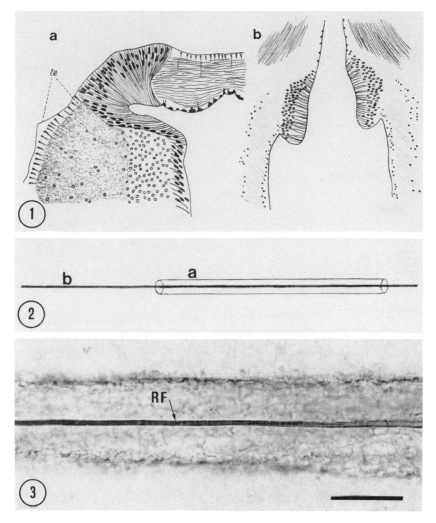

FIG. 1   (a) Ammocoetes. Area marginal to the posterior commissure (a part of the latter seen in b). Fixative: chromic acid; staining: picrocarmin. (b) *Petromyzon fluviatilis*. Area adjacent to the posterior commissure. Note fibers of the latter in the upper right- and left-hand corners. Fixative: Zenker's fluid; staining: iron hematoxylin. From Studnička (1900).

FIG. 2   Central canal of the spinal cord of *Petromyzon fluviatilis*. (a) Lumen; (b) intraluminal strand (fiber). From Reissner (1860).

FIG. 3   Reissner's fiber (RF) in the central canal of the rat spinal cord. Visualized after injection of peroxidase into the lateral ventricle. DAB-H_2O_2 reaction; vibratome section. Bar, 20 μm. From Cifuentes (1991).

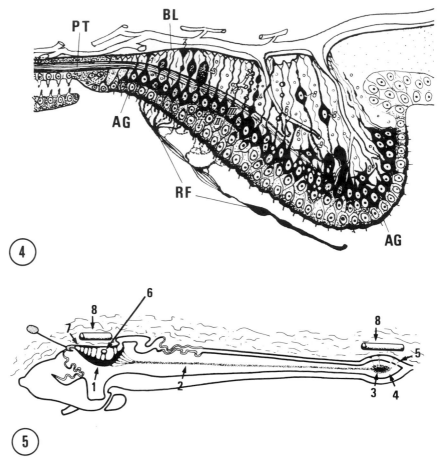

FIG. 4 Schematic representation of a paramedian sagittal section of the SCO in anuran amphibians. The columnar ependymal cells are arranged either in a simple or in a pseudo-stratified fashion. I, Ventricular pole of ependymal cells exhibiting a dense layer of apical granules (AG), threads of pre-RF material, and Reissner's fiber (RF); II, central portion of ependymal cells poorer in secretory material; III, basal pole of ependymal cells displaying long processes with their end feet adjacent to the outer basal lamina (BL). Also marked are ependymal processes showing homogeneous accumulation of secretion (1), local swellings (2), saclike accumulations of secretory granules (3), beadlike or varicose images (4), or vacuoles (5). Note vascular contacts (6) and end feet containing secretory material (7) and rare extraependymal bipolar (8) elements, representing an early manifestation of the hypendymal formation. PT, Pineal tract; K, capillary. From Oksche (1961).

FIG. 5 The SCO–RF complex of some anurans. 1, Subcommissural organ; 2, Reissner's fiber; 3, massa caudalis; 4, ampulla caudalis; 5, Reissner's fiber material escaping through dorsal wall of ampulla caudalis; 6, vascular projections of secretory cells; 7, leptomeningeal projections of secretory cells; 8, meningeal capillaries. From Oksche (1969).

(Oksche, 1962). In this connection, he gave reference to earlier work by E. Legait (1942). Oksche (1961, 1969) drew attention to possible vascular and leptomeningeal routes of the secretory material of the SCO (Figs. 4–6). In addition, he widely used histochemical, ultrastructural, and neurohistological methods in a representative number of vertebrate species (cf. Leonhardt, 1980).

In 1962, Sterba introduced a new fluorescent dye, pseudoisocyanin, for demonstration of the secretory material of the SCO. These reports were followed by numerous studies conducted by Sterba and associates (Sterba, 1977; cf. Leonhardt, 1980, and this treatise), including ultrastructural analysis, *in vitro* chemistry of RF, and immunocytochemistry.

For other achievements during the period between 1960 and 1980, including autoradiographic work (Diederen), see Leonhardt (1980). In 1982, Rodríguez and Oksche initiated their joint program on the SCO that has led to several new methodological and conceptual developments on which a considerable portion of this review is based (see E. M. Rodríguez *et al.*, 1984a,b, 1990a).

II. General Structure

The SCO is formed by two populations of secretory cells, which in many species are arranged into two distinct layers: the ependyma and the hypendyma (Krabbe, 1925, 1933; Bargmann and Schiebler, 1952; Talanti, 1958; Oksche, 1961). The degree of development of both cell populations varies greatly among the vertebrate phylum (Oksche, 1961) (Fig. 6).

The ependyma of the SCO does not undergo major evolutionary changes. The most relevant species differences are related to the arrangement of the ependymal cells into one or more layers and to the degree of development of their basal processes. The latter end on local blood vessels or, after traversing the posterior commissure, on the external basal lamina of the brain.

The term "hypendyma" was introduced by Krabbe in 1933 to depict a tissue layer located under the ependymal cells of the SCO and containing numerous blood vessels and loosely arranged parenchyma-like and neuroglial cells. Oksche (1961) characterized as "hypendymal cells" all those secretory cells of the SCO that are not located within the ependymal layer proper of this region.

Immunocytochemical studies have shown that in most vertebrate species hypendymal cells are much more abundant than originally thought (Sterba *et al.*, 1982; Rodríguez *et al.*, 1984a,b). The relative development of the hypendyma varies greatly among mammalian species, without following any apparent evolutionary trend (Rodríguez *et al.*, 1984a).

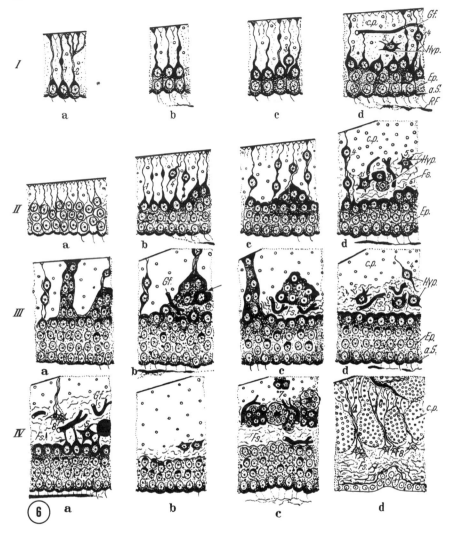

FIG. 6   Schematic comparative view of the general morphological pattern and differentiation of the subcommissural organ in different vertebrates. Secretory material is depicted as
black granules and masses. Gf., Blood vessels; Ep., ependymal cells of SCO; a. S., apical
aggregation of secretory granules; RF, Reissner's fiber. (I) Phylogenetically ancient pattern of secretory SCO cells in anurans. Note the rostrocaudal increase in the thickness
of the ventricular wall in a–d. (II) Hypendymal secretory cells (Hyp.) in the SCO of the
cat. (a) Young embryo; (b, c) newborn animal, rostral thin (b) and caudal thicker (c) portion of
the posterior commissure (*c. p.*); (d) adult cat. (III) Secretory hypendyma in the SCO of the
dog. (a) Bipolar secretory cells and wide epithelial columns leading to the outer limiting
membrane (basal lamina); (b) basal protuberance (arrow) of the ependymal formation;
(c) segregated, circumscribed hypendymal islet of SCO parenchyma (x); (d) reticular arrangement of secretory hypendymal cells. (IV) Variance in differentiation of basal ependymal and
hypendymal elements in the SCO. (a) Adult rhesus monkey. Short basal ependymal

The patterned spatial organization of the secretory hypendymal cells also varies among species. Thus, in some species, the hypendymal cells form clusters located along both sides of the SCO or at the cephalic and caudal ends of the organ (marsupials, rat, mouse); in others, they establish a distinct and continuous layer (bovine). In addition to the subependymal site of the hypendymal elements, in many species they are also located within the posterior commissure and in the vicinity of the external limiting membrane of the brain (Rodríguez et al., 1984a) (Figs. 6, 7).

Hypendymal cells may form rosette-like aggregates lining a small central cavity (Figs. 6,IV,c) (Olsson, 1958b; Talanti, 1958; Isomäki et al., 1965; Herrlinger, 1970; Ferraz de Carvalho and Prado Reis, 1977; Rodríguez et al., 1984a). Other hypendymal cells display long processes which end on the local blood vessels and the external limiting membrane of the brain (Figs. 4 and 6) (see Section V).

Pesonen (1940) was the first to report on the rich vascularization of the SCO. This was later confirmed by several authors (Bargmann and Schiebler, 1952; Wislocki and Leduc, 1952; Oksche, 1956). The only study specially designed to investigate the blood supply to the SCO was published by Duvernoy and Koritké (1964). They described in the SCO of several domestic mammals a dense hypendymal capillary network, with capillary extensions that actually penetrate the ependymal layer.

III. Ultrastructure

A. Ependymal Cells

The ependymal secretory cells of the SCO are tall cylindrical elements with the nucleus located in a basal position. The nucleus is rich in euchromatin and highly infolded (Stanka et al., 1964; Isomäki et al., 1965; Herrlinger, 1970). The indented nucleus of the secretory ependymal and hypendymal cells of the SCO allows one to distinguish them from conventional ependymal elements lining other ventricular regions.

For descriptive purposes, the cell body proper and the basal processes will be analyzed separately. In the cell body, a clear zonation of the cytoplasm can be recognized according to the distribution and density of the organelles. The following cytoplasmic areas are described below: perinuclear, intermediate, subapical, and apical. The basal processes of the ependymal cells and their endings reaching the blood vessels and

processes rich in secretory material; (b) adult mouse: low ependymal layer with a few adjacent hypendymal elements; (c) adult bovine: isletlike, tubular, and rosettelike formations of hypendymal cells; (d) adult man: isletlike remnant of the SCO (+). Fs., fibrous subependymal layer. From Oksche (1961).

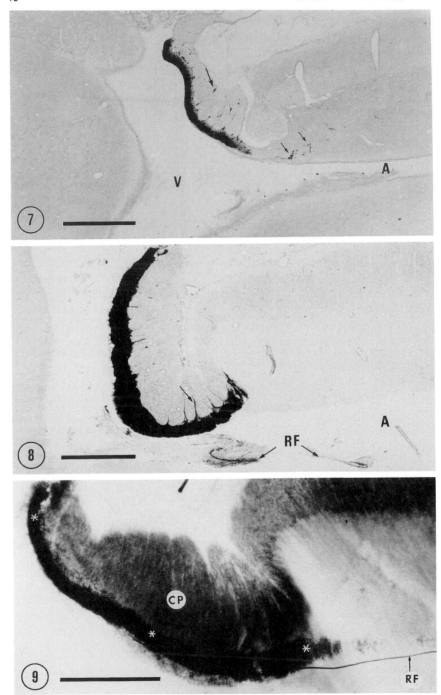

external brain surface also represent distinct cytoplasmic regions which will be described separately. The ependyma of the rat SCO will serve the basic description, with reference to certain species differences.

1. Perinuclear Region

The most striking structural feature of the secretory cells of the SCO in virtually all species is the presence of numerous and distended cisternae of the rough endoplasmic reticulum (RER), with a lumen ranging between 1 and 7 μm. In many species, the largest cisternae are concentrated in the perinuclear region and frequently they are in continuity with the nuclear envelope (Stanka et al., 1964; Herrlinger, 1970).

A distinct feature of the dilated RER cisternae of the SCO secretory cells is that only a circumscribed portion of their external surface displays attached ribosomes. A mitochondrion is frequently found at the site where ribosomes are attached to the RER membrane (Marcinkiewicz and Bouchaud, 1983). A rough estimate made from our electron micrographs of the rat SCO indicates that no more than 2 to 5% of the surface area of the RER cisternae appear studded with ribosomes. Consequently, 95% of the area enclosed by these RER cisternae is smooth-surfaced. The lumen of the RER cisternae is filled with a flocculent material of moderate electron density (Stanka et al., 1964; Müller and Sterba, 1965; Papacharalampous et al., 1968; cf. Oksche, 1969). Ultrastructural immunocytochemistry demonstrated that most of this material corresponds to secretory products (see Section IV,A,2).

Considering the above-described structural features of the RER and the general knowledge concerning protein synthesis at the cellular level (Alberts et al., 1983), two suggestions may be put forward: (1) In each individual cisterna the rate of synthesis is relatively low, although the cell as a whole, due to its excessive number of RER cisternae, may display a high synthetic activity. (2) At variance with most types of secretory cells, the RER may represent an important storage site of secretory products (see Section IV,B).

FIG. 7   Sagittal section through the rat epithalamus. Immunoperoxidase staining with anti-bovine RF serum. Indicated are hypendymal cells in posterior commissure (large arrow) and secretory cells in the roof of aqueduct (A) (small arrows); V, ventricle. Bar, 500 μm. From Rodríguez et al. (1984a).

FIG. 8   Sagittal section of the SCO of *Aotes*. Immunoperoxidase staining with anti-bovine RF serum. A, aqueduct; RF, Reissner's fiber; arrow, thin columns of immunoreactive cells in posterior commissure. Bar, 500 μm. From Rodríguez et al. (1984a).

FIG. 9   General view of the subcommissural organ of the cat (*). Cleared total preparation, stained with paraldehyde fuchsin. RF, Reissner's fiber; CP, posterior commissure. Bar, 500 μm. From Oksche (1969).

2. Intermediate Region

The cytoplasm extending from the nucleus to the ventricular surface is a tall column, which in the rat has an extension of 60 to 75 μm. The basal two-thirds of this column corresponds to the cytoplasmic region regarded as intermediate. RER, Golgi apparatus, and secretory granules are the three main structural components of this region. Although most of the RER in this region appears in the form of dilated cisternae, they are much smaller than those found in the perinuclear region (0.4 to 1.2 μm in diameter). Flattened cisternae link cisternal dilatations, giving the RER a truly reticular appearance.

The Golgi apparatus is well developed. It is an elongated structure located in the central region of the intermediate cytoplasmic region. It is formed by flattened saccules displaying numerous fenestrations and arranged in parallel arrays, as shown in a three-dimensional study using high-voltage electron microscopy (Lu and Lin, 1987). The convex aspect (*cis*-Golgi face) is lined by two to three dilated and fenestrated cisternae. Smooth-surfaced and coated vesicles are found between the *cis*-Golgi aspect and the numerous RER cisternae which surround the Golgi complex. Both types of vesicles were found to fuse with the membranes of the RER and the *cis*-Golgi cisternae. The concave aspect (*trans*-Golgi face) is lined by one to three flattened cisternae. Large vacuoles, condensing vacuoles, electron-dense material within discrete dilatations of the Golgi saccules, as usually found in most secretory cells, are missing in the SCO cells.

The involvement of the Golgi apparatus in the formation of the secretory granules, and consequently in the secretory process, has been a matter of considerable controversy. Based on electron density, many authors have described in the secretory cells of the SCO of several species the presence of two populations of secretory granules, i.e., pale and dense inclusions. Some authors postulated that both types of granules are derived directly from the RER, bypassing the Golgi apparatus (Stanka *et al.,* 1964; Oksche, 1969; Rodríguez, 1970a). Other authors, however, postulated a classical secretory process for the SCO secretory cells, whereby the initial synthesis of the secretory material occurs in the RER, followed by the addition of carbohydrates and packaging of the secretory products in the Golgi complex (Schmidt and D'Agostino, 1966; Müller and Sterba, 1965; Stanka, 1967; Diederen, 1970; Herrlinger, 1970). Still a third alternative has been postulated by Murakami and Tanizaki (1963) and Papacharalampous *et al.* (1968), according to which the dense secretory granules are formed in the Golgi apparatus and the pale secretory granules originate from the RER. This latter possibility gained a strong support from the studies by Chen *et al.* (1973), who applied ultrastructural cytochemical methods for the demonstration of carbohydrates and acid phosphatase. They concluded that

(1) the dense granules are lysosomes derived from the *trans*-Golgi cisternae, and (2) the pale granules represent secretory granules originating from the RER.

Lectin and immunochemical studies carried out during recent years have provided new valuable information concerning these crucial but controversial aspects. The problem of whether a secretory mechanism bypassing the Golgi apparatus may indeed operate within the SCO cells is discussed in detail in Section IV,B,1.

3. Subapical Region

This region is mainly occupied by microtubules, mitochondria, and smooth endoplasmic reticulum. Small spherical RER cisternae and dense secretory granules are rare. This region becomes conspicuous at the light-microscopic level when sections are immunostained using antibodies to reveal the secretory material. Indeed, this is the only cytoplasmic region displaying a very low or no immunoreactivity.

4. Apical Cell Pole

The ependymal cells project irregularly shaped protrusions into the ventricle (Stanka *et al.*, 1964; Papacharalampous *et al.*, 1968; Herrlinger, 1970). These protrusions can be clearly visualized by use of the scanning electron microscope (Krstić, 1975; Lindberg and Talanti, 1975; Weindl and Schinko, 1975; Collins and Woollam, 1979). Whereas the number of cilia of the ependymal cells varies greatly among species (Leonhardt, 1980), the numerous microvilli belong to the consistent features of these cells.

The ependymal protrusions are mainly occupied by microtubules (frequently arranged in bundles), granules of variable electron density, and dilated RER cisternae. In the rat, the secretory granules are spherical and 0.4 μm in diameter. Most of them are of moderate electron density and correspond to typical secretory granules, as has been shown by means of immunocytochemistry (Fig. 22) (Rodríguez *et al.*, 1986, 1987b); the few granules displaying a high electron density might correspond to lysosomes (Chen *et al.*, 1973). Exocytotic images of secretory granules, with flocculent material being released into the ventricle, have been observed (Stanka *et al.*, 1964; Hofer, 1986).

A most peculiar, although scarce, component of the ventricular protrusions is represented by dilated cisternae of the RER. The latter may display a few attached ribosomes and contain secretory material that can be demonstrated by immunocytochemical methods (Rodríguez *et al.*, 1987b). Although large secretory structures undergoing exocytosis have been shown (Hofer, 1986), it has not been established whether they actually correspond to RER cisternae. This is a crucial point considering the hypothesis that in the SCO a secretory mechanism bypassing the Golgi apparatus may exist (Lösecke *et al.*, 1984).

The number, size, shape, and inner structure of the secretory granules located at the ventricular cell pole exhibit considerable species variations. Such differences have been described in detail in the lamprey (Müller and Sterba, 1965; Sterba *et al.*, 1967a), toad (Murakami and Tanizaki, 1963; Rodríguez, 1970a; Wakahara, 1974), frog (Oksche, 1969; Diederen, 1970), mouse (Herrlinger, 1970; Chen *et al.*, 1973), rat (Stanka *et al.*, 1964; Collins and Woollam, 1979), guinea pig (Papacharalampous *et al.*, 1968), dog (Oksche, 1969), monkey (Hofer *et al.*, 1980; Hofer, 1986), and human embryos (Oksche, 1969).

The distinct zonation of the cytoplasm of the ependymal cells of the rat SCO that is clearly visible under the electron microscope can also be demonstrated at the light-microscopic level, especially with the use of lectin histochemistry and immunocytochemistry. This type of zonation has facilitated the investigation of the secretory process. As is shown in the following section, distinct phases of the secretory mechanism occur in discrete areas of the cell, i.e., synthesis in the perinuclear and intermediate regions, storage of precursor forms in the perinuclear area, processing and packaging in the intermediate portion, transport in the subapical zone, storage of processed forms, and ventricular release in the apical region.

A remarkable and unique feature of the SCO is that its secretory material, on release, condenses first as a film covering the ventricular surface of the organ, and then as a threadlike structure, the Reissner's fiber. This allows the microscopic visualization of the secretory products after they have been released and the investigation of the modifications undergone by the released material.

5. Basal Processes of the Ependymal Cells

In the rat SCO, the basal processes originate from the tapering portion of the infranuclear region. Since they project to and terminate on blood vessels, their length is rather variable and depends on the distance to the nearest capillary. Polyribosomes, RER, filaments, and mitochondria are the main components of this cellular portion; microtubules and secretory granules, however, are scarce. In nonmammalian species the basal processes are much longer than those in mammals, and many of them traverse the posterior commissure to end on the external limiting membrane of the brain. In the few species where the ultrastructure of these processes has been studied, the presence of numerous secretory granules and microtubles has been reported (Rodríguez, 1970b, toad; Fernández-Llebrez *et al.*, 1987a, snake, turtle; Biosca and Azcoitia, 1989, chicken).

6. Terminals of Ependymal Processes

The ultrastructural characteristics of the ependymal *endings* vary greatly among species. There are also differences between those contacting blood vessels and those establishing contact with the external basal lamina of the brain.

a. Vascular Endings   In the rat, these terminals contain (1) a mixed population of round and elongated granules of moderate electron density, ranging in size between 100 and 300 nm. Their appearance is completely different from the secretory granules found in the ventricular protrusions (see above); (2) short tubular formations about 150 nm in diameter, with a content resembling that of the granules; (3) a population of mitochondria, smooth-surfaced and coated vesicles, and polyribosomes.

The most consistent and conspicuous structural feature of the vascular ependymal endings is the presence of numerous secretory granules, with large interspecies variations with respect to their size, shape, and density (monkeys, Hofer, 1986; rabbit, Schmidt and D'Agostino, 1966; Kimble and Møllgard, 1973; opossum, Tulsi, 1983; snake and turtle, Fernández-Llebrez *et al.*, 1987a; Peruzzo *et al.*, 1990; amphibians, Oksche, 1969; Rodríguez, 1970b). A particular arrangement is found in the dog SCO. There, the secretory cells are arranged into a multilayered structure which is penetrated by numerous blood capillaries (Oksche, 1969; Rodríguez *et al.*, 1986). The ependymovascular contact areas are sites where secretory granules accumulate. In the bovine SCO, the vascular ependymal endings contain a very high number of densely packed mitochondria (Isomäki *et al.*, 1965), whereas secretory granules are rare. Dilated cisternae of the RER have been described in the vascular endfeet (primates, Hofer, 1986; snake, Fernández-Llebrez *et al.*, 1987a).

b. Leptomeningeal Endings   In most cases, in each individual species, the ultrastructure of the ependymal endings contacting the external basal lamina of the brain differs from that of the perivascular endings. A distinct feature of these endings is the presence of whorllike structures formed by concentrically arranged cisternae of RER (Oksche, 1969; Rodríguez, 1970b; Fernández-Llebrez *et al.*, 1987a; E. M. Rodríguez *et al.*, 1987b). Secretory granules resembling those occupying the vascular endings, but different from those located in the ventricular cell pole, have been described in most species investigated (E. M. Rodríguez *et al.*, 1987b).

B. Hypendymal Cells

In the rat SCO, the ultrastructure of the nucleus and cytoplasm of hypendymal cells resembles that of the ependymal elements, the main difference being a characteristic intracellular distribution of organelles. The hypendymal cells are elongated and display one or two processes which terminate either on local blood vessels or on the external limiting membrane of the brain (Rodríguez et al., 1984a,b). Hypendymal cells contain numerous spherical cisternae of the RER, which are rather small and uniform in diameter, and a well-developed Golgi apparatus.

A distinct feature of the hypendymal cells of the rat (E. M. Rodríguez et al., 1987b) and guinea pig SCO (Papacharalampous et al., 1968) is the presence of large intracellular cavities, generally located at the cell pole opposite to that giving rise to the processes; such cells often display a tennis-racket shape. The cavities described most likely correspond to deep invaginations of the plasma membrane, and may reach a diameter up to 10 μm. Microvilli and cilia project into these lumina.

Hypendymal cells are arranged into small clusters packed together by junctional complexes, thus forming rosette-shaped structures. In the rat, no dilatation of the intercellular space has been observed in these complexes. At variance, in the bovine SCO, the hypendymal rosettes display a distinct central cavity (Olsson, 1958a; Isomäki et al., 1965). Both in the rat and the bovine SCO, the cytoplasm facing the lumen of the rosette contains numerous electron-dense secretory granules.

The ultrastructure of the processes and endings of the hypendymal cells is similar to that described for the basal extensions of the ependymal cells. In the dog, the intracommissural secretory cells display signs of secretory activity similar to those shown by the ependymal cells (Oksche, 1969).

C. Blood Vessels and the Blood—SCO Barrier

The blood vessels irrigating the SCO display certain distinct features which make them a unique structure in the central nervous system.

A typical perivascular space, i.e., two basal laminae delimiting a thin layer of connective tissue, has been described in the capillaries irrigating the SCO of the toad (Murakami and Tanizaki, 1963; Rodríguez, 1970b), snake and turtle (Fernández-Llebrez et al., 1987a), opossum (Tulsi, 1983), guinea pig (Papacharalampous et al., 1968), rabbit (Schmidt and D'Agostino, 1966; Kimble and Møllgard, 1973), calf (Isomäki et al., 1965), and monkeys (Hofer, 1986). A most interesting finding is that, in the SCO of the toad (Rodríguez, 1970b), snake, and turtle (Fernández-Llebrez et al., 1987a), two types of capillaries may be found, i.e., those lacking a perivascular space and those endowed with a perivascular space. The processes of the SCO secretory cells appear to end exclusively on the capillaries displaying a perivascular space. The external perivascular basal lamina protrudes branched extensions forming complex labyrinthine structures, which penetrate the ependymal layer and increase the surface contact area between secretory cells and blood vessels (Papacharalampous *et al.*, 1968; Rodríguez, 1970b; Kimble and Møllgard, 1973; Tulsi, 1983; Fernández-Llebrez *et al.*, 1987a). The capillaries of the rat SCO, although lacking a continuous perivascular space, do exhibit areas where the single basal lamina splits into two laminae which delimit a patchlike perivascular space occupied by long-spacing collagen (see below).

A characteristic feature common to the capillaries of all circumventricular organs is the presence of a perivascular space (Gross and Weindl, 1987). The functional meaning of this space is still open to discussion. It was earlier suggested that the presence of brain capillaries surrounded by a perivascular space is indicative of a leaky blood–brain barrier within that region (Brightman *et al.*, 1970). Gross and Weindl (1987) have suggested that these spaces, separating glial endfeet from the endothelium, could be responsible for the lack of development of barrier properties by the endothelial cells of these capillaries.

Within the family of circumventricular organs, the SCO is the only organ in which the endothelium of the capillaries is not fenestrated. Furthermore, the SCO appears to have an effective blood–brain barrier, since intravenously injected peroxidase does not leak into the SCO of the rabbit (Weindl and Joynt, 1973), rat (von Bomhard *et al.*, 1974), and mouse (Broadwell and Brightman, 1976). A transendothelial vesicular transport of intravascularly injected peroxidase and peroxidase labeling of the perivascular basement lamina have been reported to occur in the blood vessels of the guinea pig SCO (Gotow and Hashimoto, 1982b). Despite this exceptional finding in the guinea pig, the presence of a tight blood–SCO barrier is generally accepted (Bouchaud, 1975; Poirier *et al.*, 1983; Gross and Weindl, 1987). This does not rule out, however, that substances released from the perivascular endfeet of the SCO secretory cells may reach the bloodstream. Indeed, Hofer (1986) has provided strong ultrastructural evidence for the transendothelial transport of secretory products in the SCO of primates.

The presence of striated bodies in widened areas of the perivascular space of the rat SCO has been reported by several authors (R. A. Naumann, 1963; R. A. Naumann and Wolfe, 1963; Wetzstein *et al.*, 1963; Stanka *et al.*, 1964; Schwink and Wetzstein, 1966). In these striated structures, lines or bands repeat approximately every 100 nm (R. A. Naumann, 1963; Wetzstein *et al.*, 1963). Each of the 100-nm periods is subdivided by four lines arranged asymmetrically (Wetzstein *et al.*, 1963). Similar structures, but with a slightly different periodicity of 50 nm, have been found in the capillaries of the guinea pig (Papacharalampous *et al.*, 1968), *Gekko japonicus* (Murakami *et al.*, 1969), and a turtle (Fernández-Llebrez *et al.*, 1987a). In the rat SCO, these periodic structures first develop during the second postnatal week (Schwink and Wetzstein, 1966). Similar perivascular broad-banded striated bodies have been found in organs and tissues other than the central nervous system, either under normal or pathological conditions (Friedman *et al.*, 1965; Kimura *et al.*, 1975; Rittig *et al.*, 1990).

The perivascular striated bodies most likely correspond to an *in situ* form of long-spacing collagen, similar to long-spacing collagen fibrils obtained *in vitro* during reconstruction of fibrils from collagen solutions (Gross *et al.*, 1954). Recent immunocytochemical evidence has shown that striated bodies in the human iris and ciliary body indeed correspond to long-spacing type VI collagen (Rittig *et al.*, 1990).

The exceptional presence of long-spacing collagen in the perivascular space of the capillaries irrigating the SCO may reflect a special type of ependymovascular interaction at this level. Thus, Schwink and Wetzstein (1966) have suggested that the striated bodies could lead to a local decrease in the efficiency of the blood–brain barrier.

The aggregation of molecules from collagen solutions in various forms of periodicity and band patterns has been shown to be dependent on several factors, such as pH, temperature, ionic composition, and addition of glycoproteins (Randall *et al.*, 1953). Since secretory glycoproteins of the SCO appear to reach the perivascular space (see Section IV,A,2), it seems likely that the presence of long-spacing collagen in this position is related to such a "basal" release of secretory products. This possibility is strongly supported by findings in the rat SCO grafted under the kidney capsule. These grafted cells not only survive well but even show an increased secretory activity (Fig. 10) (Rodríguez *et al.*, 1989). The newly formed blood vessels, which start to revascularize the grafted SCO 2 days after transplantation, are endowed with a perivascular space, but lack long-

FIG. 10   Rat SCO 1 month after transplantation under the kidney capsule; immunostained with anti-bovine RF serum. The ependymal (E) and hypendymal (H) cells are strongly immunoreactive. Secretory material is also present in the basal processes of ependymal cells (small arrows) and as a film on the free surface (large arrow). K, Kidney. Bar, 100 μm. From Rodríguez *et al.* (1989).

FIG. 11   Rat SCO 1 week after transplantation under the kidney capsule showing the pericapillary region with several extensions of the external perivascular basal lamina (arrows). Widened areas of this lamina display long-spacing collagen (asterisks). Ependymal basal processes (stars) containing glycogen and granular inclusions terminate on the basal lamina. *Insert:* Long-spacing collagen; note continuity between thick bands and thin lines of the striated structure (arrows). Bars, 1 μm. From Rodríguez *et al.* (1989).

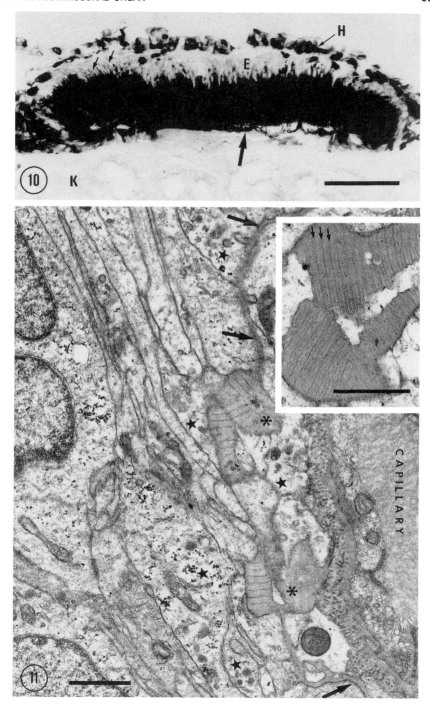

spacing collagen. One week after transplantation, long-spacing collagen starts to appear in circumscribed, mostly expanded, areas of the perivascular space (Fig. 11). This strongly suggests that the appearance of long-spacing forms of collagen adjacent to the newly formed capillaries is triggered and its presence maintained by factors provided by the grafted secretory cells. Clarification of the nature of these factors will not only help to understand this exceptional location of long-spacing collagen but may also contribute toward understanding the functional role of the SCO.

In brief, the blood vessels of the SCO exhibit certain characteristics making them a unique structure within the central nervous system; viz., they are endowed with a perivascular space containing long-spacing forms of collagen, but their nonfenestrated endothelium is tight with respect to circulating tracers of the conventional type. Transport of SCO secretory products into the blood stream, however, seems probable.

D. Junctional Complexes and the Cerebrospinal Fluid— SCO Barrier

Tight junctions and zonulae adherentes linking together the ependymal cells of the toad SCO were first observed in ultrathin sections used for electron microscopy (Rodríguez, 1970a). Rodríguez also described desmosomes along the surface contact zones of ependymal cells, including the processes and endfeet of these cells. Tight junctions and zonulae adherentes have been described in ultrathin sections of the SCO of the rabbit (Kimble *et al.*, 1973), guinea pig, and rat (Gotow and Hashimoto, 1982a,b). Tight junctions and zonulae adherentes were also found, by use of freeze-fracture electron microscopy, in the SCO of the Mongolian gerbil (Madsen and Møllgard, 1979; Lu and Peracchia, 1987). The high number of strands within the tight junction (up to 25 per cell) led Madsen and Møllgard (1979) to regard the gerbil SCO as a "tight epithelium."

The number and arrangement of strands at the tight junction vary with the species, the gerbil displaying a complex tight junction, the rat only a poorly developed junction, and the guinea pig and mouse SCO possessing an intermediate degree of complexity (Madsen and Møllgard, 1979; Gotow and Hashimoto, 1982a; Mack *et al.*, 1987; Lu and Peracchia, 1987). The existence of gap junctions is also a matter of discrepancy (see Madsen and Møllgard, 1979; Lu and Peracchia, 1987; Mack *et al.*, 1987).

Orthogonal arrays of particles, which are typical components of the astrocytic plasma membrane in mammals (Anders and Brightman, 1979), have been shown to coexist with tight junctions in the ependymal cells of the rat SCO (Gotow and Hashimoto, 1982a; Mack *et al.*, 1987).

The intraventricular administration of tracers has not led to conclusive results either. The same experiment (intraventricular administration of peroxidase) carried out by the same authors (Gotow and Hashimoto, 1982a,b) in different species showed that, whereas the ependyma of the guinea pig SCO appeared tight to the tracer, that of the rat was permeable. On the other hand, using the rabbit, Kimble *et al.* (1973) reported that intraventricularly perfused ruthenium red does not enter the intercellular space of the SCO, whereas Weindl and Joynt (1973) concluded that intraventricularly administered peroxidase does enter the intercellular space of the SCO via apical gap junctions. Thus, concerning the tightness of the SCO ependyma, there are important interspecies differences as well as contradictory results described for one and the same species.

The use of conventional tracers, such as horseradish peroxidase, for the ultrastructural evaluation of the permeability of both the blood capillaries and the ependyma of the SCO places constraints on interpreting the actual physiological properties of the blood–SCO and CSF–SCO barriers. Thus, tight junctions known to be tight to peroxidase are permeable to small solutes, like ions (Johanson, 1989). Transjunctional movement of intermediate-sized molecules such as amines and small peptides is largely unknown in general (Johanson, 1989), and completely unknown in the case of the SCO. This may be relevant when performing or interpreting physiological experiments on the SCO, since amines, neurotransmitters, and small peptides do occur in the CSF under physiological conditions (Wood, 1983). Another important aspect with respect to the blood–brain barrier concerns carrier-mediated transport across the endothelium (Fenstermacher and Rapoport, 1984). Nothing is known with respect to the transendothelial transport system operating in the SCO capillaries. This information is important, especially considering the unique characteristics of the SCO capillaries.

Although the presence of a CSF–SCO barrier needs to be investigated further, there is evidence to suggest that a barrier (to large lipid-insoluble molecules) exists, at least in certain species. There is agreement that in the circumventricular organs the blood–brain barrier is "shifted" from the blood side to the ependymal layer of these organs (Leonhardt, 1980; Gross and Weindl, 1987). This means that a "blood–CSF barrier" exists in these organs. On the other hand, all brain areas protected by a blood–brain barrier are in open communication with the ventricular CSF (cf. Leonhardt, 1980).

The SCO thus represents a unique structure within the brain, since its cells are "sequestered" within a double-barrier system whose functional significance is completely enigmatic.

IV. The Secretory Process

A. Nature of the Secretory Products

1. Histochemistry

The existence of a selectively stainable ("Gomori-positive") secretory material in the ependymal cells of the SCO was demonstrated for the first time by Stutinsky (1950). Subsequently, this observation in the frog was confirmed and extended to several other vertebrate species (Bargmann and Schiebler, 1952; Mazzi, 1952; Wislocki and Leduc, 1952; Wingstrand, 1953; Oksche, 1954, 1956; Olsson, 1958a; Talanti, 1958). The first report on selectively stainable secretory material in the SCO of a human embryo was given by Oksche (1956). The secretory material of the SCO is also stainable with the periodic acid–Schiff method (for references, see Ziegels, 1976), and it has the same distribution as the aldehyde–fuchsin stainable material (Diederen, 1970), suggesting that the secretory substance might be a glyco-protein (Bargmann and Schiebler, 1952; Wislocki and Leduc, 1952). After performing a comprehensive histochemical study of the frog (*Rana temporaria*) SCO, Oksche (1962) concluded that the secretory material is a glycoprotein or a mucopolysaccharide–protein complex lacking strong acidic groups (cf. Oksche, 1954, 1956). Leonieni (1968a), W. Naumann (1968), Diederen (1970), and Møllgard (1972) arrived at similar conclusions.

After using Bial's reagent according to Diezel for the demonstration of sialic acid, Olsson (1958a) suggested that the secretion of the bovine SCO must be rich in sialic acid perhaps in the form of a sialylated protein. Sterba and Wolf (1969) and Møllgard (1972), by use of certain histochemical tests, reported the presence of sialic acid in the SCO of the bovine and human fetus.

Wislocki and Leduc (1954) in the adult rat, Møllgard (1972) in human fetuses, and Leonieni (1968b) in some laboratory animals have reported the presence of disulfide and sulfydryl groups in the SCO secretion. W. Naumann (1968) concluded that the proteinaceous component of the secretion contains a high concentration of tyrosine, tryptophan, and cystine.

Autoradiographic investigations after the administration of radioactive cysteine and cystine have shown that the SCO and RF become strongly labeled (Sterba *et al.,* 1967b; Ermisch *et al.,* 1968; Leatherland and Dodd, 1968; Talanti, 1969, 1971; Diederen, 1972, 1973, 1975b; Ermisch, 1973; Attila and Talanti, 1973; Hess *et al.,* 1977; Diederen and Vullings, 1980a,b; Vullings and Diederen, 1985; Diederen *et al.,* 1987). Different degrees of labeling of the SCO have been reported by using other radioactive amino acids, such as methionine (Talanti, 1971), proline and histidine (Garweg and Kinsky, 1970), and leucine (Diederen *et al.*, 1987).

Enzyme histochemistry has led to the conclusion that the SCO is characterized by a high glycogenolytic and glycolytic activity and a low oxidative capacity (Shimizu *et al.*, 1957; W. Naumann, 1968; Diederen, 1970; Ziegels, 1976). Acid phosphatase and other lysosomal enzymes have been found located in the apical cytoplasm of the ependymal cells (W. Naumann, 1968; Diederen, 1970; Wenzel *et al.*, 1970). (For a complete list of references on enzyme histochemistry of the SCO, see Ziegels, 1976.)

In brief, the histochemical studies on the SCO have led to the conclusion that its secretory material is a carbohydrate–protein complex (or a complex of glycoproteins) with a high content of sialic acid, tryptophan, and cysteine.

The RF is stained by the same histochemical methods that stain the intracellular secretory material (Wingstrand, 1953; Oksche, 1961; Sterba *et al.*, 1967a; W. Naumann, 1968; S. Rodríguez *et al.*, 1985).

2. Immunohistochemistry

Sterba *et al.* (1981) raised antibodies against an aqueous extract of bovine RF and used them for immunostaining the rat brain. Only the secretory cells of the SCO were immunoreactive. From a comparative immunocytochemical study, Sterba *et al.* (1982) concluded that the ependyma and hypendyma of the SCO of all species investigated (with the exception of the hedgehog) display a positive immunoreaction. Rodríguez *et al.* (1984a) raised antibodies against RF material isolated from the bovine spinal cord completely dissolved in a buffer containing urea, dithiothreitol, and ethylenediaminetetraacetic acid. They used this antiserum for immunostaining of the SCO in 25 vertebrate species (amphibians, reptiles, birds, mammals) (Figs. 7 and 8) and extended the findings reported by Sterba *et al.* (1982).

Rodríguez *et al.* (1984a) observed that, although the SCO of several species of New- and Old-World monkeys displayed immunoreactive material, the SCOs of anthropoid apes and human fetuses were completely negative with the antisera used. An antibody raised against a 540-kDa compound (probably a precursor form; see Section IV,B,1) isolated from the bovine SCO proper (secretory ependymal–hypendymal complex) did not react with the fetal human SCO, although it stained specifically the bovine SCO (E. M. Rodríguez *et al.*, 1990b). Rodríguez *et al.* (1984a) suggested that the SCO of anthropoid apes and man, due to their lack of reactivity with various antibodies specific for the SCO of other mammals, either secretes a material not detected by the antisera used or is devoid of a true secretory activity (see Section IX).

R. Meiniel *et al.* (1988) produced monoclonal antibodies against extracts of bovine SCO. One of these antibodies recognizes a glycoprotein secreted by the bovine SCO–ependyma into the ventricle; at the same time, it is found in the hypendymal cells. Karoumi *et al.* (1990a) obtained antibodies against a crude SCO extract of 19-day-old chick embryos. This antiserum was used in an ontogenetic immunochemical study with the chick SCO (see Section IX).

Ultrastructural immunocytochemistry using anti-RF sera has shown that the immunoreactive material in the ependymal cells of the SCO is exclusively located in cisternae of the RER and in granules mainly concentrated near the ventricular surface (Lösecke *et al.*, 1984; Rodríguez *et al.*, 1986, 1987a,b; Peruzzo *et al.*, 1990) (Fig. 22) or in the vascular and leptomeningeal endfeet (E. M. Rodríguez *et al.*, 1987b; Peruzzo *et al.*, 1990) (Figs. 17, 18).

Material immunoreactive with anti-RF sera has also been detected in two extracellular locations: (1) material released into the ventricle and condensed in the form of pre-RF or RF (Lösecke *et al.*, 1984; Rodríguez *et al.*, 1986, 1987a,b) (Figs. 8, 10, 14, 19, 22); (2) material accumulated in large expansions of the intercellular space located in the vicinity of the perivascular space (Lösecke *et al.*, 1986; E. M. Rodríguez *et al.*, 1987a). This latter material most likely corresponds to the secretion-like product described in the primate SCO by use of conventional transmission electron microscopy (Hofer, 1986).

3. Combined Use of Lectin Histochemistry, Glycosidases, and Tunicamycin

The first report on the use of lectin histochemistry in the investigation of the carbohydrate component of the secretory glycoproteins of the SCO was by R. Meiniel and Meiniel (1985), who employed a set of different lectins and concluded that the SCO secretion is rich in mannosyl residues.

The combined use of immunocytochemistry (anti-RF serum), a series of lectins, and specific glycosidases applied in sequence to the same sections or to adjacent 1-μm-thick sections led to the following conclusions: (1) the secretory products within the RER are N-linked, high-mannose-type (GlcNac$_2$-Man$_n$) glycoproteins; (2) the secretions within the apical secretory granules are N-linked, complex-type glycoproteins with the sequence in the terminal chain of -GlcNac-Gal-sialic acid; (3) the apical secretory granules and the pre-RF material (see Section VI), but not RF, share the same immunocytochemical and lectin-binding properties (Rodríguez *et al.*, 1986, 1987b, 1990a; S. Rodríguez *et al.*, 1987; A. Meiniel *et al.*, 1988; Herrera and Rodríguez, 1990; R. Meiniel *et al.*, 1990; Nualart *et al.*, 1991). Ultrastructural lectin histochemistry has revealed the subcellular distribution of the lectin-binding compounds, thus further substantiating the three conclusions mentioned above (Fig. 13) (Rodríguez *et al.*, 1986; A. Meiniel *et al.*, 1988; Peruzzo *et al.*, 1990).

Core glycosylation of the secretory proteins of the rat SCO was demonstrated by the disappearance of the ConA-binding sites in the SCO secre-

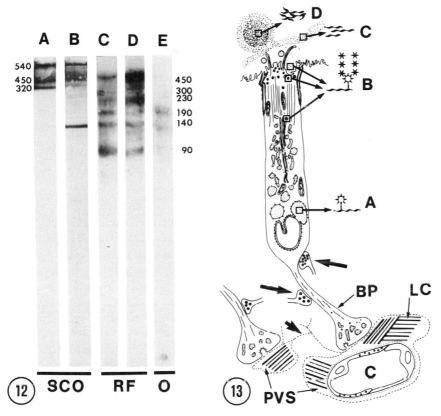

FIG. 12   Western blots of bovine SCO (A, B) and RF (C, D) extracts, and of an anti-RF immunoprecipitate of an extract of *Xenopus* oocytes microinjected with poly(A)$^+$ mRNAs purified from bovine SCO (E). A, C, and E were immunostained with anti-bovine RF serum; B, ConA binding; D, *Limax flavus* agglutinin (LFA) binding.

FIG. 13   Schematic representation of a rat SCO ependymal cell. A, N-linked, high-mannose-type glycoproteins in RER; B, complex-type glycoproteins in secretory granules and, upon release into the ventricle, C, partially packed secretory glycoproteins forming pre-RF; D, densely packed glycoproteins forming RF; BP, basal process ending on expanded areas of the perivascular space (PVS) filled with long-spacing collagen (LC); C, capillary; large arrows, axon terminals on cell body and basal process of ependymal cell; small arrow, processes of the perivascular basal lamina.

tory cells of rats injected intraventricularly with tunicamycin while these cells continued to secrete the proteinaceous component demonstrable by means of immunocytochemical reactions (Herrera and Rodríguez, 1990).

The possibility that the secretory glycoproteins of the SCO may contain O-linked sialylated oligosaccharides has been put forward by Wolf and Sterba (1972) and A. Meiniel *et al.* (1988).

Lectin histochemical studies performed at the light-microscopic level with the SCO of the dog (Rodríguez *et al.*, 1986), chick embryos (Bruel *et al.*, 1987), a snake, and a turtle (Fernández-Llebrez *et al.*, 1987a) have indicated the interesting possibility that differences may exist between the carbohydrate moiety of the secretory glycoproteins released into the ventricle and that of the secretory products accumulated in the ependymal endfeet contacting blood vessels or the subarachnoid space. However, a recent study of the ophidian SCO clearly demonstrated that the immuno-reactive secretory granules located either at the ventricular, vascular, or leptomeningeal cell poles all bind ConA and wheat germ agglutinin (Peruzzo *et al.*, 1990).

4. Quantitative Chemical Analysis

Quantitative chemical analysis has been performed using bovine RF. The water content of RF is 98.9% (Rodríguez *et al.*, 1984a), and 1.1% of dry matter consists of sialoglycoproteins (Hädge and Sterba, 1973a). According to Hädge and Sterba (1973a,b), RF consists of about 80% proteins and 16% carbohydrates. This is roughly in agreement with the reduction of the molecular weight of the glycoproteins of the bovine RF after removal of the carbohydrate component by use of endoglycosidase F (F. Nualart and E. M. Rodríguez, unpublished observation).

The amino acid composition of the bovine RF is shown in Table I. Although the data represent the amino acid composition of the entire RF, which appears to be formed by different secreted molecules, the data are valuable when designing certain experiments, such as biosynthetic label-ing. The RF possesses a large number of cysteine residues (Table I); the possibility that disulfide bonds constitute a fundamental factor in the condensation of the secretory material released into the ventricle has been considered by Wolf and Sterba (1972). The high content of hydrophobic amino acids and low solubilization of RF in aqueous media suggest that the RF polypeptides could fold, forming a hydrophobic core. A coat of sialic acid residues could be responsible for the interaction with the CSF. Con-sidering the high content of amino acids that disrupt the α-helix structure, it may be suggested that interaction between two or more polypeptides and not the formation of α-helix is predominant in the RF assembly. About 16% of carbohydrates in RF is made up of 12% neutral and basic sugars and

TABLE I

Amino Acid Composition of the Bovine Reissner's Fiber Glycoproteins

Amino acid		Percentage composition	
		I[a]	II[b]
LYS	(K)	1.7	2.1
HIS	(H)	2.5	2.2
ARG	(R)	6.9	6.3
ASP	(D)	6.5	6.3
THR	(T)	5.0	3.7
SER	(S)	8.2	5.0
GLU	(E)	7.0	11.4
PRO	(P)	10.9	6.7
GLY	(G)	12.4	5.5
ALA	(A)	9.8	4.8
VAL	(V)	5.5	3.6
MET	(M)	T[c]	0.8
ILE	(I)	1.3	1.4
LEU	(L)	8.3	6.3
TYR	(Y)	1.5	1.4
PHE	(F)	3.5	1.9
CYS–SH	(C)	9.1	6.1
CYS–S–S	—	ND[d]	9.9

[a] Determination performed in our laboratory by Nualart (1989).
[b] Data from Hädge and Sterba (1973a).
[c] Traces.
[d] Not determined.

3–4% sialic acid (Sterba and Wolf, 1969; Hädge and Sterba, 1973b). Hädge and Sterba identified the following sugar residues in the bovine RF by use of paper chromatography: galactose, glucosamine, glucose, mannose, and fucose, the two former being the most abundant. This is in rather good agreement with the N-linked complex-type nature proposed for the secretory glycoproteins of the SCO (see above).

5. Electrophoresis, Immunoblotting, and Lectin Binding

The functional role of the SCO is still open to discussion. A bioassay that could be used in the purification process of the secretory material is not available. This has led the few authors interested in purification and characterization of the SCO secretory materials to use an indirect approach.

By comparing the electrophoretic migration pattern of crude extracts of bovine SCO, pineal gland, and nonsecretory ciliated ependyma, the presence of certain bands "specific" for the SCO was established. Thus, Hein (1988) reported two bands of 240 and 350 kDa to be present in the SCO and missing from the pineal gland and the ciliated ependyma. R. Meiniel et al. (1986) reported four glycopeptides (30, 54, 72, and 100 kDa) as good candidates for glycoproteins secreted by the SCO, although only the 54-kDa component was exclusively found in the SCO extract, whereas the three others were also present, in lower concentrations, in the pineal extract.

Electrophoretic studies have also been performed using the bovine RF. Agarose and PAGE run under nondenaturating conditions led Wolf and Sterba (1972) to conclude that the bovine RF is made up by two types of glycoproteins, namely, a fibrous one of high molecular weight and a second one of low molecular weight associated to the first one by noncovalent links. Unfortunately, the actual molecular weights were not given.

Bovine SCO and RF extracted in a medium containing urea have been used for a comparative analysis using SDS-PAGE and Coomassie Blue staining. Twelve bands ranging between 320 and 23 kDa were found to be common to both extracts (Hein, 1988). Only four of these were later shown to be immunoreactive with an anti-RF serum. This complex band pattern of RF after use of urea and SDS suggests that the complex-type glycoproteins which constitute the fiber may be formed by several subunits. However, processes occurring during the collection and extraction procedures, such as degradation or aggregation and different degree of removal of sugar residues, must be kept in mind. The actual polypeptide composition of RF awaits further investigation.

Polyclonal antibodies against secretory products of the bovine SCO have been used for the identification of the secretory compounds present in bovine SCO and RF extracts analyzed by SDS-PAGE and immunoblotting (Rodríguez et al., 1987a, 1990a; Hein, 1988; Nualart et al., 1991). Four secretory glycoproteins were consistently found in the SCO proper of 540, 450, 320, and 190 kDa (Fig. 12). A fifth polypeptide of 50 kDa was detected only in some of the SCO extracts tested. The same authors described immunoreactive bands of 450, 300, 230, 190, 140, and 89 kDa obtained from RF extracts (Fig. 12). Not all of them, however, were consistently detected (Nualart et al., 1991).

R. Meiniel et al. (1986) performed a ConA-binding study on blots of sheep SCO and pineal gland. From this comparative analysis they concluded that a 54-kDa compound binding ConA corresponds to a secretory glycoprotein of the sheep SCO. This compound could be homologous to the 50-kDa polypeptide detected in our immunoblots of the bovine SCO.

Blot analyses of bovine SCO and RF using an anti-RF serum and lectins have allowed the determination of the lectin-binding properties of the immunoreactive secretory compounds when they are stored within the secretory cells (SCO extract) and after they have been released (RF extract) (E. M. Rodríguez *et al.*, 1987a; Nualart, 1989; Nualart *et al.*, 1991). Thus, the four immunoreactive compounds present in the SCO (540, 450, 320 and 190 kDa) bind ConA (Fig. 12), but only two of them (450 and 190 kDa) bind *Limax flavus* agglutinin (LFA; affinity for sialic acid). All immunoreactive compounds of RF bind LFA (Fig. 12) and none of them binds ConA (see Table II). These findings allow several relevant conclusions to be discussed with reference to the process of biosynthesis (see following section).

6. Immunoaffinity Chromatography

Immunoaffinity chromatography has been used for the purification and identification of the SCO secretory products in the bovine SCO (E. M. Rodríguez *et al.*, 1987a; Nualart *et al.*, 1991) and in the SCO of chick embryos (Karoumi *et al.*, 1990b).

Extracts of bovine SCO have been filtered through a column containing anti-RF serum (IgG fraction), covalently attached to Sepharose. The bound and unbound fractions were processed for immunoblot analysis using an anti-RF serum. The bound fractions contained three of the four immunoreactive compounds (540, 450, and 320 kDa) detected in crude extracts of the SCO (Hein, 1988; Nualart *et al.*, 1991).

Karoumi *et al.* (1990b) used a column containing an antiserum against an extract of chick SCO immobilized on Ultragel to purify the secretory products in the SCO of 19-day-old chick embryos. The bound fraction contained 10 compounds binding ConA, ranging from 240 to 42 kDa. Three of these compounds bound wheat germ agglutinin (affinity for glucosamine, sialic acid). Immunoblotting of the bound fraction was not performed in this study.

The apparent discrepancy with respect to the findings obtained with the bovine and chicken SCO could reflect species differences in the migration properties of the secretory products when processed for SDS-PAGE. As is discussed in the following section, the 540-, 450-, and 320-kDa compounds present in the bovine SCO appear to be missing in the SCO of reptiles. Thus, it would not be surprising that they were also missing from the chick SCO.

7. Translation of SCO mRNA in *Xenopus laevis* Oocytes

Nualart (1989) microinjected *Xenopus laevis* oocytes with mRNA poly(A)$^+$ purified from bovine SCO. Extracts of the injected oocytes were processed for immunoprecipitation using an anti-RF serum. The immunoprecipitates were analyzed by SDS-PAGE and immunoblotting techniques using an anti-RF serum. Three immunoreactive bands of 190, 140, and 89 kDa were found (Fig. 12) similar to three of the compounds detected in immunoblots of bovine RF. As is discussed in the following section, only one of the two precursors in the bovine SCO secretion, the putative 320-kDa precursor, had been translated in the injected oocytes.

B. Biosynthesis

1. Evidence for the Presence of Precursor and Processed Forms: Processing

Results obtained with the use of monoclonal antibodies raised against the secretory proteins of the bovine SCO speak in favor of the presence of precursor and processed forms in this organ (O. Garrido, G. Eller, and E. M. Rodríguez, unpublished observations). Indeed, antibodies exist which exclusively react either with the glycoproteins stored in the RER or with those present in the secretory granules. Lectin histochemical studies of the SCO have revealed that the secretory glycoproteins are N-linked, high-mannose-type while stored in the RER, but become transformed into complex-type glycoproteins when they are packed into secretory granules (see previous section). This conclusion has been further substantiated by radioautographic studies using radioactive amino acids and sugars (Diederen *et al.*, 1987).

Thus, immunocytochemical, lectin histochemical, and radioautographic studies have provided evidence to postulate that both components of the SCO secretory glycoproteins, i.e., the protein and carbohydrate moieties, undergo a maturation process (processing) along the secretory pathway.

The application of complementary methods, i.e., the parallel use of immunostaining and lectin binding to blots of bovine SCO and RF (Table II), the raising of specific antibodies against each of the three major compounds of the bovine SCO (540, 450, and 320 kDa), and the use of these antibodies in a comparative immunocytochemical study allow one to draw the following conclusions (E. M. Rodríguez *et al.*, 1987a; Hein, 1988; Nualart *et al.*, 1991): (1) The 540-kDa glycoprotein may represent a precursor form since it is present in the SCO proper and missing in RF. Its high affinity for ConA and lack of affinity for LFA characterize this compound as a core-glycosylated protein which has not yet passed the Golgi apparatus (Table II). Furthermore, the antiserum against the 540-kDa compound immunoreacts with the secretion stored in the RER of the dog SCO, but not with the apical secretory granules. (2) Based on the same line of reasoning, it was suggested that the 450-kDa glycoprotein is a processed form. This compound is present in the SCO *and* RF, thus indicating that it has been released into the CSF; its affinity for ConA and LFA (Table II) indicates that this glycoprotein has passed the Golgi apparatus. The antiserum against the 450-kDa compound reacts with material occurring in the RER and in the secretory granules of the dog SCO. (3) The absence of the 320-kDa glycoprotein from RF and its lack of affinity to LFA characterize this fraction as a precursor form. In immunoblots of the SCO the anti-320-kDa serum revealed a 190-kDa polypeptide. This latter compound, which binds ConA and LFA and is present in both the SCO and RF, might correspond to a processed form of the putative 320-kDa precursor (Table II).

The possibility that the bovine SCO harbors two precursor forms of secretory material (540 and 320 kDa) is strongly supported by immunocytochemical findings using antisera against the 540, 450, and 320 kDa glycoproteins, respectively (E. M. Rodríguez et al., 1990a; Nualart et al., 1991). The first two antisera produced a strong reaction with the bovine and dog SCO, but did not react with the rat and ophidian SCO. The anti-320-kDa serum reacted with the bovine, dog, and rat SCO, but did not stain the ophidian SCO. Taking into account that the three antisera used are poly-

TABLE II

Immunoblotting and Lectin Binding of Secretory Material in Bovine Subcommissural Organ and Reissner's Fiber[a]

Subcommissural organ			Reissner's fiber		
AFRU[b] (kDa)	ConA[c] (kDa)	LFA[d] (kDa)	AFRU[b] (kDa)	Con A[c] (kDa)	LFA[d] (kDa)
540	540	—[e]	—	—	—
450	450	450	450	—	450
320	320	—	—	—	—
—	—	—	300	—	300
—	—	—	230	—	230
190[f]	190	190	190	—	190
—	—	—	140	—	140
—	—	—	89	—	89
50[g]	NE[h]	NE[h]			

[a] From Nualart et al. (1991).
[b] Anti-Reissner's fiber serum.
[c] Concanavalin A (affinity for mannose, glucose).
[d] Limax flavus agglutinin (affinity for sialic acid).
[e] —, Negative values obtained.
[f] Immunostained in the SCO only with the anti-320-kDa serum.
[g] Seen only in some blots.
[h] Not established.

clonal and that, due to the large molecular size and the glycosylated nature of the secretory proteins, the number of their reactive epitopes must be relatively high, the above-mentioned immunocytochemical findings might be explained as follows: (1) the bovine and the canine SCO contain two precursor forms; (2) in the rat only one of the precursors (or a similar compound) is present; (3) none of these fractions occurs in the SCO of the snake. Apparently, both precursors found in the bovine SCO would also be missing in the SCO of chick embryos. In this SCO, Karoumi et al. (1990b) have described 10 ConA-positive glycoproteins, which bind to an immunoaffinity column containing an antiserum against chick SCO. Three of these glycoproteins (98, 88, and 52 kDa) bound wheat germ agglutinin and were regarded by Karoumi et al. (1990b) as complex-type glycoproteins that have passed the Golgi apparatus. According to their M_r, none of these four glycopeptides corresponds to those found in the bovine SCO.

The fact that the bovine RF contains glycoproteins which are not present in the bovine SCO proper, such as the 300- and 230-kDa compounds (see previous section), might suggest that processing after release takes place. Autoradiographic studies carried out in our laboratory have shown that the released secretory material first assembles into a film on the surface of the SCO (pre-RF material) and about 1 hr later becomes packed into RF. This may favor the idea of a postrelease, prepackaging processing (Nualart et al., 1991). The possibility that the compounds present in RF and missing in the SCO represent artifacts depending on postmortem changes has also been considered (Nualart et al., 1991).

The findings obtained by injecting mRNA poly(A)$^+$ from the bovine SCO into Xenopus laevis oocytes (see previous section) are not in contradiction to the existence of two precursors in the bovine SCO; in this case, the mRNA of only one of them would have been translated in the oocytes.

2. Intracellular Transport

In the rat SCO, the apical secretory granules containing processed secretory material are located distal to a well-defined cytoplasmic region (subapical region; see Section III,A,3) containing a large number of microtubules. This suggests the interesting possibility that processing and transport of granules to the ventricular cell pole are interrelated phenomena. In fact, no apical secretory granules occur in the SCO of rats after colchicine has been injected into the CSF (unpublished observations by the authors). Studies designed to clarify the routing and sorting of the secretory compounds within the secretory pathways are to date missing. This is largely due to the fact that, so far, a cell line of SCO secretory cells has not been established.

3. Ventricular Release

Exocytotic figures involving secretory granules and the apical plasma membrane have been shown by use of conventional electron microscopy (see Section III,A,4) and immunocytochemistry (Lösecke *et al.*, 1984). These authors also described saclike structures which they interpreted as a component of the RER system undergoing exocytosis; they suggested two secretory mechanisms in the SCO, one of them bypassing the Golgi apparatus. Based on the observation in the bovine RF of ConA-positive, wheat germ agglutinin-negative compounds, E. M. Rodríguez *et al.* (1987a) suggested that core-glycosylated glycoproteins that have not passed the Golgi apparatus could also be released into the ventricle. This assumption was not confirmed by more recent work conducted in our laboratory (Nualart *et al.*, 1991).

The possibility of the existence in the SCO of a secretory mechanism bypassing the Golgi apparatus, in addition to the conventional mechanism involving this organelle, long debated by most investigators involved in SCO research, has to be reinvestigated using new adequate techniques.

4. Dynamic Studies of the Secretory Process

Although several investigations using radioactive precursors have been performed (see Section IV,A,1), only a few of them were designed to study the dynamics of the secretory process. Ermisch *et al.* (1971) and Diederen (1973) found that, in *Rana esculenta,* radiolabeled material is released by the SCO into the ventricle 2–3 hr after the administration of [^{35}S]cysteine, as shown by the labeling of the proximal end of RF. Incorporation of the label into the supranuclear and perinuclear regions of the frog SCO occurs ~30 min after the injection (Diederen, 1972, 1973). Then the secretory material located subnuclearly becomes progressively labeled until ~20 hr after the injection, when both cell compartments are equally strongly labeled (Diederen, 1972). Talanti (1969) performed a radioautographic study of the rat SCO at various intervals, ranging between 10 min and 3 days, after the administration of [^{35}S]cysteine, and concluded that the incorporation of the label into the SCO is rapid, reaching its maximum within 2–4 hr. He also pointed out that, 3 days after the injection of the precursor, a considerable amount of radioactivity was still present in the ependyma of the SCO.

A prerequisite for the use of radioactive precursors as markers of the secretory activity of the SCO is that these labeled precursors actually become part of the secretory product. Biosynthetic labeling of the SCO secretory products, their separation by immunoaffinity procedures, and their analysis by SDS-PAGE and fluorography or by HPLC appear to be the adequate procedures complying with that prerequisite. This type of analysis, however, has not yet been performed.

The combined use of immunocytochemistry (anti-RF serum) and auto-radiography after the administration of [^{35}S]cysteine into the ventricular CSF of the rat has allowed us to establish the coexistence, within the same cell compartment or structure, of the secretory material and the label (Figs. 14 and 15). The progressive appearance of the label along the secretory pathway, as well as at extracellular sites containing the secre-tion, such as the pre-RF and RF, could be taken as an indication that the

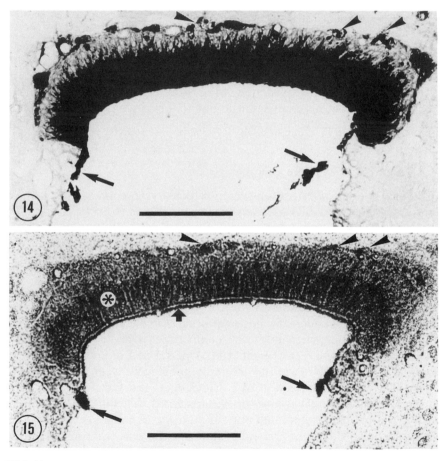

FIGS. 14 and 15   Frontal sections of the SCO of a rat sacrificed 48 hr after injection of [^{35}S[cysteine into the lateral ventricle. Immunoperoxidase staining using an anti-bovine RF serum (Fig. 14) and radioautography (Fig. 15) show strong reaction in the supranuclear region of the ependymal cells (asterisk), film of pre-RF material on the apical surface of the organ (small arrow), and threads of pre-RF material emanating from the lateral aspects of the SCO (long arrows). Hypendymal cells (arrowheads) also concentrate the radioactive label. Bars, 100 μm.

label has actually been incorporated into the secretory products. This experimental approach has shown that materials located within the RER were labeled 20 min after the intraventricular injection of [^{35}S]cysteine. RER, apical secretory granules, and pre-RF material were labeled 1 hr after the injection of the precursor, thus indicating that synthesis, processing, transport to the ventricular cell pole, and ventricular release occur approximately within 1 hr (Herrera, 1988; E. M. Rodríguez et al., 1990a). The first labeling of RF takes place ~2 hr after the injection, indicating that "packaging" of pre-RF material into RF requires approximately 1 hr.

This time course of events has been further substantiated in our laboratory by using the same combined approach of immunocytochemistry and autoradiography, but employing [^{3}H]galactose as a precursor. One hour after the intraventricular administration of the precursor, it was already possible to detect the label in the Golgi area, apical secretory granules, and pre-RF material (shorter time intervals were not investigated). The RER was free of the label (unpublished observations by the authors).

Several days after the administration of [^{35}S]cysteine the label can still be detected in relatively large amounts in the secretory cells of the SCO (Leatherland and Dodd, 1968; Talanti, 1969; Diederen et al., 1987; E. M. Rodríguez et al., 1990a). In the rat, 3 days after the intraventricular injection of [^{35}S]cysteine, the label is still found in the RER and in the most proximal portion of the RF. Since the bulk of ventricular CSF is renewed approximately every 6–8 hr, it may be assumed that the radioactive label is no longer available to the SCO after it has been cleared from the CSF and blood. This leads to the following conclusions: (1) the label observed within the RER several days after its administration has been incorporated into the secretory products during the first few hours after its injection; (2) certain secretory compounds are stored within the RER for several days after their synthesis (Diederen et al., 1987). This deposit might be slowly and continuously transported (and processed?) to the apical surface and released into the ventricle since the proximal end of RF continues to be labeled at these delayed postinjection intervals (Diederen et al., 1987).

The presence in the SCO of a large storage pool of secretory material, which is slowly released into the ventricle, is also demonstrated in experiments blocking core glycosylation of the secretory proteins by the use of tunicamycin (Herrera and Rodríguez, 1990). Twelve hours after the intraventricular administration of tunicamycin, the large storage pool of secretory material within the RER continues to bind ConA. This suggests (1) that most of the glycosylated secretory proteins existing in the SCO before the treatment are still present 12 hr after the injection; (2) that the newly synthetized secretory proteins devoid of carbohydrates represent only a small proportion. At longer postinjection intervals, the "old" secretory glycoproteins would continue to be released and, simultaneously, the newly synthetized, nonglycosylated proteins would accumulate in the SCO. This results in a progressive disappearance of the ConA-binding sites, a process becoming completed 60 hr after the injection of tunicamycin (Herrera and Rodríguez, 1990).

The above-discussed observations point toward the possibility that two pools of secretory material exist in the SCO. One of them is rapidly released into the ventricle after its synthesis. The other, however, remains stored within the RER and is released into the ventricle over a time period of several days. Whether these two pools are related to the two types of secretions discussed above (Section IV,B,1) is an unsolved problem.

The fact that, in the SCO, the main storage site of the secretion appears to be the RER, and the observation that the storage time within this cellular compartment may last for several days, characterize the SCO as a rather unique type of gland.

The amount of radioactivity marking the RER and the secretory granules of the SCO at defined postinjection intervals after application of [^{35}S]cysteine has been taken as a measure of the turnover rate of the secretory material in the SCO (Diederen and Vullings, 1980b; Diederen *et al.*, 1987). Changes in this turnover rate have been reported after osmotic stimulation (Leatherland and Dodd, 1968; Vullings and Diederen, 1985), manipulation of the photoperiod (Diederen, 1972, 1973), changes in the environmental temperature (Diederen, 1975a), and experimental variations in the composition of the CSF (Hess *et al.*, 1977).

V. Sites of Release of the Secretory Products

The fate of the secretory products of the SCO has been the subject of numerous investigations. The ventricular release of one or more secretory products is a well-established fact. On the other hand, a "basal pathway of secretion" has been suggested because of the presence of secretory material in the distal processes of the ependymal cells and their endfeet, a conspicuous feature in lower vertebrates (Oksche, 1954, 1956, 1961, 1962, 1964; Okada, 1956; Hofer, 1958; Rodríguez *et al.*, 1984a,b, 1987b; Fernández-Llebrez *et al.*, 1987a). Ultrastructural evidence also supports the hypothesis of a basal route of secretion in the SCO (Murakami and Tanizaki, 1963; Oksche, 1969; Rodríguez, 1970b; Kimble and Møllgard, 1973; Tulsi, 1983; Fernández-Llebrez *et al.*, 1987a; Schafer and Blum, 1988; Peruzzo *et al.*, 1990). Two sites are to be considered as target areas for the basal processes of the SCO secretory cells, viz., the local blood vessels and the topographically corresponding portions of the external limiting membrane (basal lamina) of the brain, adjacent to the leptomeninx primitiva in nonmammalian species, and to the pia mater and the subarachnoid space in mammals.

A. Subarachnoid (Leptomeningeal) Space

The cell processes of the SCO terminating on the brain surface originate from its ependymal (nonmammalian species) and hypendymal (mammals) elements. A recent ultrastructural immunocytochemical and lectin histochemical study of the SCO of a snake (Peruzzo *et al.*, 1990) has shown that secretory granules stored in the terminal endfeet and at the ventricular cell pole display the same immunocytochemical and lectin-binding properties. These authors concluded that, with reference to the lectin-binding affinities, the immunoreactive granules observed in the terminal of the process are true secretory granules derived from the Golgi apparatus. They also indicated that, since the Golgi complex is only present in the cell soma, a transport from the cell body to the endfeet of the basal process must occur.

Transport of secretory granules along the cell process and their accumulation in the enlarged endfeet support the possibility of a local release of the granule content into the leptomeningeal space (Rodríguez *et al.*, 1984a; Peruzzo *et al.*, 1990) (Figs. 16 and 17). After lesion of the basal processes, establishing the leptomeningeal connection of the anuran SCO, an accumulation of stainable secretory material could be observed in the proximal portion of the transected process (Oksche, 1962).

B. Local Blood Vessels

A close association between secretory cells of the SCO, either ependymal or hypendymal, and local blood capillaries has been repeatedly reported by use of classical staining and electron-microscopic procedures (E. Legait, 1949; Murakami *et al.*, 1957; Oksche, 1961, 1962, 1969; Murakami and Tanizaki, 1963; Palkovits, 1965; Schmidt and D'Agostino, 1966; Leonieni, 1968a; Papacharalampous *et al.*, 1968; Rodríguez, 1970b; Kimble and Møllgard, 1973; Tulsi, 1983; Hofer, 1986; Ramkrishna and Saigal, 1986; Fernández-Llebrez *et al.*, 1987a; Biosca and Azcoitia, 1989).

Light- and electron-microscopic immunocytochemistry by use of an anti-RF serum has shown that the perivascular endfeet contain immunoreactive material stored in secretory granules (Rodríguez *et al.*, 1984b; Peruzzo *et al.*, 1990) (Fig. 18). The lectin-binding properties of these granules indicate that they are derived from the Golgi apparatus (Peruzzo *et al.*, 1990), and therefore transport of these granules from the cell body to the vascular endfeet can be postulated. This is strong support for the

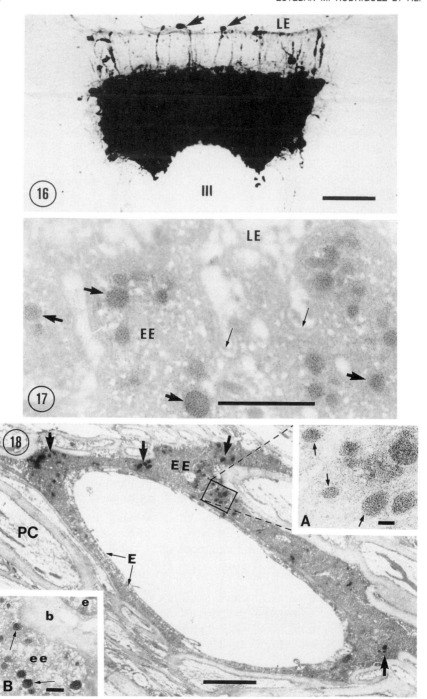

assumption that the content of these granules might be released to the perivascular space. RF-immunoreactive material has, indeed, been detected in expanded areas of the intercellular (extracellular) space located in the vicinity of blood capillaries of the SCO (Lösecke *et al.*, 1986; E. M. Rodríguez *et al.*, 1987a) and the perivascular space proper (Lösecke *et al.*, 1986). In the canine SCO, there are perivascular endfeet containing secretory granules which do not react with the anti-RF serum (E. R. Rodríguez *et al.*, 1987a), indicating that, in this species, a secretory material different from that forming the RF is gaining access to the bloodstream.

C. Third Ventricle

A substantial body of evidence indicates that the ependymal cells of the SCO release at least a major part of their secretory material into the ventricular CSF. In turn, some of this released material condenses to form RF (Sterba, 1969). In this respect, the most convincing evidence has been provided by immunocytochemical investigations. Thus, antibodies raised against RF extracts react specifically with the secretory material of the SCO (Sterba *et al.*, 1982; Rodríguez *et al.*, 1984a), and antibodies against secretory products extracted from the SCO immunoreact with RF (S. Rodríguez *et al.*, 1985; R. Meiniel *et al.*, 1988; Karoumi *et al.*, 1990a). Recent evidence obtained in our laboratory indicates that some of the secretion released into the ventricle remains soluble in the CSF (H. Richter, F. Nualart, P. Peña, and E. M. Rodríguez, unpublished observation).

VI. Reissner's Fiber

At least a part of the material secreted by the SCO into the ventricle is not soluble in the CSF, but condenses first on the surface of the organ, and

---

FIG. 16 A 100-μm-thick vibratome section of the rat SCO immunostained with anti-bovine RF serum. Immunoreactive basal processes of secretory cells reach the leptomeningeal space (LE) by means of terminal swellings (arrows). III, Third ventricle. Bar, 100 μm.

FIGS. 17 and 18 Ultrastructural immunocytochemistry of *Natrix maura* SCO using methacrylate embedding and immunoperoxidase–silver methenamine staining by use of anti-bovine RF serum. Fig. 17: Immunoreactive granules (large arrows) in ependymal endings (EE). Small arrows indicate external basal lamina of the brain; LE, leptomeningeal space. Bar, 1 μm. Fig. 18: Capillary located in the posterior commissure (PC) and surrounded by endfeet of ependymal processes. Arrows show immunoreactive secretory granules. E, Endothelium. Bar, 2 μm. (Insert A) Detailed magnification of secretory granules (arrows). Bar, 0.1 μm. (Insert B) Endfeet (ee) contacting a capillary (e) and containing immunoreactive granules (arrows); b, perivascular basal lamina. Bar, 0.2 μm. Both figures from Peruzzo *et al.* (1990).

only then forms a RF (Figs. 19 and 20). RF grows in a rostrocaudal direction by addition of newly released secretory material shifted continuously to its distal end. At this distal end located in the filum terminale, RF material becomes unpacked to form an amorphous mass, known as massa caudalis, that partially fills a terminal dilation of the central canal (Figs. 5 and 23) (for review, see Leonhardt, 1980).

Thus, RF has to be regarded as the result of a dynamic process involving various mechanisms such as (1) shift of the secretory material along the fiber structure; (2) prepackaging, packaging, and unpackaging of the secretory molecules; (3) chemical modification of the RF molecules when arriving at the distal end of the fiber; (4) hydrodynamic factors having influence on the formation and the shape of the fiber.

There is substantial evidence for distinguishing three different spatial and temporal stages of the CSF-insoluble secretion released by the SCO: (1) pre-RF material, (2) RF proper, (3) massa caudalis (E. M. Rodríguez *et al.*, 1987b).

A. Pre-Reissner's Fiber Stage

On release by the ependymal cells, the secretory material capable of forming RF first becomes aggregated in the form of a film on top of microvilli and cilia (Fig. 22). This film, recently regarded as "pre-RF material" (Rodríguez *et al.*, 1986, 1987b), was first demonstrated at the light-microscopic level with the same staining and histochemical procedures that revealed the intracellular secretory material (Wingstrand, 1953; Olsson, 1958a,b; Oksche, 1961; Sterba *et al.*, 1967a; W. Naumann, 1968; Rodríguez *et al.*, 1986, 1987b; Herrera and Rodríguez, 1990). The layer of pre-RF material has also been visualized by transmission electron microscopy (Müller and Sterba, 1965; Sterba *et al.*, 1967a; Wenger *et al.*, 1969; Rodríguez, 1970a; Krstić, 1973; Hofer *et al.*, 1980) and scanning electron microscopy (Krstić, 1975; Lindberg and Talanti, 1975; Weindl and Schinko, 1975; Collins, 1983; Sturrock, 1984). At the ultrastructural level, the pre-RF material appears in the form of loosely arranged bundles of thin filaments (Wenger *et al.*, 1969; Rodríguez, 1970a).

It seems most likely that both the material stored in the apical secretory

---

FIG. 19 Sagittal section through the bovine cervical spinal cord. PAP reaction using an antibody against bovine RF. Open arrow, fine threads in core of Reissner's fiber (RF); small arrows, immunoreactive spherical structures on surface of RF; large arrow, immunoreactive ependymal cell; E, ependyma. Bar, 100 μm. From S. Rodríguez *et al.* (1985).

FIG. 20 Scanning electron micrograph of the isolated bovine RF obtained after perfusing the central canal with saline. Bar, 10 μm.

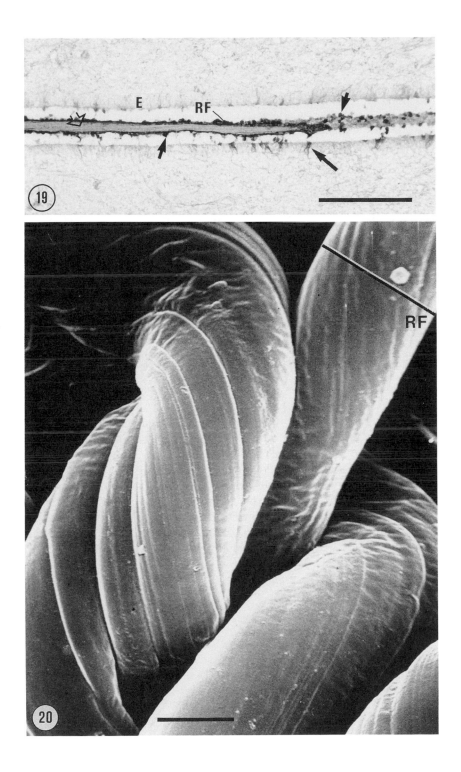

granules and that transiently arranged as pre-RF material correspond to the same processed forms of the secretory glycoproteins (Rodríguez *et al.*, 1986, 1987a,b; Lösecke *et al.*, 1984; Herrera and Rodríguez, 1990).

That the existence of the pre-RF stage of the material released into the ventricle is a vital phenomenon may be concluded from the following observations: (1) During ontogeny of the chick and duck SCO, pre-RF appears 4 days earlier than RF proper (Schoebitz *et al.*, 1986). (2) Biosynthetic labeling of the secretory material by use of radioactive precursors indicates that pre-RF becomes labeled 1 hr after the administration of the precursor, whereas RF shows labeling only 1 hr later (E. M. Rodríguez *et al.*, 1990a, see Section IV,B,4). (3) Convincing evidence for the existence of the pre-RF stage is provided by the *in vivo* use of an antiserum against RF (Fig. 21). The intraventricular administration of this antiserum, followed by the fixation of the brain and processing of the tissue sections for demonstration of IgG revealed that the injected antiserum labeled specifically the layer of pre-RF and RF, thus confirming that both structures are present in the living animal (S. Rodríguez *et al.*, 1990).

Very little is known of the nature of the modifications of the pre-RF material which finally lead to the formation of RF. As indicated by the biosynthetic labeling studies, transformation of pre-RF material into RF may occur within 1 hr. However, in the rat, secreted material may remain in the pre-RF stage for up to 3 days (S. Rodríguez *et al.*, 1990). The following findings indicate that chemical changes must occur during passage from the pre-RF stage to the RF stage: (1) The minor or even missing immunoreactivity of RF to the respective antisera is a transient phenomenon that should be ascribed to masking of epitopes (see following section). Masking of epitopes could reflect a higher degree of packaging. (2) The bovine SCO proper is characterized by two processed forms of secretory glycoproteins, whereas the bovine RF appears to contain at least six glycoproteins. This difference has been interpreted as a postrelease processing that takes place during the pre-RF stage (Nualart *et al.*, 1991).

FIG. 21   Ultrastructural immunocytochemistry of the rat SCO after *in vivo* administration of anti-bovine RF serum into the lateral ventricle CSF. Incomplete immunogold procedure omitting the first antibody reveals the sites of reaction of the *in vivo* injected antibodies. Reaction occurs exclusively in extracellular material, i.e., pre-RF (long arrows), and masses in between it and the apical surface of the ependymal cells. The injected antibodies did not reach intracellular structures in the SCO, as revealed by the negative reaction in apical secretory granules (short arrows). Bar, 2 μm.

FIG. 22   Ultrastructural immunocytochemistry. Complete immunogold procedure using anti-bovine RF serum on a thin section of normal rat SCO. Intracellular (secretory granules, short arrows) and extracellular (pre-RF, long arrows) secretory material displays strong and specific immunoreactivity. Bar, 2 μm.

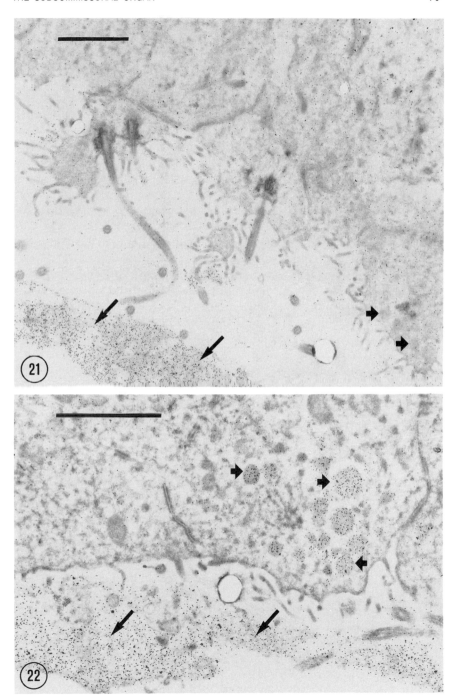

(3) When bovine RF is dissolved in a medium lacking disulfide-bond-blocking agents, such as 2-mercaptoethanol, a perplexing phenomenon is observed: after a few hours of being in solution, the RF components reaggregate and form a fiber resembling the native RF (Wolf and Sterba, 1972; unpublished observation by the authors). Formation of disulfide bonds may be an essential part of the mechanism of packaging of RF components and could be related to the high content of cysteine in RF glycoproteins (see Table I). However, the intimate chemical mechanism underlying the final assembly of the glycoproteins into such a geometrically regular structure (Fig. 20) (uniform diameter, fixed growing rate, etc.) remains enigmatic.

CSF-related factors appear to participate in RF formation. In chick embryos pre-RF material is first seen at the seventh day of incubation, and a RF proper is found in the Sylvian aqueduct only from the eleventh day of incubation (Schoebitz *et al.*, 1986). Thus, factors other than the ventricular release of secretory material are required for the formation of RF. Olsson (1958a) and Oksche (1961) have postulated that hydrodynamic factors of the CSF circulation play a role in the aggregation of the secreted glycoproteins in the form of a typical RF. Two recent findings strongly support this assumption. The rat SCO grafted under the kidney capsule continues to secrete into newly organized cavities (Fig. 10). The secreted material fills these cavities but does not form a RF-like structure (Rodríguez *et al.*, 1989). Rats with induced postnatal hydrocephalus do not possess a RF although their SCO continues to release secretory material into the ventricle (Irigoin *et al.*, 1990).

B. The Reissner's Fiber Proper

In 1860, Reissner described for the first time the presence within the central canal of *Petromyzon fluviatilis* of a "string (Fig. 2), which in cross sections has a diameter of 1.5 μm and is characterized by its high refringence, its extremely regular shape and by lying free within the central canal." Kutschin (1866) named the structure described by Reissner as "Reissner's fiber." Studnička (1899) described RF in a series of vertebrate species and regarded it as a secretory product of the adjacent ependymal cells. Sargent (1900) indicated that RF originates in the region of the third ventricle. Dendy and Nicholls (1910) established the relationship between RF and a highly differentiated ependyma covering the posterior commissure, the "subcommissural organ."

RF-like structures already exist, although in a different topographical position, in *Oikopleura* (Holmberg and Olsson, 1984) and amphioxus (Wolff, 1907; Olsson and Wingstrand, 1954). In most vertebrate species

RF is a well-developed, even spectacular structure (for references, see Olsson, 1958a,b; Oksche, 1961; Palkovits, 1965; Leonhardt, 1980; Tulsi, 1982). RF has been reported to be missing in the spinal cord of the bat (Wislocki *et al.*, 1956; E. M. Rodríguez *et al.*, 1987b), camel (Afifi, 1964), chimpanzee (Dendy and Nicholls, 1910), and man (for references, see Leonhardt, 1980). The diameter of RF varies from 0.3 μm in the amphioxus (Olsson and Wingstrand, 1954) to 20–50 μm in the bovine (Figs. 19 and 20). In large animals, RF may be 1 m or more in length. In the snake (*Python* sp.), it may reach a length of 3 m (see E. Legait, 1942).

At the ultrastructural level, RF was first studied with the transmission electron microscope by Afzelius and Olsson (1957) in the hagfish. RF is formed by filaments with a diameter ranging between 5 and 15 nm (Kohno, 1969; Rodríguez, 1970a; Hoheisel *et al.*, 1971). That RF is made up of densely packed filaments agrees well with observations made under the polarizing microscope (Sterba *et al.*, 1967c).

By using an antibody against bovine RF and comparing the immunoreactivity of RF in different species it becomes evident that there is a distinct interspecific difference in the intensity of the immunostaining of RF (S. Rodríguez *et al.*, 1987). Since the secretory glycoproteins are strongly immunoreactive at the site of their release into the CSF (pre-RF) and at the distal site of accumulation (massa caudalis), the lack of immunoreactivity in the "intermediate" portion of the material (RF proper) should be interpreted as a transitory stage in which the accessibility of the antibodies to the epitopes is decreased or prevented. Findings on the bovine RF indicate that the degree of packaging of RF material and the extent of immunoreactivity might be interrelated phenomena (E. M. Rodríguez *et al.*, 1987b). An alternative possibility to explain the pattern of immunoreactivity of RF refers to the spatial distribution of sialic acid residues within the fiber (S. Rodríguez *et al.*, 1987).

The observation of a RF in the regenerated lizard tail led Kolmer (1921) to suggest that the fiber moves in a rostrocaudal direction. Surgical transection of the spinal cord of the pike followed by the microscopic analysis of the proximal and distal segments of the spinal cord at different postsurgical intervals led Olsson (1957) to state that "fibre-substance is transported caudally where it accumulates as a caudal mass." Ontogenetic studies (Wingstrand, 1953; Schoebitz *et al.*, 1986) have led to a similar conclusion.

Two to 3 hr after the injection of [^{35}S]cysteine, radioactively labeled secretory material appears in the proximal portion of RF (see Section IV,B,4). With increasing time periods after the injection, the label is detected in more distal segments of RF. The progression of the labeled molecules along the fiber has been used to determine the growth rate of RF in different species. Thus, it has been estimated that, in the mouse, RF grows 10% of its total length per day (Ermisch, 1973); in the rat, 7% per day (Herrera, 1988); in the lamprey, 1% per day; and in the carp, 0.5 to 2% per day (Sterba et al., 1967b; Ermisch et al., 1968).

C. Massa Caudalis

In amphioxus and in lower vertebrates, the central canal extends only for a short distance beyond the caudal end of the spinal cord, whereas in higher vertebrates it continues for a long distance along a terminal filum. At the caudal end of the spinal cord proper the central canal is encompassed by an ependymal tube, directly surrounded by meninx primitiva (Studnička, 1899; Olsson, 1955; Wislocki et al., 1956). The distal end of the central canal displays a dilation known as the terminal ventricle or ampulla caudalis (Figs. 5 and 23), (Olsson, 1958b). When reaching the ampulla caudalis, RF ends as an irregular mass, the so-called massa caudalis (Studnička, 1899; Kolmer, 1905; Olsson, 1955, 1958a,b; Wislocki et al., 1956; Hofer, 1964; Fährmann, 1963; Heuschneider, 1968; Oksche, 1969).

Only a few ultrastructural studies of the massa caudalis have been published. These detailed reports deal with the massa and ampulla caudalis of lampreys (Sterba and Naumann, 1966; Hofer et al., 1984; S. Rodríguez et al., 1987). In these cyclostomes, the massa caudalis appears as a conglomerate of fibrillar material, less densely packed than in RF and in close apposition to the dorsal wall of the ampulla. The only scanning electron-microscopic study of the massa caudalis (Castenholz, 1984) deals with the situation in the cat and rabbit. Castenholz described the massa caudalis as a dense network of fine filaments, with finger-shaped processes extending from the core to the ependymal surface.

Since RF grows constantly in a rostrocaudal direction, RF material constantly arrives at the ampulla caudalis. In addition, since the volume of the massa caudalis does not undergo conspicuous variations, the mechanism(s) responsible for the "escape" of RF material from the ampulla must also operate in a continuous manner. Therefore, an equilibrium between the rate of supply of RF material to the ampulla and the rate of disappearance of this material may be considered. It may also be suggested that the massa caudalis should be regarded as a transitory stage in the process leading to the decomposition of RF.

Does RF material undergo certain changes when shifting from the RF state to the massa caudalis state? The transmission and scanning electron-microscopic studies mentioned above and the strong immunoreaction of the massa caudalis with antisera against RF (S. Rodríguez et al., 1987; Yulis et al., 1990) indicate a restructuring or unpackaging of the filaments forming RF. Using histochemical methods, W. Naumann (1968) detected in the massa caudalis a number of enzymes, several of them hydrolytic. A lectin histochemical study of the lamprey massa caudalis (S. Rodríguez et al., 1987) led S. Rodríguez et al. to conclude that the glycoproteins located at the periphery of the massa caudalis have lost their sialic acid residues and possess galactose as the terminal residue. Indeed, it is known that sialoglycoproteins become degradable after the loss of their sialic acid residue. Due to this chemical modification, they become available to macrophages as galactose remains as the terminal sugar residue (see Sharon and Lis, 1982).

D. Fate of Reissner's Fiber Material

The continuous arrival of RF material in the ampulla caudalis implies that mechanisms leading to the discharge of this secretory substance from the ampulla must operate continuously (Peruzzo et al., 1987). What is the fate of RF material beyond the level of the ampulla caudalis? As early as in 1899 Studnička suggested that RF gradually fades from the ampulla into the surrounding connective tissue that is well supplied with "lymphatic vessels."

In lamprey larvae, where a detailed ultrastructural analysis of the ampulla caudalis and adjacent structures has been performed (Fig. 23) (Hofer et al., 1984; S. Rodríguez et al., 1987; Peruzzo et al., 1987) it has been clearly shown that (1) the openings in the dorsal wall of the ampulla establish a direct communication between the latter and large cavities or lacunae lined by slender processes of cells of unknown nature ("lympathic vessels"? of Studnička, 1899); (2) the lacunae are in open communication with blood capillaries devoid of a basal lamina ("lymphatic vessels"?) which, in turn, communicate with typical blood capillaries endowed with an endothelium and a basal lamina; (3) in all these compartments a RF-immunoreactive fibrillar material identical to that forming the massa caudalis can be seen (Peruzzo et al., 1987). The fact that no immunoreactive material was found outside these structures led Peruzzo et al. (1987) to conclude that the local blood capillaries appear to represent the only final target of RF material arriving at the ampulla caudalis.

Thus, the passage of RF material from the ampulla caudalis of lower vertebrates into blood vessels, once suggested at the light-microscopic level (Olsson, 1955; Hofer, 1964; W. Naumann, 1968), later postulated on the basis of conventional electron-microscopic evidence (Sterba and Naumann, 1966; Oksche, 1969), and further supported by two detailed ultrastructural studies (Hofer et al., 1984; S. Rodríguez et al., 1987), has been largely substantiated by the use of ultrastructural immunocytochemistry (Peruzzo et al., 1987). The definitive proof, i.e., the detection of RF

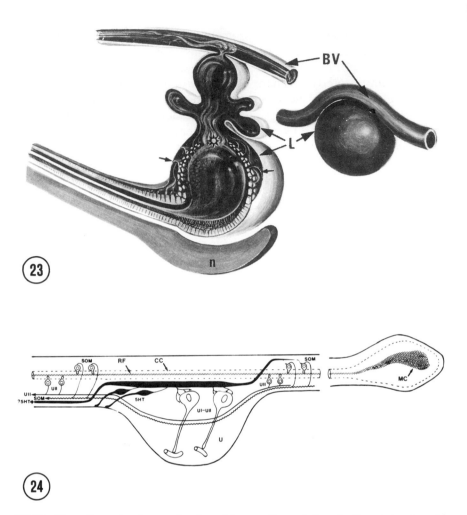

FIG. 23 Three-dimensional reconstruction of the ampulla caudalis and adjacent structures in lamprey (*Geotria australis*) larvae. Note Reissner's fiber in the central canal and its massa caudalis occupying the ampulla caudalis. Immunoreactive RF material is shown to escape via wide intercellular spaces in the dorsal wall and via a large central opening (open star), to fill the lumen of lacunae (L), and to penetrate into the lumen of blood vessels (BV). Small arrows point to the thin wall of lacunae; *n*, notochord. The drawing on the right represents the type of spatial relationship found between a blood capillary and a large lacuna. From Peruzzo *et al.* (1987).

FIG. 24 Diagram summarizing putative neuronal interrelationships in the caudal spinal cord of coho salmon. 5-Hydroxytryptamine-immunoreactive (5HT-IR), thin, beaded fibers inner-vate caudal secretory neurons and somatostatin-immunoreactive (SOM) CSF-contacting neurons in the filum terminale. These fibers probably originate from local (caudal) 5HT-IR neurons, serotoninergic neurons located in upper regions of the CNS, or both. Dendrites of urotensin II-immunoreactive (UII) and SOM-IR CSF-contacting neurons contact Reissner's fiber (RF) inside the central canal (CC). Thin fibers arising from these CSF-contacting neurons form apparently ascending pathways. U, urophysis; MC, massa caudalis. From Yulis *et al.* (1990).

material in circulating blood, awaits further investigation by means of adequate micromethods.

Studies in higher vertebrates (Wislocki *et al.*, 1956; Tulsi, 1982; Hofer *et al.*, 1987; unpublished observations by the authors) indicate that the architecture of the caudal end of the central canal shows marked differences with regard to the situation in lower vertebrates. In several mammalian species investigated, the dorsal wall of the filum terminale displays preterminal openings which connect the central canal with the subarachnoid space. Via these gaps, a passage of RF material from the central canal has been observed (Wislocki *et al.*, 1956). It has been proposed that ventricular CSF may also escape from the central canal following this route (Nakayama, 1976). Thus, a soluble form of RF material may gain access from this location to the subarachnoidal CSF or to the local blood vessels, as suggested by Wislocki *et al.* (1956).

There is evidence that in a number of submammalian species the meninx primitiva does not fully differentiate into pia and arachnoid. Consequently, these animals may lack a continuous subarachnoid space and free-circulating external CSF (cf. Bargmann *et al.*, 1982). In contrast, mammals possess a continuous external leptomeningeal compartment where CSF can circulate freely (Davson, 1967). In mammals, CSF extends from the central cavities (ventricles and central canal) to the subarachnoid space to be finally filtered to venous blood. In submammalian species, the ampulla caudalis and its close association with blood vessels (see above) may represent the most important site of drainage of CSF into blood. Due to the special structural features of the dorsal wall of the lamprey ampulla caudalis and of the associated blood vessels, it may be postulated that the lamprey ampulla caudalis represents a primitive form of arachnoid villus. Indeed, there appears to be a certain degree of similarity between the ultrastructure of the ampulla caudalis–vasculature interphase in the lamprey (Hofer *et al.*, 1984; Peruzzo *et al.*, 1987) and the structural pattern in arachnoid villi (Butler *et al.*, 1983). It might be postulated that, in mammals, the secretory products of the SCO that enter the subarachnoid CSF either rostrally via the leptomeningeal projections of the SCO parenchyma (see Section V), or caudally via the openings of the central canal at the level of the filum terminale, are finally drained into the venous system of the brain circulation. The observation that the cisternal CSF contains soluble, assayable RF material (H. Richter, F. Nualart, P. Peña, and E. M. Rodríguez, unpublished observation) may support this hypothesis.

VII. Origin of Reissner's Fiber Material

It has been postulated that, depending on the species and the developmental stage, RF-like material may be discharged by three different secretory structures of the chordate central nervous system: the SCO, the flexural organ, and the infundibular organ (Olsson, 1958b).

The central nervous system of the acranian chordate *Branchiostoma lanceolatum* (amphioxus) resembles mainly the spinal cord and possesses only a minor brainlike extension. The latter is devoid of structures that could be regarded as homologs of the epithalamic circumventricular organs. However, in the absence of a SCO, this primitive "brain" displays a RF-like differentiation. This RF has a similar light-microscopic appearance, staining properties (Gomori, PAS), and caudal termination as the RF of vertebrates (Olsson and Wingstrand, 1954; Olsson, 1955, 1958b; Hofer, 1959). The peculiar fiber of *Branchiostoma* originates in a group of secretory cells located at the ventral aspect of the entrance to the central canal (Olsson and Wingstrand, 1954; Hofer, 1959) (Fig. 25), corresponding to the so-called "infundibular organ" (Boecke, 1902). Ultrastructural studies of the infundibular organ clearly indicated that its cells are secretory in nature and that they release their secretory material into the ventricle (Olsson, 1962; Obermüller-Wilén, 1976).

In early stages of the ontogenetic development of *Salmo salar, Essox lucius, Xenopus laevis* (Olsson, 1956), and *Etmopterus spinax* (Altner, 1964), a RF-like structure was found to originate not at the SCO, but from a secretory ependyma located on the frontal tip of the ventral brain fold (plica ventralis encephali), called the "flexural organ" by Olsson (1956) (Fig. 25). The ependymocytes of the flexural organ closely resemble the parenchymal elements of the SCO; they are elongated, extend basal processes to the external limiting membrane, and contain granules of secretory material displaying affinity for chrome–alum hematoxylin–phloxine (Gomori) and periodic acid–Schiff (PAS) stains (Olsson, 1958b). Concomitantly with the flexural organ, the SCO is also present at early stages of salmon development and shows signs of active secretion. As the ontogenic development proceeds, fibrillar strands arising from the flexural organ join a characteristic RF originating in the SCO. Later on, the contribution of the flexural organ ceases and the function of RF production is assumed exclusively by the SCO. According to Olsson (1956), "the flexural cells of embryonic vertebrates, the infundibular cells of amphioxus and the subcommissural cells of vertebrates all have positive chrome haematoxylin secretion. This does not mean that the cells produce identical secretions, but it seems unlikely that the composition of the different Reissner's fibers should vary within great limits."

A recent immunocytochemical study performed in our laboratory using an antiserum against bovine RF has shown that the SCO and flexural organ of salmon embryos and the infundibular organ of amphioxus are strongly immunoreactive (Olsson *et al.,* 1991) (Fig. 26). It was concluded from these observations that all three ependymal organs secrete RF-like mate-

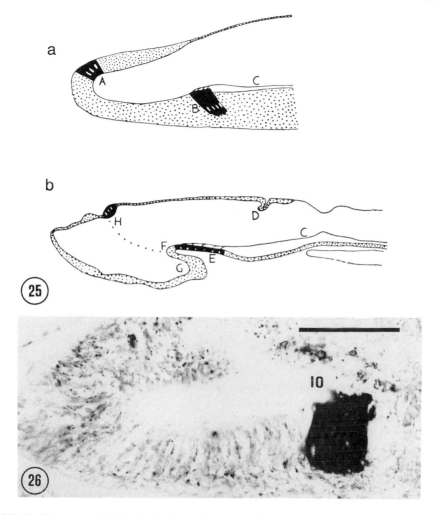

FIG. 25   Diagrams of (a) the brain of amphioxus and (b) the brain of a 14-day-old (5.6 mm) salmon embryo. A, Secretory ependyma; B, infundibular organ. C, Reissner's fiber; D, plica rhombomesencephalica; E, flexural organ; F, plica ventralis; G, infundibulum; H, sub-commissural organ. Amphioxus, ×290; *Salmo,* ×60. From Olsson (1956).

FIG. 26   Section of the brain vesicle of a specimen of amphioxus (12 mm) immunostained with anti-bovine RF serum. Note the strong reaction in the cells of the infundibular organ (IO). Bar, 5 μm.

rial and that the nature of at least one of the RF components is highly conservative. The lack of immunoreactivity of the amphioxus infundibular organ to antisera raised against secretory compounds extracted from the bovine SCO, and its strong reactivity with antisera raised against the bovine RF, led Olsson *et al.* (1991) to the assumption that (1) the SCO and infundibular organ obviously secrete materials of different chemical nature; (2) RF of adult vertebrates might be formed by compounds discharged from two different sources, i.e., the SCO and a second—so far unknown—secretory structure. The latter, and not the SCO, would have to be regarded as the homolog of the infundibular organ of amphioxus. In this respect, some ultrastructural observations in the opossum (Tulsi, 1982) and the immunocytochemical findings in the bovine central canal (S. Rodríguez *et al.*, 1985) (Fig. 19) may be of interest. These authors have suggested that ependymal cells of the central canal might release materials to be added to RF.

VIII. Mechanism(s) Controlling the Secretory Activity of the SCO

A. Innervation of the SCO

Several reports contain information concerning the innervation of the SCO in a variety of species, such as fishes (W. Naumann, 1968; Hafeez and Zerihun, 1974; Korf, 1974), amphibians (Oksche, 1955; Murakami and Tanizaki, 1963; Mautner, 1965; Oksche and Vaupel-von Harnack, 1965; Altner, 1968; Diederen, 1970; Rodríguez, 1970b; Paul *et al.*, 1971; Paul, 1972), reptiles (Ariëns Kappers, 1967), birds (Palkovits and Wetzig, 1962), and mammals (Stanka, 1964; Isomäki *et al.*, 1965; Fuxe, 1965; Leonieni and Rechardt, 1972; Tulsi, 1983). However, in spite of the peculiar structural features described, none of the above-mentioned reports provides an overall interpretation of the nature of the innervation of the SCO. On the basis of rather indirect evidence, the existence of catecholaminergic and cholinergic systems related to the SCO has been proposed (Miline *et al.*, 1969; Leonieni and Rechardt, 1972; Rechardt and Leonieni, 1972; Kimble and Møllgard, 1973; Miline, 1974; Wake *et al.*, 1974; Kohl, 1975; Meunier and Bouchaud, 1978). The existence of peptidergic fibers among the SCO ependymocytes, described early on by use of the Gomori stain (H. Legait and Legait, 1956), has been supported by recent immunocytochemical or radioimmunoassay studies. According to these reports, luteinizing hormone-releasing hormone (LHRH) (Kizer *et al.*, 1976; Matsuura and Sano, 1987), vasopressin (Buijs and Pévet, 1980; Matsuura and Sano,

1987), arginine vasotocin (Rosenbloom and Fisher, 1975), oxytocin (Dogterom *et al.*, 1979; Buijs and Pévet, 1980), oxytocin–neurophysin (Fernández-Llebrez *et al.*, 1987b), mesotocin (Fernández-Llebrez *et al.*, 1987b), somatostatin (Pelletier *et al.*, 1975), substance P (Hökfelt *et al.*, 1978), and α-neo-endorphin (Zamir *et al.*, 1983) are present in the area of the SCO.

The most abundant and consistent evidence on the nature of SCO innervation indicates a monoaminergic input to the SCO cells. Formaldehyde-induced fluorescence, radioautography, and immunocytochemical studies have established the existence of a dense plexus of serotoninergic fibers along the basal portion of the SCO (Fuxe, 1965; Fuxe *et al.*, 1968; Björklund *et al.*, 1972; Wiklund, 1974; Fuxe and Jonsson, 1974; Bouchaud and Arluison, 1977; Wiklund *et al.*, 1977; Calas *et al.*, 1978; Bouchaud, 1979; Møllgard and Wiklund, 1979; Y. K. Takeuchi and Sano, 1983; Bouchaud and Bosler, 1986; Matsuura and Sano, 1987; Matsuura *et al.*, 1989). Electron-microscopic studies after administration of $5[^3H]$-HT or specific neurotoxic destruction of the serotoninergic innervation by application of 5,6-dihydroxytryptamine (5,6-DHT) or 5,7-DHT have revealed that the SCO of the rat is densely innervated by serotonin-containing nerve fibers which establish well-differentiated axoglandular synapses on the basal processes and at the laterobasal aspects of the ependymal cells (Bouchaud and Arluison, 1977; Møllgard *et al.*, 1978; Calas *et al.*, 1978; Bouchaud, 1979; Møllgard and Wiklund, 1979).

Bouchaud (1979), using high-resolution radioautography after administration of $5[^3H]$-HT, observed populations of labeled and unlabeled terminals establishing synaptic contacts with ependymocytes of the rat SCO. This finding, together with morphological differences between labeled (serotoninergic) and unlabeled terminals, allowed Bouchaud to propose a multiple innervation of the rat SCO. Results obtained by Matsuura and Sano (1987) in the dog are also in line with a multiple innervation of the SCO.

The pattern of the serotoninergic innervation found in the rat and Japanese squirrel (Matsuura *et al.*, 1989) cannot be regarded as a general morphological feature present in all mammals or even in rodents. In the mouse, guinea pig, Mongolian gerbil, rabbit, and monkey, this plexus appears to be poorly developed or even missing (Wiklund *et al.*, 1977; Tramu *et al.*, 1983; Matsuura *et al.*, 1989). These differences regarding the pattern of serotoninergic innervation of the SCO may indicate a substantial, species-dependent heterogeneity in the mode of the neural control of SCO cells.

Ontogenetic studies indicate an absence of serotoninergic innervation in neonatal specimens of the rat, gerbil, mouse, rabbit, and cat (Wiklund, 1974; Wiklund *et al.*, 1977; Marcinkiewicz and Bouchaud, 1986; Matsuura *et al.,* 1989). There is evidence that serotonin-immunoreactive fibers reach the rat SCO as early as the third postnatal day. During the first weeks of development a process of maturation of presynaptic and postsynaptic elements takes place until they reach the mature state observed in adult specimens, approximately at the end of the first postnatal month (Wiklund, 1974; Wiklund *et al.,* 1977; Marcinkiewicz and Bouchaud, 1986; Matsuura *et al.,* 1989). In the cat and dog the serotoninergic innervation of the SCO also develops postnatally (Matsuura *et al.,* 1989). Therefore, it appears that certain (so far unknown) events occurring during the early postnatal stage of development are essential for the completion of the serotoninergic innervation.

In mammals, the main population of serotonin-immunoreactive perikarya in the CNS is located in the raphe nuclei of the brain stem (Dahlström and Fuxe, 1964). The results obtained by Leger *et al.* (1983), performing electrolytic lesions of different raphe nuclei, revealed that the serotoninergic innervation of the rat SCO is mainly derived from nuclei raphe centralis superior and raphe dorsalis, each nucleus contributing about one-third of the input; the remainder is suggested to originate from the nucleus raphe pontis. Ueda *et al.* (1988) reported that, in rats treated with 5,6-DHT, the serotoninergic innervation of the SCO may be reestablished by transplantation of fetal raphe nuclei into the third ventricle. This evidence supports the view that SCO cells represent a target for serotoninergic fibers originating in neurons of raphe nuclei. To date, the pathways of serotoninergic fibers from the mesencephalic raphe nuclei to the SCO have not been precisely established.

Autoradiographic studies performed in rats after intraventricular administration of [^{3}H]GABA (γ-aminobutyric acid), measurements of glutamate decarboxylase (GAD) activity, and immunocytochemical localization of GAD show accumulation of the labeled amino acid and presence of GAD in SCO ependymocytes and in numerous nerve terminals and fibers contacting the SCO (Gamrani *et al.,* 1981; Weissmann-Nanopoulos *et al.,* 1983). Two populations of labeled nerve terminals were found: one that is morphologically and pharmacologically similar to the serotoninergic terminals, and another that resembles closely the population described as nonserotoninergic by Bouchaud (1979) (see above). Furthermore, the first population disappears after 5,7-DHT treatment and lesions of mesencephalic raphe nuclei, whereas the second group survives both procedures. These results suggest that an important part of the SCO afferents might be serotoninergic and GABAergic, and that they originate in perikarya located in the mesencephalic raphe nuclei. The origin of the population of putative GABAergic terminals surviving after lesions of the raphe nuclei remains to be determined.

It has been postulated by several authors that the serotoninergic innervation influences the secretory activity of SCO cells and that the metabolic state of the GABAergic elements of the SCO is also under serotoninergic control. Møllgard and co-workers (1978, 1979) have shown that neurotoxic destruction of the serotoninergic input to the SCO of the adult rat leads to an increase in the histochemically detectable amount of the secretory material; this effect is not reversed even after 8 months, when the innervation of the SCO has been substituted by nonserotoninergic fibers (Wiklund and Møllgard, 1979). Møllgard and co-workers concluded that the secretory activity of the SCO is under a strong inhibition by serotoninergic fibers. This hypothesis was further supported by lesions of the raphe nuclei causing an intense secretory activity in the SCO cells detectable by histochemical methods (Leger *et al.*, 1983). It is important to consider that all the publications reporting an inhibitory effect of 5-HT on SCO activity are based on morphological techniques that only provide a rough estimate of the amount of secretory material in SCO secretory cells. Studies on the turnover of the SCO secretion under conditions of differential serotoninergic innervation are required in order to raise a more definitive functional conclusion.

B. Interrelationship between Reissner's Fiber and the Walls of the Central Canal

Some of the hypotheses postulated by different authors about the functional role of the RF have been based on a possible interaction between RF and constituents of the central canal walls (cf. Leonhardt, 1980). Ependymal cells and also CSF-contacting neurons have been postulated as likely candidates for an interrelationship with RF.

It has been shown that the RF, while passing throughout the central canal, establishes a close relationship with cilia and microvilli projecting from ependymal cells (Woolam and Collins, 1980; Kohno, 1969; Sturrock, 1984; Erhardt and Meinel, 1983). The apparent attraction exerted by RF on ependymal cilia and microvilli, in addition to the demonstration of cells and cell debris attached to RF (Tulsi, 1982; Sturrock, 1984), suggests that the fiber possesses an intrinsic adhering capacity, which has been regarded as a manifestation of a highly adhesive surface or to the presence of electric charges (sialic acid residues?) (Weindl and Schinko, 1975; Sturrock, 1984; Woolam and Collins, 1980).

By means of scanning electron microscopy, strands of apparent RF material have been shown extending from the apical pole of the ependyma of the central canal to the RF (Tulsi, 1982; Sturrock, 1984). These filaments are not stained by neurosecretory stains, suggesting that they may differ in composition from the secretory products of the SCO (Tulsi, 1982).

S. Rodríguez *et al.* (1985) reported that, at lower segments of the bovine central canal, some ependymal cells and a flocculent material surrounding the RF display an intense immunoreactivity with an antiserum against bovine RF (Fig. 19), but are not reactive with the Gomori and PAS stains. Two possibilities arise from these observations: (1) ependymal cells located in discrete regions of the central canal release into the CSF a nonglycosylated material resembling SCO secretory products; or (2) part of the RF material, after a loss of its carbohydrate component and solubilization within the central canal CSF, is absorbed by the above-mentioned ependymal cells (S. Rodríguez *et al.,* 1985).

A population of spinal CSF-contacting neurons, analogous to the CSF-contacting neuron systems found in the hypothalamus, has been described in the medulla oblongata and spinal cord of several species (for review, see Vigh and Vigh-Teichman, 1973). The general features of this group of neurons are rather uniform among the vertebrates. The bipolar cells project dendritelike processes toward the central canal. These processes display a terminal enlargement which projects an atypical cilium and numerous stereocilia extending radially into the CSF, where they touch the surface of the RF (Vigh and Vigh-Teichman, 1973; Vigh *et al.,* 1977).

Acetylcholinesterase (AChE)-positive nerve cells (Vigh-Teichman *et al.,* 1970; Vigh *et al.,* 1970; Vigh and Vigh-Teichman, 1971) and monoamine-fluorescent perikarya (Baumgarten *et al.,* 1970; Vigh *et al.,* 1970; Vigh and Vigh-Teichman, 1971) were demonstrated around the central canal of several species. Immunocytochemical studies (Yulis and Lederis, 1986, 1988a,b; Yulis *et al.,* 1990) in teleosts have shown the presence of urotensin II- and somatostatin-immunoreactive CSF-contacting neurons along the central canal, from the medulla oblongata to the filum terminale. The CSF-contacting neurons containing immunoreactive urotensin II are arranged as a longitudinal row in the ventromedial wall of the central canal. These neurons project (1) dendritelike processes into the CSF, where they contact RF; (2) beaded axons toward the lateroventral external surface of the spinal cord and also to more remote brain regions.

Axosomatic and axodendritic synapses on spinal CSF-contacting neurons have been reported (Vigh and Vigh-Teichman, 1973). Urotensin II-immunoreactive CSF-contacting neurons appear to be innervated by somatostatin-containing axons (Yulis and Lederis, 1988b), probably originating in somatostatin-immunoreactive CSF-contacting neurons located in the vicinity (Yulis *et al.,* 1990). In the filum terminale of the salmon, somatostatin-containing CSF-contacting neurons, in turn, appear to be innervated by serotoninergic fibers originating from a "caudal serotoninergic system" (Fig. 24) (Yulis *et al.,* 1990).

C. Probable Feedback Mechanism

In spite of increasing evidence suggesting the existence of a vascular route of release of SCO secretion (see above) and also a discharge of soluble secretory products into the CSF, it is reasonable to consider the production of RF material as the main function of the SCO. At present, the physiological role of RF is still a matter of speculation. Thus, assumptions concerning the control mechanisms involved in regulation of the production rate and the composition of the SCO secretion are mostly based on structural findings and not on responses elicited by a biological action of the RF material on central and peripheral target sites.

CSF-contacting neurons present along the entire length of the central canal are located in a very "strategic" position in relation to RF (Fig. 24). Their dendrites contacting the RF could be engaged in a feedback mechanism controlling the activity of the SCO (Yulis *et al.*, 1990). Beaded axons projecting from spinal urotensin II- and somatostatin-immunoreactive CSF-contacting neurons form ascending bundles that branch off widely in several regions of the teleost brain (Yulis and Lederis, 1986, 1988a). The existence of CSF- and RF-contacting neuronal systems in teleosts may establish the structural basis of a control of SCO activity in response to signals originating in the compartment of the central canal. It appears noteworthy that the SCO of teleosts lacks a serotoninergic innervation. In addition, a system of CSF-contacting serotoninergic neurons occurs in the paraventricular and posterior recess organs of the fish brain (Meurling and Rodríguez, 1990); by means of a release of serotonin into the ventricular CSF it could exert a control over the activity of the SCO.

IX. Ontogenetic Development of the SCO

The embryonic SCO has been the subject of numerous investigations carried out in several species, i.e., petromyzontids (Sterba *et al.*, 1965, 1967a), a shark (Altner, 1964), an oviparous teleost (Marini *et al.*, 1978), trout (Olsson, 1956), several species of amphibians (Mazzi, 1954; Marini, 1957, 1962; Oksche, 1961; W. Naumann, 1986), chicken (Schumacher, 1928; Wingstrand, 1953; Ziegels, 1977; Schoebitz *et al.*, 1986; Bruel *et al.*, 1987; W. Naumann *et al.*, 1987; Karoumi *et al.*, 1990a), duck (Schoebitz *et al.*, 1986), canary (Kux, 1929), and several mammalian species (Talanti, 1959; Rakic and Sidman, 1968; Kohl and Linderer, 1973; Castañeyra-Perdomo *et al.*, 1983b; Marcinkiewicz and Bouchaud, 1983, 1986; Taniguchi *et al.*, 1985; W. Naumann *et al.*, 1987; R. Meiniel *et al.*, 1990). The human fetal SCO will be dealt with separately.

The ontogeny of the chick SCO has been the subject of the most detailed and multimethodological studies. Using classical stains for neurosecretion, Wingstrand (1953) and Ziegels (1977) reported that the secretory material of the SCO first appears in 7-day-old chick embryos. However, immunocytochemical studies have shown a much earlier onset of the synthesis of the secretory material, viz., during the third day of incubation (Schoebitz *et al.*, 1986; W. Naumann *et al.*, 1987; Karoumi *et al.*, 1990a) (Fig. 27). Concanavalin A- and wheat germ agglutinin-positive material was detected at days 5 and 7 of incubation, respectively (Bruel *et al.*, 1987). This indicates a delayed glycosylation, especially the steps occurring in the Golgi apparatus, in contrast to the surprisingly early synthetized secretory proteins. The ventricular release of secretory material observed at day 7 of incubation (Wingstrand, 1953; Schoebitz *et al.*, 1986; Karoumi *et al.*, 1990a) (Fig. 28) agrees well with the appearance of wheat germ agglutinin-positive (post-Golgi) material. Despite this early ventricular release of secretory material, a RF proper becomes visible in the Sylvian aqueduct only at day 11 of the incubation (Schoebitz *et al.*, 1986). At day 12, RF has already reached the lumbar spinal cord, indicating a growth rate of about 70–80% of total length per day. Thus, the developing RF appears to grow much faster than RF in adult specimens (see Section VI,B). Immunoreactive hypendymal cells first appear on day 7 of incubation. From days 7 to 21, they proliferate, develop processes, and migrate in a dorsal direction.

In an investigation of the morphogenesis of the mouse SCO with the use of [^{3}H]thymidine and autoradiography, Rakic and Sidman (1968) established that the cells of the SCO arise between days 11 and 19 of embryonic life. They also observed that (1) DNA synthesis in the SCO cells started at least 1 day earlier than in the adjacent ependymal elements; (2) SCO cells formed during the embryonic period persisted through the adult life, thus indicating that SCO cells belong to the "nonrenewal" category; (3) the first signs of secretory activity appeared at day 14 of gestation. Full differentia-

---

FIG. 27   Three-day-old chick embryo immunostained with anti-bovine RF. The SCO (arrow) is the only positive structure. Tc, Telencephalon; Dc, diencephalon; Mc, mesencephalon; Mt, metencephalon; Ist, isthmus; vt, velum transversum; or, optic recess. Bar, 500 μm. From Schoebitz *et al.* (1986).

FIG. 28   Seven-day-old chick embryo immunostained with anti-bovine RF. Immunoreactive material appears in the SCO and on floor (arrows) of the Sylvian aqueduct (sa). OL, Optic lobe; P, pineal. Bar, 500 μm. From Schoebitz *et al.* (1986).

FIG. 29   Sagittal section of the primordium of the pineal (P) and subcommissural (SCO) organs in a 3-month-old human fetus Bouin's fixative and chrome–alum hematoxylin–phloxin stain. V, Ventricle; HC, habenular commissure; PC, posterior commissure; SA, Sylvian aqueduct. Bar, 500 μm. From Oksche (1961).

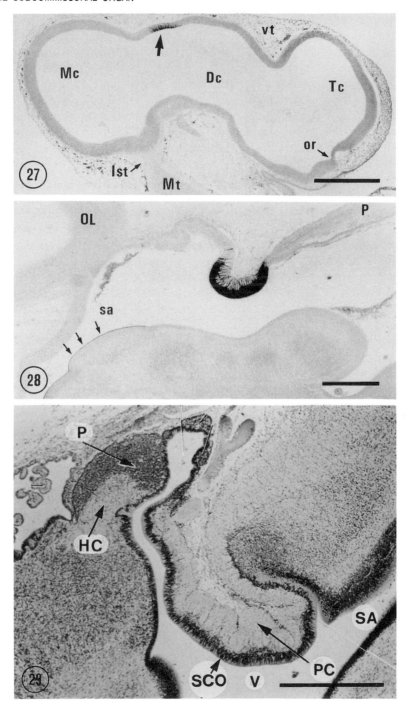

tion of the cells of the mouse SCO is, however, attained during the first postnatal month (Castañeyra-Perdomo *et al.,* 1983b). Similar findings have been reported for the rat SCO, with the onset of the secretory activity around prenatal day 15–16, with a progressive completion of differentiation during the four postnatal weeks (Kohl and Linderer, 1973; Marcinkiewicz and Bouchaud, 1983, 1986; W. Naumann *et al.,* 1987). In the bovine, with a gestational period of 9 months, 2-month-old embryos already display a morphologically differentiated SCO containing immunoreactive secretory material (R. Meiniel *et al.,* 1990).

In the case of the chicken and bovine, the SCO appears to represent one of the first secretory structures of the CNS to differentiate. This latter statement also holds true for poikilothermic vertebrates, as has been shown by Oksche (1961) in his studies of frog larvae. Considering the early onset of the secretory activity of the SCO in ontogeny, and the maintenance of this activity throughout the embryonic life, the possibility that the material elaborated by the SCO might participate in basic developmental mechanisms of the CNS must be kept in mind.

The human SCO reaches its maximum development during embryonic life (Dendy and Nicholls, 1910; Puusepp and Voss, 1924; Pesonen, 1940; E. Legait, 1942; Oksche, 1956, 1961, 1964; Wislocki and Roth, 1958; Olsson, 1961; Palkovits, 1965; Møllgard, 1972; Castañeyra-Perdomo *et al.,* 1985) (Fig. 29). After birth, the SCO undergoes regressive changes. In the adult human only remnants of SCO parenchyma can be detected (Puusepp and Voss, 1924; Pesonen, 1940; Oksche, 1961, 1964) (Fig. 6). According to Oksche (1969), the regression of the human SCO has already begun during the last period of fetal life.

On the other hand, Palkovits (1965) concluded that, in the human, the SCO remains highly differentiated during the first postnatal year; then it undergoes a progressive involution until the age of 30 years. In the human, a RF is apparently missing (see Leonhardt, 1980). However, material stainable with the periodic acid–Schiff (PAS) and aldehyde–fuchsin (Gomori) techniques has been shown to be present in the fetal human SCO, concentrated at the ventricular pole of the ependymal cells (Oksche, 1954, 1961, 1969; Wislocki and Roth, 1958; Olsson, 1961; Palkovits, 1965), and, according to some authors, also in the hypendymal elements (Palkovits, 1965; Møllgard *et al.,* 1973). Most authors have interpreted the PAS- and Gomori-positive material as a sign of the secretory activity of the fetal human SCO.

This possibility has been challenged by Rodríguez *et al.* (1984a), who found that the SCO of anthropoid apes and human embryos does not react with a series of different antibodies raised against secretory products extracted from the bovine RF and SCO (see also E. M. Rodríguez *et al.,* 1990b). The same antisera react clearly with the SCO secretion of most vertebrate species. Also, at the ultrastructural level the findings are contradictory. Murakami *et al.* (1970) interpreted their ultrastructural observation as an indication that the human fetal SCO is rudimentary, not secretory in nature. On the other hand, Oksche (1969, and unpublished observations) described in the SCO of a 3-month-old fetus (Fig. 32): (1) numerous cisternae of the RER filled with flocculent material, similar to those in the SCO of other vertebrates; (2) a well-developed Golgi apparatus formed by numerous dictyosomes; (3) images indicative of ventricular release of material; (4) ependymal processes contacting the external brain surface with endfeet containing electron-dense profiles resembling those found in the vascular and leptomeningeal processes of SCO cells of most vertebrate species.

A recent lectin histochemical study gives support to Oksche's ultrastructural findings and favors the view of a secretory activity of the human fetal SCO (E. M. Rodríguez *et al.*, 1990b) (Figs. 30 and 31). These authors postulated that the SCO of human fetuses secretes glycoproteins with a carbohydrate chain similar to the carbohydrate material elaborated by the SCO of other vertebrate species; only the protein backbone appears to be different.

X. Phylogenetic Development of the SCO

In comparative terms, the SCO is one of the most constant structures of the vertebrate brain (Oksche, 1961, 1969; Rodríguez *et al.*, 1984a; for earlier observations, see Bargmann, 1943). Irrespective of certain species-dependent variations in the shape and the differentiation of its ependymal and hypendymal components, there is a high degree of conformity in the location and fine structure of the SCO (Fig. 6). For this reason, in the present context, there is no need for a detailed, systematically based, comparative description of the SCO. The SCO occupies the entrance to the Sylvian aqueduct, and it shows a close spatial relationship to the pineal organ. A further remarkable, although more remote structure of this epithalamic complex of circumventricular organs is the choroid plexus of the third ventricle.

Already, the earliest phylogenetic manifestations of the SCO display the general morphological pattern of a SCO–RF complex (see Section VII), as can be observed in the cyclostomes (lampreys and hagfish). Both orders of cyclostomes possess a well-developed, secretory active SCO (Adam, 1957, 1963) producing RF (see also Sterba, 1962; Sterba *et al.,* 1967a). The problem of similar ependymal structures producing RF-like material in the acranian chordate *Branchistoma lanceolatum* (infundibular organ) and a

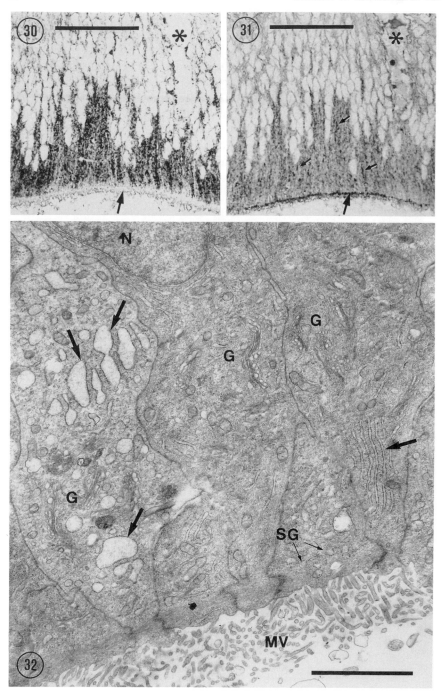

transient differentiation (flexural organ) during early developmental stages of teleosts has already been discussed in Section VII (see for literature) of this review. Very obviously, intrinsic cavities, or at least wide spaces containing circulating fluid (CSF), are a crucial prerequisite for the formation of RF-like structures (cf. Section VII). Thus, the rudimentary ventricular compartment adjacent to the SCO of *Myxine* still fulfills the basic structural and functional requirements leading to the formation of a RF (Adam, 1963), but there is no evidence that a RF-like structure exists in the compact ganglia of invertebrates.

It is of interest that, in the hagfish, *Myxine glutinosa*, which possesses a well-developed SCO, the pineal organ is missing. Also, in the armadillo, *Dasypus novemcinctus*, a species in which a typical pineal body is absent and the existence of scattered pinealocyte-like elements within the posterior commissure represents an unsolved problem, the SCO displays a high degree of secretory activity (Rodríguez *et al.*, 1984a; cf. Hofer *et al.*, 1976).

Only in a very limited number of vertebrate species are the anatomical differentiation and secretory activity of the SCO poor, e.g., in the European hedgehog. However, this is not a general feature of the insectivore brain since the SCO of the European mole is rather extended and rich in secretion (cf. Hofer *et al.*, 1976).

A peculiar situation can be observed in primates, with particular reference to the development in man (see Section IX). The human SCO displays its highest degree of development during the third and fourth months of fetal life, followed by a gradual decline during the second half of pregnancy and a more conspicuous regressive development after the birth. On the other hand, in a representative number of Old and New World monkeys the SCO is conspicuous and rich in secretion even in adult animals (Rodríguez *et al.*, 1984a). However, in very old anthropoid apes (e.g., chimpanzees) the SCO, although still present, shows a decrease in height and regularity of its parenchyma.

FIGS. 30 and 31   Semithin sagittal sections through the SCO of a 3-month-old human fetus processed for ConA (Fig. 30) and wheat germ agglutinin (WGA) (Fig. 31) binding. Both lectins react with different structures in the supranuclear cytoplasm: ConA with RER and WGA with Golgi complexes (small arrows), Apical secretory granules are ConA-negative and WGA-positive (large arrows). Asterisks show the same blood vessel in both sections. Bars, 50 μm. From E. M. Rodríguez *et al.* (1990b).

FIG. 32   Subcommissural organ of a human embryo (estimated age 109 days) showing the supranuclear region of ependymal cells. The cells display a well-developed Golgi apparatus (G), numerous RER cisternae, many of them distended and close to the apical plasma membrane (arrows), and profiles of secretory granules (SG). N, Cell nucleus; MV, microvilli. Bar, 1 μm.

XI. Location of Immunoreactive Reissner's Fiber Material in Areas Different from the SCO—RF Complex

In addition to the flexural organ of vertebrate embryos, the infundibular organ of amphioxus (Section VII) (Fig. 26), and certain ependymal cells surrounding the central canal of the bovine spinal cord (Section VIII,B) (Fig. 19), two other cell types have been reported to display immunoreactivity to antisera against RF material.

A. Somatostatin-Secreting Cells

In all mammalian, amphibian, and some of the reptilian species investigated, the somatostatin-immunoreactive neurons located in the periventricular and arcuate (infundibular) nuclei of the hypothalamus are strongly reactive to an antiserum raised against bovine RF extracted with DTT, EDTA, and urea. Immunoreactive material is also found in the axons projecting to the median eminence and in axon terminals contacting the portal capillaries. Incubation of the somatostatin and RF antisera with 14-28-somatostatin abolishes the immunoreaction of the somatostatinergic neurons with the former but not with the latter serum (Rodríguez *et al.,* 1987a). Preadsorption of the RF antiserum with a peptidic extract of the bovine SCO and with a peptidic extract of the bovine median eminence abolishes the reactivity of this antiserum with the somatostatinergic neurons (unpublished observations by the authors). Somatostatin-secreting cells in other locations, such as those of the endocrine pancreas, also immunoreact with the antiserum against RF. On the other hand, anti-somatostatin does not react with the SCO–RF complex. All these findings suggest the interesting possibility that cells secreting somatostatin may also secrete a material of low molecular weight displaying at least a partial chemical similarity to a compound present in the bovine SCO and RF. The identification of this putative compound both in the SCO and median eminence appears most relevant mainly for two reasons: (1) it might correspond to a peptide secreted by the SCO; (2) its location within hypothalamic neurons, and most importantly, in the axon terminals contacting portal capillaries, strongly points toward the possibility that such a compound could be released into the portal circulation.

B. Pinealocytes

In all classes of vertebrates, pinealocytes secrete melatonin, an indolamine which is rhythmically synthesized and released (Quay, 1986). The

presence in pinealocytes of an additional secretory substance with proteinaceous or peptidic properties has been suggested (Pévet, 1977, 1981;
Collin, 1979, 1981; Ebels, 1979; cf. Oksche, 1988; for phylogenetic
aspects, see Collin, 1971; Oksche, 1971; Korf and Oksche, 1986).

Recently, Rodríguez et al. (1988) applied antibodies raised against secretory proteins extracted from the bovine RF (anti-RF) to the pineal
organ of different species. Double or sequential immunostaining using
anti-RF and antibodies against pinealocyte markers (rod-opsin and S-
antigen) showed that, in the salmon and rat, pineal RF-immunoreactive
material is located in pinealocytes of the sensory line. RF-immunoreactive
pinealocytes were detected in the lamprey, salmon, toad, two species of
lizards, domestic fowl, and rat. In the salmon, the RF-immunoreactive
material is mostly confined to the vascular endfeet of the photoreceptive
pinealocytes. Interestingly, in the rat, the material immunostained with the
anti-RF serum was missing from pinealocytes forming the main body of the
pineal, but present in the S-antigen-immunoreactive pinealocytes located
in the deep portion of the pineal organ and in the medial habenular nucleus
(Rodríguez et al., 1988). These latter findings support the view that several
types of pinealocytes exist, differing in their chemical nature and synthetic
capacity. The nature and biological significance of the RF-immunoreactive
material in pinealocytes remain open to discussion. Still, one may speculate that these immunoreactions reflect the existence of a secretory protein
or peptide produced by a particular population of pinealocytes.

XII. Reactivity of the SCO to Neuronal and Glial Markers

Although the SCO is a remarkably stable glandular structure throughout
evolution, especially with respect to the nature of its secretory products
and its capacity to form RF, there are certain characteristics of its secretory cells that undergo large species variations. Two of the most variable
features are the innervation pattern (see Section VIII,A) and the synthesis
of certain molecules regarded as markers of neurons and glial cells.

Thus, whereas the secretory ependymal cells of the rat and cat SCO do
not react with antisera against two typical glial markers, such as the
fibrillary acidic protein (GFAP) and S-100 protein (S-100) (Didier et al.,
1986; Chouaf et al., 1989; Rodríguez et al., 1989), the SCO of another
rodent species, the Mongolian gerbil, displays a strong reaction with antisera against GFAP (Redecker, 1989). Furthermore, the SCO cells of the
mouse and rabbit exhibit a positive immunoreaction with antisera against
S-100 (Chouaf et al., 1989). The rat SCO possesses α,α-enolase, also

regarded as a glial marker, and lacks γ,γ-enolase, a specific neuron enolase (Weissmann-Nanopoulos *et al.*, 1983). Antibodies against the three poly-peptides of neurofilaments do not immunostain the SCO of the rat, mouse, cat, and rabbit (Chouaf *et al.*, 1989). Furthermore, while the SCO epen-dyma of the mouse and rabbit take up and accumulate GABA, those of the mouse and rabbit do not (Gamrani *et al.*, 1981; Chouaf *et al.*, 1989; Didier-Bazes *et al.*, 1989).

The expression by the SCO cells of one or other cell marker may depend on factors external to the SCO itself, such as the extent and nature of the neural input (Holton and Weston, 1982; Chouaf *et al.*, 1991). Indeed, the rat SCO grafted under the kidney capsule is not reinnervated; it continues to lack GFAP and S-100, but exhibits a positive immunoreaction against neuron-specific enolase (Rodríguez *et al.*, 1989). When the SCO of new-born rats is grafted in the fourth ventricle of adult rats, its cells display GFAP and S-100 but not neuron-specific enolase. The rat SCO expresses GFAP only transiently during a short period of development, from the nineteenth prenatal day to the third postnatal day (Chouaf *et al.*, 1991). Interestingly, in the rat SCO, synaptoid contacts with the ependymal cells start to be established from the third postnatal day onward (Marcinkiewicz and Bouchaud, 1986).

XIII. Experimental Studies of the SCO

A. Early Hypotheses

Soon after the first publications describing the existence of the SCO, numerous hypotheses with respect to the function of the SCO were pro-posed. Most of these hypotheses were based on indirect evidence mainly obtained by use of morphological procedures. This resulted in a number of contradictory and unrelated hypotheses (see Leonhardt, 1980). Three of these hypotheses are discussed in some detail since they are based on a variety of experimental designs and methodologies. Before this, a brief survey of contributions dealing with the function of the SCO is presented.

Sargent (1904) proposed that the RF of fishes functions as an optical reflex apparatus. For Nicholls (1912, 1917) the RF is part of a sense organ recording body movements. According to Kolmer (1921), the SCO, RF, and the CSF-contacting neurons of the central canal form a functional complex, the "sagittal organ." According to this concept, RF would stimulate the dendrites of the CSF-contacting neurons after changes in the body position, resembling the situation in certain receptive areas of the inner ear. Marburg (1920) postulated that the SCO–RF complex partici-pates in the regulation of the CSF pressure.

E. Legait (1942), Oksche (1955, 1962), and Leatherland and Dodd (1968) discussed a possible role of the SCO in color change mechanisms, especially in anurans and teleosts (eels). Sathyanesan (1965, 1966) reported an increase in the aldehyde–fuchsin-stainable secretory material in the SCO of some teleosts after exposure to continuous illumination. Diederen (1975b) and Vullings et al. (1983) observed in frogs that light causes the pineal complex and the lateral eyes to exert an inhibitory influence on the secretory activity of the SCO, suggesting a functional relationship between the SCO, the pineal complex, and the retina. It has been assumed that in lower vertebrates (Oksche, 1955; Ariëns Kappers, 1967) and in the rat (Scepovic, 1963; Ziegels and Devecerski, 1976) some kind of enigmatic relationship may exist between the SCO and the pineal organ.

A possible correlation has been investigated between the SCO and: (1) the hypothalamo-hypophysial system (Kivalo et al., 1961), (2) the thyroid gland (Yamada, 1961; Talanti, 1967; Talanti and Pasanen, 1968; Ferres-Torres et al., 1985), (3) the adrenal gland (see below), and (4) the gonads. D'Uva et al. (1977, 1978) postulated that the secretory activity of the SCO, together with other factors, participates in the regulation of the seasonal sexual cycle, and that the SCO itself may be under the influence of some of the factors influencing the sexual cycle. This hypothesis is supported by findings obtained by Ziegels (1979) and Castañeyra-Perdomo et al. (1983a). Radioimmunoassay of luteinizing hormone (LH) and follicle-stimulating hormone (FSH) in rats bearing a lesion in the SCO led Limonta et al. (1982) to suggest that the SCO might participate in the regulation of the hypersecretion of FSH during the proestrous day.

1. Morphogenetic Functions

Hauser (1969) proposed that in amphibians the normal growth of the regenerating tail tissue depends on the secretory activity of the SCO and on the presence of RF in the regenerating neural tube (see also Kirsche, 1956; Winkelmann, 1960). Rühle (1971) has claimed that elimination of the SCO during the postmetamorphic period of the newt Pleurodeles waltlii causes characteristic deformations of the entire body axis within a few weeks after the operation. Hauser (1972) reported that the selective destruction of the SCO results in disturbed tail regeneration in Xenopus larvae; the latter appears to depend on the reconstitution of RF and not on the restoration of the continuity of the central canal as such. Also, a marked distortion of the body axis was observed after destruction of the SCO in amphibians (Murbach and Hauser, 1974; Hauser, 1976). In such cases, however, a possible damage to the posterior commissure and adjacent reticular formations should be discussed. Other studies showed that, in young nursery-bred graylings exhibiting marked distortion of the body axis, an active SCO producing RF was found, but RF never reached the end of the neural tube (Murbach, 1976).

Transection of the distal portion of the rat RF (filum terminale) during the first two postnatal weeks does not interfere with normal tail growth (Sterba and Wolf, 1970). Thus, the question whether RF might play a role in normal tail growth during the prenatal life of the rat remains open to discussion. Concerning a possible role of the SCO in basic morphogenetic mechanisms of the CNS, see Section IX.

2. The RF and the Composition of the CSF

The chemical nature of the RF material (presence of sialic acid), especially its capacity to adhere particulate materials and to bind substances present in the CSF, has given support to an early hypothesis by Olsson (1958a) that RF may act as a detoxicator, keeping the CSF free from waste material.

The adhesive properties of RF have been suggested mainly on the basis of findings obtained by transmission and scanning electron microscopy (Sterba *et al.,* 1967a; Kohno, 1969; Weindl and Schinko, 1975; Schinko and Weindl, 1977; Schober and Sterba, 1977). Binding experiments carried out *in vitro* and *in vivo* have shown that RF specifically binds tyrosine and some biogenic amines such as adrenaline, noradrenaline, and serotonin (Sterba and Ermisch, 1969; Sterba *et al.,* 1969; Ermisch *et al.,* 1970; Hess and Sterba, 1972, 1973; Hess *et al.,* 1973).

Olsson (1958a) was the first investigator to describe the presence of sialic acid in the secretion of the SCO and RF; he suggested that it may participate in the detoxifying function of RF. Sterba and Wolf (1969) and Arnold (1969) indicated that, because of its high content of sialic acid, RF might balance the CSF concentration of biogenic amines by cation exchange. In an *in vivo* study performed with frogs, Diederen *et al.* (1983) demonstrated that, at the level of the Sylvian aqueduct and fourth ventricle, RF binds adrenaline and noradrenaline, and then transports these catecholamines to the central canal of the spinal cord. The same authors suggested that the RF might be involved in the regulation of the composition of the CSF, by releasing biogenic amines to or removing them from the CSF. In this respect, the observation is of interest that an increase of noradrenaline in the CSF appears to stimulate the secretory activity of the SCO (Diederen *et al.,* 1977; Hess *et al.,* 1977). Systemic administration of adrenaline had been reported to "deplete" the secretory material of the SCO of frogs (E. Legait, 1942) and rats (Gilbert and Armstrong, 1966).

3. Participation of the SCO in Water and Electrolyte Metabolism

The probable involvement of the SCO in the fluid electrolyte balance represents the experimental aspect most widely investigated. Thus, the

SCO has been linked to various and different aspects of water and electrolyte metabolism, such as volume reception, thirst, sodium excretion, diuresis, and aldosterone secretion.

Gilbert (1956, 1957, 1960, 1963) reported that the electrocoagulative ablation of the SCO in rats results in an immediate and drastic fall in water consumption, while subcutaneous injections of aqueous extracts of bovine and rat SCO into rats are followed by a clear-cut alteration in water intake and a decrease in the urinary sodium excretion. Gilbert suggested that the SCO may be regarded as a volume receptor of the body, regulating the total body fluid content. In this context, Palkovits (1968) suggested that the function of the SCO decreases in hypervolemia and increases in hypovolemia.

After ablation of the rat SCO, Upton et al. (1961) reported a fall in the urine output and sodium excretion, but did not find any change in water intake. At variance with this finding, Brown and Afifi (1965), who also achieved a complete ablation of the rat SCO, did not detect any change in urinary sodium excretion, but recognized a decrease in water intake and urine output. Bugnon et al. (1965), Lenys (1965), and Crow (1967) did not find any change in water intake, urine output, and urinary excretion of sodium after ablation of the SCO or injection of SCO extracts.

There are also very contradictory results with respect to the biological properties of SCO extracts. Wingstrand (1953), who was the first to prepare a SCO extract, found that the SCO of chick embryos extracted in acetic acid had no effect on diuresis in the rat. Palkovits and Földvári (1960) and Palkovits (1965) prepared an aqueous extract of rat SCO and described a potent antidiuretic activity of this extract. Arginine–vasotocin (AVT), detected by bioassay (Pavel, 1971) and radioimmunoassay (Rosenbloom and Fisher, 1975), has been reported to be present in the rat, bovine, and rabbit SCO. However, Dogterom et al. (1979), by using a specific radioimmunoassay for AVT, came to the conclusion that the SCO obviously contains a peptide related to but not identical with AVT.

Rottman (1962) reported an increased urinary sodium excretion in rats injected with an aqueous extract of SCO. Hein (1988) filtered an ammonium bicarbonate extract of bovine SCO through a Sephadex G-50 column and tested the eluted fractions on water-loaded, alcohol-anesthetized rats placed in a device for automatic recording of urine flow and sodium and potassium excretion. This author found that two of the collected fractions had a moderate diuretic activity and a potent natriuretic effect.

A critical analysis of the major problems concerned with most of the SCO studies involving the use of SCO extracts and lesion experiments has been made by Severs et al. (1987).

Ghiani et al. (1988) postulated the presence of specific receptors for angiotensin II in the SCO cells. According to Ghiani et al. (1988), these receptors are modulated by changes in plasma volume, which in turn very likely influences CSF volume. Binding sites for ^{125}I-labeled atrial natriuretic factor were absent in the SCO (Bianchi et al., 1986) or they occurred only in moderate concentrations (Mantyh et al., 1987).

Also contradictory are the effects of osmotic stimulation on the SCO. No changes in the SCO after osmotic stimulation have been observed by Fridberg and Olsson (1959), Kivalo et al. (1961), and Ziegels and Devecerski (1973). Osmotic-triggered changes in the amount of stainable secretory material, enzymatic activity, and ultrastructure of the SCO have been described in the rat (Gilbert, 1958; Bugnon et al., 1965; Lenys, 1965; Palkovits, 1968; Schütte, 1971; Leonieni and Rechardt, 1972), frog (Oksche, 1962, 1969), and eels (Leatherland and Dodd, 1968). The latter report represents the only study in which the turnover of the secretory material has been investigated by using [35S]cysteine. The authors reported a faster turnover indicative of an increased secretory activity of the SCO as a result of changes in ambient salinity. A semiquantitative analysis of the secretory material of the SCO and the detection of changes in the turnover rate of this material by use of radioactive precursors, and of the growth rate of RF, led Vullings and Diederen (1985) to conclude that, in the frog, Rana temporaria, water deprivation enhances the secretory activity of the SCO.

The following observations led some investigators to suggest that the SCO produces an "endocrine factor" that stimulates the production and release of aldosterone (see Severs et al., 1987): (1) a drastic decrease of water resorption and increase of water excretion after destruction of the SCO (Gilbert, 1956; Brown and Afifi, 1965); (2) a significant inhibition of urine flow in rats after intravenous injection of SCO extracts (Palkovits and Földvári, 1960; Palkovits et al., 1965); (3) a decreased functional activity of the zona glomerulosa and fall in aldosterone production in vitro after electrocoagulation of the SCO in rats (Palkovits, 1965); (4) changes in the amount of stainable material and in the ultrastructure of the SCO after adrenalectomy and ACTH administration (Varano et al., 1978; Srebro and Szirmai, 1989); (5) higher and lower rates of incorporation of [35S]cysteine into the rat SCO after hydrocortisone treatment and adrenalectomy, respectively (Attila and Talanti, 1973); (6) in vitro incorporation of [3H]leucine into the SCO of frogs treated in vivo with aldosterone or Ringer's solution, showing that the SCO of aldosterone-treated animals contains an increased amount of radioactively labeled material (Vullings and Diederen, 1985). However, other authors did not detect any changes either in the zona glomerulosa of the adrenal cortex after destruction of the SCO or in the SCO after administration of aldosterone or bilateral adrenalectomy (Bugnon et al., 1965, 1966; Lenys, 1965; Bugnon and Lenys, 1966). In a study performed with [3H]aldosterone in mice, Ermisch and Rühle (1977) concluded that the SCO does not possess receptors for aldosterone.

By applying a more reliable experimental approach, Dundore *et al.* (1984, 1987) infused aldosterone, via a cannula inserted in the pineal recess above the rostral pole of the SCO, and made the following observations: (1) an increased loss of urinary sodium and changes in the sodium/ potassium ratio; (2) a decreased adrenal medullary cross-sectional area; (3) an increased adrenal corticosterone content; (4) an elevation of plasma levels of adrenaline; (5) no changes in the adrenal cortical cross-sectional area or in the plasma levels of corticosterone. The authors concluded that their data support the concept that the SCO area interacts with physiological systems related to the adrenal cortex and medulla.

In our laboratory, we have studied the rat SCO by use of the immunoperoxidase procedure employing anti-RF sera and with conventional transmission electron microscopy under the following experimental conditions: (1) acute and chronic salt loading, (2) water deprivation, (3) perfusion of hypertonic saline into CSF, (4) pinealectomy, (5) hypophysectomy, (6) bilateral adrenalectomy, (7) bilateral gonadectomy, (8) continuous illumination for 10 days, (9) continuous darkness for 10 days. In each of these experiments at least 10 rats were used. No changes were detected either at the ultrastructural level or in the amount of immunoreactive material in any of the nine experimental conditions (unpublished observations).

It is likely that the SCO is involved in some aspects of fluid/electrolyte balance, but the conflicting data from the literature do not give enough support to the hypothesis that the SCO participates in osmoregulation.

B. Current Experimental Models

1. SCO Transplants

The rat SCO grafted under the kidney capsule survives transplantation for several months (Rodríguez *et al.*, 1989) (Fig. 10). The immunocytochemical and ultrastructural study of the grafted SCO revealed an absence of nerve fibers within the graft and suggested a state of enhanced secretory activity. The cells secrete into newly formed cavities lined by secretory and conventional ependymal cells. The capillaries revascularizing the graft start to display long-spacing collagen in their perivascular space 1 week after transplantation (Fig. 11). This finding and the numerous ependymovascular contacts, with ependymal endfeet containing secretory material, strongly suggest the possibility that compounds secreted by the grafted SCO cells may reach the bloodstream (Rodríguez *et al.*, 1989).

Transplantation of the SCO into richly vascularized areas devoid of serotonin fibers such as the kidney capsule may provide a useful experimental model for investigation of putative hormonal properties of the SCO. The SCO of newborn rats grafted in the fourth brain ventricle of adult rats also survives well for at least 3 months (Chouaf *et al.*, 1991). This site of transplantation may also become a useful tool in order to analyze various aspects of the SCO–RF complex.

2. Tissue Culture of the SCO

Tissue cultures of the SCO of adult rats survive for about 3 months (Lehmann and Sterba, 1989). Immunocytochemistry using anti-RF serum and transmission electron microscopy suggest that the cultured SCO cells continue to secrete glycoproteins. Similar results were obtained using mouse SCO. In this case, evidence for a release of secretory material into the culture medium was obtained (Lehmann *et al.*, 1989). Cell cultures of ependymal and hypendymal cells of the rat SCO, even after 1 to 2 years of culture, remain differentiated, are loaded with immunoreactive secretory material, and are capable of releasing this material to the culture medium (E. M. Rodríguez, R. Cariedes, O. Garrido, and A. Oksche, unpublished observations).

Tissue culture is thus a potent tool, but has only been recently applied to the investigation of the SCO; it will obviously contribute to revealing many relevant aspects of the cell biology of the SCO.

3. Immunological Blockage of RF Formation

A single injection of an antiserum against RF into the lateral ventricle of the rat triggers several events: (1) The injected antibody binds selectively to the pre-RF material and to RF (Fig. 21). (2) Approximately 4 hr after the injection, RF detaches from the SCO and undergoes fragmentation. (3) One day after the injection, a new RF starts to grow from the rostral end of the SCO. (4) Eight days after the administration of the antibody, the new RF reaches the fourth ventricle. (5) At the end of the first month, RF is found at the entrance of the central canal of the spinal cord where RF material accumulates, resembling a massa caudalis. (6) RF is missing from the central canal from the second postinjection day and for the entire observation period, which lasted for 5 months (S. Rodríguez *et al.*, 1990; S. Rodríguez, 1991).

Thus, single injections of an antiserum against RF into the CSF provide an experimental model displaying several distinct phases: (1) animals completely lacking normal RF structure (first 24 hr); (2) animals with a "short" RF only extending along the aqueduct (first week); (3) animals with a RF extending up to the entrance of central canal (starting first month); (4) animals lacking RF in the central canal.

It also seems likely that, while the injected antibodies were available in the CSF, they also blocked, by forming antigen–antibody complexes, soluble secretory materials released into the CSF. Antibodies cleared from the CSF approximately 8 hr after their injection (S. Rodríguez *et al.*, 1990).

This experimental model provides a useful tool to investigate functional aspects of the SCO–RF complex. Indeed, during the 5 days following a single injection of an anti-RF serum into the CSF, the rats showed a decrease in urine flow and water intake when compared to rats injected with IgG (S. Rodríguez, 1991).

4. Hydrocephalus and the SCO

In 1954, Overholser *et al.* suggested that the secretion of the SCO released into the CSF during fetal life prevents the closure of the Sylvian aqueduct, thus allowing the CSF to circulate freely between the third and fourth ventricles. If this were the case, a disturbance in the morphogenesis of the SCO should lead to a stenosis of the aqueduct and a congenital hydrocephalus (Overholser *et al.*, 1954; Newberne, 1962). Recent findings have given support to this hypothesis. Maldevelopment of the SCO induced by X-irradiation during fetal life precedes stenosis of the aqueduct and leads to congenital hydrocephalus (I. K. Takeuchi and Takeuchi, 1986).

A mouse strain (MT/HOK Idr) with spontaneous congenital hydrocephalus lacks a SCO (I. K. Takeuchi *et al.*, 1987). A drastic reduction in the size of the SCO has been reported in mouse and rats with congenital hydrocephalus (I. K. Takeuchi *et al.*, 1988; Jones and Bucknall, 1988) and in rats with a postnatally induced hydrocephalus (Irigoin *et al.*, 1990). Irigoin *et al.* also reported that in hydrocephalic rats immunoreactive secretory material continues to be released into the ventricle, probably at an enhanced rate, but a RF is not formed.

Current studies in our laboratory reveal that CSF from hydrocephalic children contains a glycoprotein not detected in normal human CSF. Antibodies raised against this glycoprotein immunoreact with the rat SCO. So far, the source of this glycoprotein is not known, but the SCO of the hydrocephalic children appears to be a possible candidate (unpublished observations).

The investigation of the SCO in the hydrocephalic state may lead not only to a better understanding of certain dynamic aspects of the SCO–RF complex, but also to an elucidation of at least a part of the mechanisms leading to congenital hydrocephalus.

XIV. Conclusions and Perspectives

The SCO is an ancient and persistent glandular structure of the vertebrate brain interposed between the neural tissue and the CSF. During ontogeny, this unique ependymal derivative represents one of the first secretory structures of the brain to differentiate.

Five secretory, core-glycosylated glycoproteins of 540, 450, 320, 190, and 50 kDa have been identified in the bovine SCO. Conspicuous species differences appear to exist with respect to the molecular weight of the glycoproteins secreted by the SCO. An extreme case is the human fetal SCO, which appears to secrete glycoproteins with a carbohydrate moiety similar to that elaborated by the SCO of other vertebrate species, but endowed with a different protein backbone.

Two of the glycoproteins detected in the bovine SCO (540 and 320 kDa) may represent two different precursor forms. Specific antibodies raised against these compounds showed that both compounds are present in the bovine and canine SCO. Only one of these precursors is found in the rat SCO, and neither of them occurs in the SCO of human fetuses and lower vertebrates. While stored in the RER the precursors are N-linked, high-mannose-type ($GlcNac_2$-Man_n) glycoproteins. The processed forms (450, 190, and 50 kDa) stored in secretory granules are N-linked, complex-type glycoproteins with the following sequence in the terminal chain: -GlcNac-Gal-sialic acid.

In the rat SCO, synthesis, processing, transport to the ventricular cell pole, and release into the ventricle occur within approximately 1 hr; in addition to this rapidly releasable pool, there is a second pool of secretory glycoproteins that remains stored in the RER and is released into the ventricle over several days. A distinct feature of the SCO is that the RER represents the main storage site of its secretory products.

The blood vessels of the SCO represent a highly specialized vascular structure within the CNS. The SCO is "sequestered" within a double-barrier system (blood–SCO, CSF–SCO) of unknown functional significance, indicating the unique character of this circumventricular organ.

Secretory material is shifted along basal processes of the ependymal and hypendymal cells of the SCO and stored in their terminals contacting blood vessels or the external limiting membrane of the brain. Although there is evidence indicating a vascular mode of release, this pathway still remains open to discussion. The crucial question as to whether an identical product(s) is discharged to the perivascular space and into the CSF-containing compartments awaits further investigation.

The principal route of secretion is directed into the CSF. On release, and for a period of about 1 hr, the glycoproteins first condense into a film covering the surface of the SCO (pre-RF). During this stage, the secreted glycoproteins appear to undergo a postrelease processing. Then, by a process apparently involving a higher degree of packaging and formation of disulfide bonds, the pre-RF material becomes assembled into a thread-like structure, the Reissner's fiber (RF). As new molecules are continuously added to the proximal end of RF, the fiber grows in a rostrocaudal direction at a rather fixed rate. Consequently, RF material continuously arrives at the caudal-most end of the central canal of the spinal cord, the ampulla caudalis, where it forms the massa caudalis, a transient deposit of the secretion. This mass reflects an equilibrium between the rate of supply of SCO secretion to the ampulla and the rate of escape of this material. Certain, only partly understood changes, e.g., an unpackaging and loss of sialic acid residues, may occur in the glycoproteins when shifted via RF to the massa caudalis.

In lower vertebrates, the blood capillaries surrounding the ampulla represent the final target of the RF material arriving at the ampulla.

In mammals, a soluble fraction of the SCO secretion appears to circulate in the CSF, including the subarachnoid space as the outer CSF-containing compartment. From there, it might be possible for dissolved secretory material to access venous leptomeningeal vessels, under particular consideration of the arachnoid villi.

The mammalian SCO is supplied with serotoninergic and GABAergic afferents originating in the raphe nuclei. The serotoninergic innervation appears to exert an inhibitory influence on the secretory activity of the SCO. In addition, certain observations indicate a more general serotonin-mediated control of the SCO, possibly also involving the pineal organ, and, in submammalian vertebrates, the system of the CSF-contacting neurons, especially those located in the central canal and paraventricular organ.

Two unsolved aspects of the SCO–RF complex appear to be crucial: (1) the exact chemical structure of the secretory glycoproteins and of their encoding genes, and (2) the functional significance of the SCO.

The secretory material produced by the bovine SCO has been purified and is currently used for microsequencing studies. Polyclonal and monoclonal antibodies to different components of the SCO secretion are available. Thus, experiments using recombinant DNA technology and cDNA probes appear to be in sight and offer a promising alley of further analysis.

A major challenge is, however, to discover the biological function of the SCO–RF complex. Although several hypotheses have been advanced over almost a century, they are very contradictory, and, in many cases, lacking substantial evidence.

Recently, some experimental models, such as transplantation of the SCO and the immunological blockage of the secretory material released into the CSF, and also new methodological tools, viz., organ and cell culture of SCO tissue, as well as sensitive enzyme immunoassays, have been developed. These and future developments, including radioimmunoassays, DNA probes for *in situ* hybridization, and an appropriate bioassay may all contribute to meeting the challenge.

The interesting possibility that the SCO might, somehow, participate in the mechanism leading to congenital hydrocephalus, has recently been reinforced, and places the SCO in the perspective of clinical studies.

Acknowledgments

We are grateful to the Volksgenwerk-Stiftung Federal Republic of Germany, the National Research Council of Chile (FONDECYT), and the Research Department (D. I.) of Universidad Austral de Chile for their support of our research on the subcommissural organ during the past 10 years.

References

Adam, H. (1957). *Anat. Anz.* **103,** 173–188.
Adam, H. (1963). *In* "The Biology of Myxine," (A. Brodal and R. Fange, eds.), pp. 137–149. Universitetsforlaget, Oslo.
Afifi, A. K. (1964). *J. Comp. Neurol.* **123,** 139–146.
Afzelius, B. A., and Olsson, R. (1957). *Z. Zellforsch. Mikrosk. Anat.* **46,** 672–685.
Alberts, B., Bray, D., Lewis, J., Raff, M., Roberts, K., and Watson, J. D. (1983). "Molecular Biology of the Cell." Garland Publishing, New York.
Altner, H. (1964). *Zool. Anz., Suppl.* **27,** 441–452.
Altner, H. (1968). *Z. Zellforsch. Mikrosk. Anat.* **84,** 102–140.
Anders, J. J., and Brightman, M. W. (1979). *J. Neurocytol.* **8,** 777–795.
Ariëns Kapers, J. (1967). *Z. Zellforsch. Mikrosk. Anat.* **81,** 581–618.
Arnold, W. (1969). *Z. Zellforsch. Mikrosk. Anat.* **101,** 152–166.
Attila, U., and Talanti, S. (1973). *Acta Physiol. Scand.* **87,** 422–424.
Bargmann, W. (1943). *In* "Handbuch der mikroskopischen Anatomie des Menschen" (V. W. Möllendorff, ed.), Vol. VI, pp. 309–502. Springer-Verlag, Berlin.
Bargmann, W., and Schiebler, T. H. (1952). *Z. Zellforsch. Mikrosk. Anat.* **37,** 583–596.
Bargmann, W., Oksche, A., Fix, J. D., and Haymaker, W. (1982). *In* "Histology and Histopathology of the Nervous System" (W. Haymaker and R. D. Adams, eds.), pp. 560–641. Thomas, Springfield, Illinois.
Baumgarten, H. G., Falck, B., and Wartenberg, H. (1970). *Z. Zellforsch. Mikrosk. Anat.* **107,** 479–498.
Bianchi, C., Gutkowska, J., Ballak, M., Thibault, G., Garcia, R., Genest, J., and Cantin, M. (1986). *Neuroendocrinology* **44,** 365–372.
Biosca, A., and Azcoitia, I. (1989). *J. Hirnforsch.* **30,** 273–279.
Björklund, A., Owman, C., and West, K. A. (1972). *Z. Zellforsch. Mikrosk. Anat.* **127,** 570–579.
Boecke, J. (1902). *Anat. Anz.* **21,** 411–414.
Bouchaud, C. (1975). *J. Microsc. (Paris)* **24,** 45–58.
Bouchaud, C. (1979). *Neurosci. Lett.* **12,** 253–258.
Bouchaud, C., and Arluison, M. (1977). *Biol. Cell.* **30,** 61–64.
Bouchaud, C., and Bosler, O. (1986). *Int. Rev. Cytol.* **105,** 283–327.
Brightman, M. W., Reese, T. S., and Feder, N. (1970). *In* "Capillary Permeability" (C. Crone and N. A. Larsen, eds.), pp. 483–490. Academic Press, New York.
Broadwell, R. D., and Brightman, M. W. (1976). *J. Comp. Neurol.* **166,** 257–284.
Brown, D. D., and Afifi, A. K. (1965). *Anat. Rec.* **153,** 255–263.
Bruel, M. T., Meiniel R., Meiniel, A., and David, D. (1987). *J. Neural Transm.* **70,** 145–168.
Bugnon, C., and Lenys, D. (1966). *C. R. Assoc. Anat.* **134,** 199–212.
Bugnon, C., Lenys, R., and Lenys, D. (1965). *Ann. Sci. Univ. Besancon* [3] **1,** 43–60.
Bugnon, C., Lenys, D., and Lenys, R. (1966). *C. R. Assoc. Anat.* **131,** 219–234.

Buijs, R.-M., and Pévet, P. (1980). *Cell Tissue Res.* **205**, 11–17.

Butler, A. B., Mann, J. D., Maffeo, C. J., Dacey, R. G., Johnson, R. N., and Bass, N. H. (1983). *In* "Neurobiology of Cerebrospinal Fluid" (J. H. Wood, ed.), Vol. II, pp. 707–726. Plenum, New York.

Calas, A., Bosler, O., Arluison, M., and Bouchaud, C. (1978). *In* "Brain-Endocrine Interaction. III. Neural Hormones and Reproduction" (D. E. Scott, G. P. Kozlowski, and A. Weindl, eds), pp. 238–250. Karger, Basel.

Castañeyra-Perdomo, A., Ferres-Torres, R., and Meyer, G. (1983a). *Neurosci. Lett.* **39**, 27–31.

Castañeyra-Perdomo, A., Meyer, G., and Ferres-Torres, R. (1983b). *J. Hirnforsch.* **24**, 368–370.

Castañeyra-Perdomo, A., Meyer, G., and Ferres-Torres, R. (1985). *J. Anat.* **143**, 195–200.

Castenholz, A. (1984). *Cell Tissue Res.* **237**, 181–183.

Chen, I. L., Lu, K. S., and Lin, H. S. (1973). *Z. Zellforsch. Mikrosk. Anat.* **139**, 217–236.

Chouaf, L., Didier-Bazes, M., Aguera, M., Tardy, M., Sallanon, M., Kitahama, K., and Belin, M. F. (1989). *Cell Tissue Res.* **257**, 255–262.

Chouaf, L., Didier-Bazes, M., Hardin, H., Aguera, M., Fèvre-Montange, M., Voutsinos, B., and Belin, M. F. (1991). *Cell Tissue Res.* (in press).

Cifuentes, M. (1991). Thesis of Doctor in Biological Sciences, University of Malaga, Spain.

Collin, J. P. (1971). *In* "The Pineal Gland" (G. E. W. Wolstenholme and J. Knight, eds.), pp. 79–125. Churchill-Livingstone, Edinburgh and London.

Collin, J. P. (1979). *Prog. Brain Res.* **52**, 271–296.

Collin, J. P. (1981). *In* "The Pineal Organ: Photobiology—Biochronometry—Endocrinology" (A. Oksche and P. Pévet, eds.), pp. 187–210. Elsevier, Amsterdam.

Collins, P. (1983). *J. Anat.* **137**, 665–673.

Collins, P., and Woolam, D. (1979). *J. Anat.* **129**, 623–631.

Crow, L. T., (1967). *Anat. Rec.* **157**, 457–464.

Dahlström, A., and Fuxe, K. (1964) *Acta Physiol. Scand., Suppl.* **62**, 1–55.

Davson, H. (1967). "Physiology of the Cerebrospinal Fluid." Churchill, London.

Dendy, A. (1902). *Proc. R. Soc. London* **69**, 485.

Dendy, A., and Nicholls, D. V. (1910). *Proc. R. Soc. London* **82**, 515–592.

Didier, M., Harandi, M., Aguera, M., Bancel, B., Tardy, M., Fages, C., Calas, A., Stagaard, M., Møllgard, K., and Belin, M. F. (1986). *Cell Tissue Res.* **245**, 343–351.

Didier-Bazes, M., Aguera, M., Chouaf, L., Harandi, M., Calas, A., Meiniel, A., and Belin, M. F. (1989). *Brain Res.* **489**, 137–145.

Diederen, J. H. B. (1970). *Z. Zellforsch. Mikrosk. Anat.* **111**, 379–403.

Diederen, J. H. B. (1972). *Z. Zellforsch. Mikrosk. Anat.* **129**, 237–255.

Diederen, J. H. B. (1973). *Z. Zellforsch. Mikrosk. Anat.* **139**, 83–94.

Diederen, J. H. B. (1975a). *Cell Tissue Res.* **156**, 267–271.

Diederen, J. H. B. (1975b). *Cell Tissue Res.* **158**, 37–60.

Diederen, J. H. B., and Vullings, H. G. B. (1980a). *Comp. Biochem. Physiol. A* **66**, 593–597.

Diederen, J. H. B., and Vullings, H. G. B. (1980b). *Cell Tissue Res.* **212**, 383–394.

Diederen, J. H. B., Vullings, H. G. B., and Hess, J. (1977). p. 9 *Conf. Eur. Comp. Endocrinol., Giessen.*

Diederen, J. H. B., Vullings, H. G. B., Rombout, J. H. W. M., and de Gunst-Schoonderwoerd, A. T. M. (1983). *Acta Zool. (Stockholm)* **64**, 47–53.

Diederen, J. H. B., Vullings, H. G. B., and Legerstee-Oostveen, G. G. (1987). *Cell Tissue Res.* **248**, 215–222.

Dogterom, J., Snijdewint, F. G. M., Pévet, P., and Buijs, R. M. (1979). *In* "The Pineal Gland of Vertebrates Including Man" (J. Ariëns Kappers and P. Pévet, eds.), pp. 465–470. Elsevier, Amsterdam.

Dundore, R. L., Wurpel, J. N. D., Balaban, C. D., Kiel, L. C., and Severs, W. B. (1984). *Neurosci. Res.* **1,** 341–351.

Dundore, R. L., Wurpel, J. N. D., Ballaban, C. D., Harrison, T. S., Kiel, L. C., Seaton, J. F., and Severs, W. B. (1987). *Brain Res.* **401,** 122–131.

D'Uva, V., Ciarcia, G., Ciarletta, A., and Angelini, F. (1977). *Monit. Zool. Ital.* [N. S.] **11,** 193–210.

D'Uva, V., Ciarcia, G., Ciarletta, A., and Angelini, F. (1978). *J. Exp. Zool.* **205**(2), 285–292.

Duvernoy, H., and Koritké, J. G. (1964). *Arch. Biol.* **75,** Suppl., 693–748.

Ebels, I. (1979). *Prog. Brain Res.* **52,** 309–321.

Erhardt, H., and Meinel, W. (1983). *Gegenbaurs Morphol. Jahrb., Leizig* **129**(6), 783–798.

Ermisch, A. (1973). *Wiss. Z.—Karl-Marx-Univ. Leipzig, Mat.-Naturwiss. Reihe* **22,** 297–336.

Ermisch, A., and Rühle, H. J. (1977). *Nova Acta Leopold., Suppl.* **9,** 133–138.

Ermisch, A., Sterba, G., Hartmann, G., and Frezer, K. (1968). *Z. Zellforsch. Mikrosk. Anat.* **91,** 220–235.

Ermisch, A., Sterba, G., and Hess, J. (1970). *Experientia* **26,** 1319–1321.

Ermisch, A., Sterba, G., Mueller, A., and Hess, J. (1971). *Acta Zool. (Stockholm)* **52,** 1–21.

Fährmann, W. (1963). *Z. Zellforsch. Mikrosk. Anat.* **58,** 820–836.

Fenstermacher, J. D., and Rapoport, S. I. (1984). *In* "Handbook of Physiology" (E. M. Renkin and C. C. Michel, eds.), Sect. 2, Vol. IV, pp. 969–999. Am. Physiol. Soc., Bethesda, Maryland.

Fernández-Llebrez, P., Pérez, J., Nadales, A. E., Pérez-Figares, J. M., and Rodríguez, E. M. (1987a). *Histochemistry* **87,** 607–614.

Fernández-Llebrez, P., Pérez, J., Cifuentes, M., Alvial, G., and Rodríguez, E. M. (1987b). *Cell Tissue Res.* **248,** 473–478.

Ferraz de Carvalho, C. A., and Prado Reis, F. (1977). *Anat. Anz.* **141,** 372–390.

Ferres-Torres, R., Castañeyra-Perdomo, A., and Ramos-Navarro, J. (1985). *Brain Res.* **331,** 348–352.

Fridberg, G., and Olsson, R. (1959). *Z. Zellforsch. Mikrosk. Anat.* **49,** 531–540.

Friedmann, I., Cawthorne, T., and Bird, E. S. (1965). *Nature (London)* **207,** 171–174.

Fuxe, K. (1965). *Acta Physiol. Scand.* **64,** 37–85.

Fuxe, K., and Jonsson, G. (1974). *Adv. Biochem. Pharmacol.* **10,** 1–12.

Fuxe, K., Hökfelt, T., and Ungerstedt, U. (1968). *Adv. Pharmacol.* **6A,** 235–251.

Gamrani, H., Belin, M. F., Aguera, M., Calas, A., and Pujol, J. F. (1981). *J. Neurocytol.* **10,** 411–424.

Garweg, G., and Kinsky, I. (1970). *Naturwissenschaften* **57,** 253.

Ghiani, P., Uva, B., Vallarino, M., Mandich, A., and Masini, M. A. (1988). *Neurosci. Lett.* **85,** 212–216.

Gilbert, G. J. (1956). *Anat. Rec.* **126,** 253–266.

Gilbert, G. J. (1957). *Am. J. Physiol.* **191,** 243–247.

Gilbert, G. J. (1958). *Anat. Rec.* **132,** 563–567.

Gilbert, G. J. (1960). *Neurology* **10,** 138–142.

Gilbert, G. J. (1963). *Neurology* **13,** 43–55.

Gilbert, G. J., and Armstrong, E. P. (1966). *Neurology* **16,** 236–241.

Gotow, T., and Hashimoto, P. H. (1982a). *J. Neurocytol.* **11,** 363–379.

Gotow, T., and Hashimoto, P. H. (1982b). *J. Neurocytol.* **11,** 447–462.

Gross, J., Highberger, J. H., and Schmitt, F. O. (1954). *Proc. Natl. Acad. Sci. U.S.A.* **40,** 679–684.

Gross, P. M., and Weindl, A. (1987). *J. Cereb. Blood Flow Metab.* **7,** 663–672.

Hädge, D., and Sterba, G. (1973a). *Acta Biol. Med. Ger.* **30,** 581–585.

Hädge, D., and Sterba, G. (1973b). *Acta Biol. Med. Ger.* **30,** 587–592.

Hafeez, M. A., and Zerihun, L. (1974). *Cell Tissue Res.* **154**, 485–511.
Hauser, R. (1969). *Wilhelm Roux' Arch. Entwicklungsmech. Org.* **163**, 221–247.
Hauser, R. (1972). *Wilhelm Roux' Arch. Entwicklungsmech. Org.* **169**, 170–184.
Hauser, R. (1976). *Rev. Suisse Zool.* **83**(4), 898–903.
Hein, S. (1988). Thesis of Doctor in Biological Sciences, Universidad de Chile, Chile.
Herrera, H. (1988). Thesis of Magister in Sciences, Universidad Austral de Chile, Chile.
Herrera, H., and Rodríguez, E. M. (1990). *Histochemistry* **93**, 607–615.
Herrlinger, H. (1970). *Ergeb. Anat. Entwicklungsgesch.* **42**, 1–73.
Hess, J., and Sterba, G. (1972). *Acta Biol. Med. Ger.* **28**, 849–851.
Hess, J., and Sterba, G. (1973). *Brain Res.* **58**, 303–312.
Hess, J., Hoheisel, G., and Sterba, G. (1973). *J. Hirnforsch.* **14**, 257–260.
Hess, J., Diederen, J. H. B., and Vullings, H. G. B. (1977). *Cell Tissue Res.* **185**, 505–514.
Heuschneider, J. K. H. (1968). Inaug.-Diss., Med. Fak., Univ. Munchen.
Hofer, H. (1958). *Zool. Anz., Suppl.* **22**, 202–251.
Hofer, H. (1959). *Zool. Jahrb., Abt. Anat. Ontog. Tiere* **77**, 465–490.
Hofer, H. (1964). *Zool. Anz., Suppl.* **27**, 430–440.
Hofer, H. (1986). *Gegenbaurs Morphol. Jahrb, Leipzig* **132**, 205–230.
Hofer, H., Merker, G., and Oksche, A. (1976). *Anat. Anz.* **140**, Suppl., 97–102.
Hofer, H., Meiniel, W., and Erhardt, H. (1980). *Cell Tissue Res.* **205**, 295–301.
Hofer, H., Meiniel, W., Erhardt, H., and Wolter, A. (1984). *Gegenbaurs Morphol. Jahrb., Leipzig* **130**, 77–110.
Hofer, H., Meinel, W., and Erhardt, H. (1987). *Gegenbaurs Morphol. Jahrb., Leipzig* **133**, 869–887.
Hoheisel, G., Sterba, G., and Ermisch, A. (1971). *J. Hirnforsch.* **13**, 33–38.
Hökfelt, T., Ljungdahl, A., Steinbusch, H., Verhofstad, A., Nilsson, G., Brodin, E., Pernow, B., and Goldstein, M. (1978). *Neuroscience* **3**, 517–538.
Holmberg, K., and Olsson, R. (1984). *Vidensk. Medd. Dan. Naturh. Foren.* **145**, 43–52.
Holton, B., and Weston, J. A. (1982). *Dev. Biol.* **89**, 72–81.
Irigoin, C., Rodríguez, E. M., Heinrichs, M., Frese, K., Herzog, S., Oksche, A., and Rott, R. (1990). *Exp. Brain Res.* **82**, 384–392.
Isomäki, A. M., Kivalo, E., and Talanti, S. (1965). *Ann. Acad. Sci. Fenn., Ser. A5* **111**, 1–64.
Johanson, C. E. (1989). *In* "Implications of the Blood-Brain Barrier and its Manipulation" (E. A. Neuwelt, ed.), Vol. I, pp. 223–260. Plenum, New York.
Jones, H. C., and Bucknall, R. M. (1988). *Neuropathol. Appl. Neurobiol.* **14**, 263–274.
Karoumi, A., Meiniel, R., Croisille, Y., Belin, M. F., and Meiniel, A. (1990a). *J. Neural Transm. (Gen. Sect.)* **79**, 141–153.
Karoumi, A., Croisille, Y., Croisille, F., Meiniel, R., Belin, M. F., and Meiniel, A. (1990b). *J. Neural Transm.* **80**, 203–212.
Kimble, J. E., and Møllgard, K. (1973). *Z. Zellforsch. Mikrosk. Anat.* **142**, 223–239.
Kimble, J. E., Sorensen, S. C., and Møllgard, K. (1973). *Z. Zellforsch. Mikrosk. Anat.* **137**, 375–386.
Kimura, K., Koizumi, F., Kihara, I., and Kitamura, S. (1975). *Lab. Invest.* **32**, 279–285.
Kirsche, W. (1956). *Z. Mikrosk.-Anat. Forsch.* **62**, 521–586.
Kivalo, E., Talanti, S., and Rinne, U. K. (1961). *Anat. Rec.* **139**, 357–361.
Kizer, J. S., Palkovits, M., and Brownstein, M. J. (1976). *Endocrinology (Baltimore)* **98**, 311–317.
Kohl, W. (1975). *Prog. Histochem. Cytochem.* **7**(4), 1–50.
Kohl, W., and Linderer, T. (1973). *Histochemie* **33**, 349–368.
Kohno, K. (1969). *Z. Zellforsch. Mikrosk. Anat.* **94**, 565–573.
Kolmer, W. (1905). *Anat. Hefte., Abt. 2* **29**, 165–214.
Kolmer, W. (1921). *Z. Anat. Enwicklungsgesch.* **60**, 652–717.

Korf, H. W. (1974). *Cell Tissue Res.* **155,** 475–489.
Korf, H. W., and Oksche, A. (1986). *In* "Vertebrate Endocrinology. Fundamentals and Biomedical Implications" (P. K. T. Pang and M. P. Schriebman, eds.), Vol. I, pp. 105–145. Academic Press, Orlando, Florida.
Krabbe, K. H. (1925). *Biol. Medd.—K. Dan. Vidensk. Selsk.* **5,** 1–83.
Krabbe, K. H. (1933). *Nord. Med. Tidskr.* **6,** 1030–1035.
Krstić, R. (1973). *Z. Zellforsch. Mikrosk. Anat.* **139,** 237–252.
Krstić, R. (1975). *Z. Mikrosk.-Anat. Forsch.* **89,** 1157–1165.
Kutschin, O. (1866). *Arch. Mikrosk. Anat.* **2,** 525–530.
Kux, E. (1929). *Z. Mikrosk.-Anat. Forsch.* **16,** 141.
Leatherland, J. F., and Dodd, J. M. (1968). *Z. Zellforsch. Mikrosk. Anat.* **89,** 533–549.
Legait, E. (1942). Thèse Docteur en Médicine, Université de Nancy, France.
Legait, E. (1949). *Bull. Soc. Sci. Nancy* **1,** 1–12.
Legait, H., and Legait, E. (1956). *C. R. Seances Soc. Biol. Ses Fil.* **150,** 1982–1984.
Leger, L., Degueurce, A., Lundberg, J. I., Pujol, J. F., and Møllgard, K. (1983). *Neuroscience* **10,** 411–423.
Lehmann, W., and Sterba, G. (1989). *Biomed. Res.* **10,** Suppl. 3, 169–176.
Lehmann, W., Sterba, G., and Wobus, A. M. (1989). *Acta Zool. (Stockholm)* **70**(4), 199–203.
Lenys, R. (1965). Thèse Docteur en Médicine, Université de Nancy, France.
Leonhardt, H. (1980). *In* "Handbuch der mikroskopischen Anatomie des Menschen" (A. Oksche and L. Vollrath, eds.), Vol. IV/10, pp. 177–665. Springer-Verlag, New York.
Leonieni, J. (1968a). *Folia Morphol.* **27,** 11–20.
Leonieni, J. (1968b). *Folia Histochem. Cytochem.* **6**(4), 485–498.
Leonieni, J., and Rechardt, L. (1972). *Z. Zellforsch. Mikrosk. Anat.* **133,** 377–387.
Limonta, P., Maggi, R., Martini, L., and Piva, F. (1982). *J. Endocrinol.* **95,** 207–213.
Lindberg, L. A., and Talanti, S. (1975). *Cell Tissue Res.* **163,** 125–132.
Lösecke, W., Naumann, W., and Sterba, G. (1984). *Cell Tissue Res.* **235,** 201–206.
Lösecke, W., Naumann, W., and Sterba, G. (1986). *Cell Tissue Res.* **244,** 449–456.
Lu, K.-S., and Lin, H.-S. (1987). *Proc. Natl. Sci. Counc., Repub. China, Part B* **11,** 332–340.
Lu, K.-S., and Peracchia, C. (1987). *Proc. Natl. Sci. Counc., Repub. China, Part B* **11,** 289–296.
Mack, A., Neuhaus, J., and Wolburg, H. (1987). *Cell Tissue Res.* **248,** 619–625.
Madsen, J. K., and Møllgard, K. (1979). *J. Neurocytol.* **8,** 481–491.
Mantyh, C. R., Kruger, L., Brecha, N. C., and Mantyh, P. W. (1987). *Brain Res.* **412,** 329–342.
Marburg, O. (1920). *Arb. Neurol. Inst. Univ. Wien.* **23,** 3–37.
Marcinkiewicz, M., and Bouchaud, C. (1983). *Biol. Cell.* **48,** 47–52.
Marcinkiewicz, M., and Bouchaud, C. (1986). *Biol. Cell.* **56,** 57–65.
Marini, M. (1957). *Atti Accad. Naz. Lincei, Cl. Sci. Fis., Mat. Nat. Rend.* **23,** 96–101.
Marini, M. (1962). *Atti Accad. Naz. Lincei, Cl. Sci. Fis., Mat. Nat. Rend.* **33,** 170–175.
Marini, M., Laguardia, M. T., and Gentile, R. (1978). *Atti Accad. Naz. Lincei, Cl. Sci. Fis., Mat. Nat. Rend.* **64,** 226–230.
Matsuura, T., and Sano, Y. (1987). *Cell Tissue Res.* **248,** 287–295.
Matsuura, T., Kumamoto, K., Ebara, S., and Sano, Y. (1989). *Biomed. Res.* **10,** Suppl. 3, 177–186.
Mautner, W. (1965). *Z. Zellforsch. Mikrosk. Anat.* **67,** 234–270.
Mazzi, V. (1952). *Arch. Zool. Ital.* **37,** 445–464.
Mazzi, V. (1954). *Monit. Zool. Ital.* **62,** 78–82.
Meiniel, A., Molat, J. L., and Meiniel, R. (1988). *Cell Tissue Res.* **253,** 383–395.
Meiniel, R., and Meiniel, A. (1985). *Cell Tissue Res.* **239,** 359–364.

Meiniel, R., Molat, J. L., and Meiniel, A. (1986). *Cell Tissue Res.* **245**, 605–613.

Meiniel, R., Duchier, N., and Meiniel, A. (1988). *Cell Tissue Res.* **254**, 611–615.

Meiniel, R., Molat, J. L., Duchier-Liris, N., and Meiniel, A. (1990). *Dev. Brain Res* **55**, 171–180.

Meunier, M. T., and Bouchaud, C. (1978). *Arch. Anat. Microsc. Morphol. Exp.* **67**, 81–98.

Meurling, P., and Rodríguez, E. M. (1990). *Cell Tissue Res.* **259**, 463–473.

Miline, R. (1974). *In* "Ependyma and Neurohormonal Regulation" (A. Mitro, ed.), pp. 77–103. Veda Publishing House of the Slovak Academy of Sciences, Bratislava.

Miline, R., Krstić, R., and Devecerski, V. (1969). *In* "Zirkumventrikulare Organe und Liquor" (G. Sterba, ed.), pp. 53–57. Fischer, Jena.

Møllgard, K. (1972). *Histochemie* **32**, 31–48.

Møllgard, K., and Wiklund, L. (1979). *J. Neurocytol.* **8**, 445–467.

Møllgard, K., Møller, M., and Kimble, J. (1973). *Histochemie* **37**, 61–74.

Møllgard, K., Lundberg, J. J., Wiklund, L., Lochenmajer, L., and Baumgarten, H. G. (1978). *Ann. N.Y. Acad. Sci.* **305**, 262–288.

Møllgard, K., Jacobsen, M., and Jacobsen, G. K. (1979). *Neurosci. Lett.* **14**, 85–90.

Müller, H., and Sterba, G. (1965). *Zool. Anz., Suppl.* **29**, 441–453.

Murakami, M., and Tanizaki, T. (1963). *Arch. Histol. Jpn.* **23**, 337–358.

Murakami, M., Ban, F., and Aiura, S. (1957). *Kurume Med. J.* **4**, 8–17.

Murakami, M., Nakayama, N., and Tanaka, H. (1969). *Experientia* **25**, 522–523.

Murakami, M., Nakayama, Y., Shimada, T., and Amagase, B. (1970). *Arch. Histol. Jpn.* **31**, 529–540.

Murbach, V. (1976). *Rev. Suisse Zool.* **83**(4), 903–908.

Murbach, V., and Hauser, R. (1974). *Rev. Suisse Zool.* **81**(3), 678–684.

Nakayama, Y. (1976). *J. Neurocytol.* **5**, 449–458.

Naumann, R. A. (1963). *Anat. Rec.* **145**, 266.

Naumann, R. A., and Wolfe, D. E. (1963). *Nature (London)* **198**, 701–703.

Naumann, W. (1968). *Z. Zellforsch. Mikrosk. Anat.* **87**, 571 591.

Naumann, W. (1986). *Acta Histochem., Suppl.* **32**, 265–272.

Naumann, W., Müller, G., and Kloss, P. (1987). *Wiss. Z.—Karl-Marx-Univ. Leipzig., Math.-Naturwiss. Reihe* **36**, 17–20.

Newberne, P. M. (1962). *J. Nutr.* **76**, 393–414.

Nicholls, G. E. (1912). *Anat. Anz.* **40**, 409–432.

Nicholls, G. E. (1917). *J. Comp. Neurol.* **2**, 117–191.

Nualart, F. (1989). Thesis of Magister in Science, Universidad Austral de Chile, Chile.

Nualart, F., Hein, S., Rodríguez, E. M., and Oksche, A. (1991). *Mol. Brain Res.* **11**, 227–238.

Obermüller-Wilén, H. (1976). *Acta Zool. (Stockholm)* **57**, 211–216.

Okada, M. (1956). *Arch. Histol. Jpn.* **9**, 199–204.

Oksche, A. (1954). *Verh. Anat. Ges.* **52**, Anat. Anz. 101, 88–96.

Oksche, A. (1955). *Morphol. Jahrb.* **95**, 393–425.

Oksche, A. (1956). *Anat. Anz.* **102**, 404–419.

Oksche, A. (1961). *Z. Zellforsch. Mikrosk. Anat.* **54**, 549–612.

Oksche, A. (1962). *Z. Zellforsch. Mikrosk. Anat.* **57**, 240–326.

Oksche, A. (1964). *Verh. Anat. Ges.* **58**, Anat. Anz. 112, 373–383.

Oksche, A. (1969). *J. Neuro.-Visc. Relat., Suppl.* **9**, 111–139.

Oksche, A. (1971). *In* "The Pineal Gland" (G. E. W. Wolstenholme and J. Knight, eds.), pp. 127–146. Churchill-Livingstone, Edinburgh and London.

Oksche, A. (1988). *Acta Anat.* **132**, 216–224.

Oksche, A., and Vaupel-von Harnack, M. (1965). *Z. Zellforsch. Mikrosk. Anat.* **68**, 389–426.

Olsson, R. (1955). *Acta Zool. (Stockholm)* **36**, 167–198.

Olsson, R. (1956). *Acta Zool. (Stockholm)* **37**, 235–250.

Olsson, R. (1957). Z. Zellforsch. Mikrosk. Anat. **46,** 12–17.

Olsson, R. (1958a). Acta Zool. (Stockholm) **39,** 71–102.

Olsson, R. (1958b). "The Subcommissural Organ," Thesis, University of Stockholm, Sweden.

Olsson, R. (1961). Gen. Comp. Endocrinol. **1,** 117–123.

Olsson, R. (1962). Ark. Zool. **15,** 347–355.

Olsson, R., and Wingstrand, K. G. (1954). Univ. Bergen Arbok, Naturvitensk. Rekke **14,** 1–14.

Olsson, R., Yulis, C. R., and Rodríguez, E. M. (1991). Submitted.

Overholser, M. D., Whitley, J. R., O'Dell, B. L., and Hogan, A. G. (1954). Anat. Rec. **120,** 917–934.

Palkovits, M. (1965). Stud. Biol. Acad. Sci. Hung. **4,** 1–103.

Palkovits, M. (1968). Z. Zellforsch. Mikrosk. Anat. **84,** 59–71.

Palkovits, M., and Földvári, I. P. (1960). Acta Biol. Acad. Sci. Hung. **11,** 91–102.

Palkovits, M., and Wetzig, H. (1962). Z. Mikrosk.-Anat. Forsch. **68,** 612–626.

Palkovits, M., Monos, E., and Fachet, J. (1965). Acta Endocrinol. (Copenhagen) **48,** 169–176.

Papacharalampous, N. X., Schwink, A., and Wetzstein, R. (1968). Z. Zellforsch. Mikrosk. Anat. **90,** 202–229.

Paul, E. (1972). Z. Zellforsch. Mikrosk. Anat. **128,** 504–511.

Paul, E., Hartwig, H. G., and Oksche, A. (1971). Z. Zellforsch. Mikrosk. Anat. **112,** 466–493.

Pavel, S. (1971). Endocrinology (Baltimore) **89,** 613–617.

Pelletier, G., Leclerc, R., Dube, D., Labrie, F., Puviani, R., Arimura, A., and Schally, A. V. (1975). Am. J. Anat. **142,** 397–401.

Peruzzo, B., Rodríguez, S., Delannoy, L., Hein, S., Rodríguez, E. M., and Oksche, A. (1987). Cell Tissue Res. **247,** 367–376.

Peruzzo, B., Pérez, J., Fernández-Llebrez, P., Pérez-Figares, J. M., Rodríguez, E. M., and Oksche, A. (1990). Histochemistry **93,** 269–277.

Pesonen, N. (1940). Acta Soc. Med. Fenn. Duodecim, Ser. A **22,** 53–78.

Pévet, P. (1977). Cell Tissue Res. **182,** 215–219.

Pévet, P. (1981). In "The Pineal Organ: Photobiology—Biochronometry—Endocrinology" (A. Oksche and P. Pévet, eds.), pp. 211–235. Elsevier, Amsterdam.

Poirier, J., Fleury, J., and Gherardi, R. (1983). Rev. Méd. Int. **4**(2), 131–144.

Puusepp, L., and Voss, H. E. V. (1924). Folia Neuropathol. **2,** 13–21.

Quay, W. B. (1986). In "Pineal and Retinal Relationships" (P. O'Brien and D. C. Klein, eds.), pp. 107–118. Academic Press, Orlando, Florida.

Rakic, P., and Sidman, R. L. (1968). Am. J. Anat. **122,** 317–336.

Ramkrishna, V., and Saigal, R. P. (1986). Acta Vet. Hung. **34**(1–2), 3–9.

Randall, J. T., Brown, G. L., Jackson, S. F., Kelly, F. C., North, A., Seeds, N. E., and Wilkinson, G. R. (1953). In "Nature and Structure of Collagen" (J. T. Randall, ed.), pp. 213–225. Butterworth, London.

Rechardt, L., and Leonieni, J. (1972). Histochemie **30,** 115–121.

Redecker, P. (1989). Cell Tissue Res. **255,** 595–600.

Reissner, E. (1860). Arch. Anat. Physiol. **77,** 545–588.

Rittig, M., Lütjen-Drecoll, E., Rauterberg, J., Jander, R., and Mollenhauer, J. (1990). Cell Tissue Res. **259,** 305–312.

Rodríguez, E. M. (1970a). Z. Zellforsch. Mikrosk. Anat. **111,** 15–31.

Rodríguez, E. M. (1970b). Z. Zellforsch. Mikrosk. Anat. **111,** 32–50.

Rodríguez, E. M., Oksche, A., Hein, S., Rodríguez, S., and Yulis, C. R. (1984a). Cell Tissue Res. **237,** 427–441.

Rodríguez, E. M., Oksche, A., Hein, S., Rodríguez, S., and Yulis, C. R. (1984b). *Cell Tissue Res.* **237,** 443–449.

Rodríguez, E. M., Herrera, H., Peruzzo, B., Rodríguez, S., Hein, S., and Oksche, A. (1986). *Cell Tissue Res.* **243,** 545–559.

Rodríguez, E. M., Hein, S., Rodríguez, S., Herrera, H., Peruzzo, B., Nualart, F., and Oksche, A. (1987a). *In* "Functional Morphology of Neuroendocrine Systems" (B. Scharrer, H.-W. Korf, and H.-G. Hartwig, eds.), pp. 189–202. Springer-Verlag, Berlin.

Rodríguez, E. M., Oksche, A., Rodríguez, S., Hein, S., Peruzzo, B., Schoebitz, K., and Herrera, H. (1987b). *In* "Circumventricular Organs and Body Fluids" (P. M. Gross, ed.), Vol. II, pp. 3–41. CRC Press, Boca Raton, Florida.

Rodríguez, E. M., Korf, H.-W., Oksche, A., Yulis, C. R., and Hein, S. (1988). *Cell Tissue Res.* **254,** 469–480.

Rodríguez, E. M., Rodríguez, S., Schoebitz, K., Yulis, C. R., Hoffmann, P., Manns, V., and Oksche, A. (1989). *Cell Tissue Res.* **258,** 499–514.

Rodríguez, E. M., Oksche, A., Rodríguez, S., Nualart, F., Irigoin, C., Fernández-Llebrez, P., and Pérez-Figares, J. M. (1990a). *In* "Progress in Comparative Endocrinology" (A. Epple, ed.), pp. 282–292. Wiley-Liss, New York.

Rodríguez, E. M., Garrido, O., and Oksche, A. (1990b). *Cell Tissue Res.* **262,** 105–113.

Rodríguez, S. (1991). Thesis of Doctor in Biological Sciences, University of Malaga, Spain.

Rodríguez, S., Hein, S., Yulis, C. R., Delannoy, L., Siegmund, I., and Rodríguez, E. M. (1985). *Cell Tissue Res.* **240,** 649–662.

Rodríguez, S., Rodríguez, P. A., Banse, C., Rodríguez, E. M., and Oksche, A. (1987). *Cell Tissue Res.* **247,** 359–366.

Rodríguez, S., Rodríguez, E. M., Jara, P., Peruzzo, B., and Oksche, A. (1990). *Exp. Brain Res.* **81,** 113–124.

Rosenbloom, A. A., and Fisher, D. A. (1975). *Endocrinology (Baltimore)* **96,** 1038–1039.

Rottman, A. R. (1962). Thesis of Science, Chicago, Illinois.

Rühle, H. J. (1971). *Acta Zool. (Stockholm)* **52,** 23–68.

Sargent, P. E. (1900). *Anat. Anz.* **17,** 33–44.

Sargent, P. E. (1904). *Bull. Mus. Comp. Zool.* **45,** 129–258.

Sathyanesan, A. G. (1965). *Neuroendocrinology* **1,** 178–183.

Sathyanesan, A. G. (1966). *Am. J. Anat.* **118,** 1–10

Scepovic, M. (1963). Thèse, University of Sarajevo.

Schafer, H., and Blum, V. (1988). *J. Morphol.* **196,** 345–351.

Scharrer, B. (1978). *Gen. Comp. Endocrinol.* **34,** 50–62.

Scharrer, B. (1990). *Am. Zool.* **30,** 887–895.

Schinko, I., and Weindl, A. (1977). *Nova Acta Leopold., Suppl.* **9,** 169–172.

Schmidt, W. R., and D'Agostino, A. M. (1966). *Neurology* **16,** 373–379.

Schober, F., and Sterba, G. (1977). *Nova Acta Leopold., Suppl.* **9,** 145–149.

Schoebitz, K., Garrido, O., Heinrichs, M., Speer, L., and Rodríguez, E. M. (1986). *Histochemistry* **84,** 31–40.

Schumacher, S. (1928). *Z. Mikrosk.-Anat. Forsch.* **13,** 269–327.

Schütte, B. (1971). *Acta Histochem.* **41,** 210–228.

Schwink, A., and Wetzstein, R. (1966). *Z. Zellforsch. Mikrosk. Anat.* **73,** 56–88.

Severs, W. B., Dundore, R. L., and Balaban, C. D. (1987). *In* "Circumventricular Organs and Body Fluids" (P. M. Gross, ed.), Vol. II, pp. 43–58. CRC Press, Boca Raton, Florida.

Sharon, N., and Lis, H. (1982). *Mol. Cell. Biochem.* **42,** 167–187.

Shimizu, N., Morikawa, N., and Ishi, Y. (1957). *J. Comp. Neurol.* **108,** 1–14.

Srebro, Z., and Szirmai, E. (1989). *Agressologie* **30,** 85–88.

Stanka, P. (1964). *Z. Mikrosk.-Anat. Forsch.* **71,** 1–9.

Stanka, P. (1967). *Z. Zellforsch. Mikrosk. Anat.* 77, 404–415.
Stanka, P., Schwink, A., and Wetztein, R. (1964). *Z. Zellforsch. Mikrosk. Anat.* 63, 277–301.
Sterba, G. (1962). *Zool. Jahrb., Abt. Anat. Ontog. Tiere* 80, 135–158.
Sterba, G. (1969). *In* "Zirkumventrikuläre Organe and Liquor" (G. Sterba, ed.), pp. 17–32. Fischer, Jena.
Sterba, G. (1977). *Nova Acta Leopold., Suppl.* 9, 103–114.
Sterba, G., and Ermisch, A. (1969). *Acta Biol. Med. Ger.* 22, K9–K14.
Sterba, G., and Naumann, W. (1966). *Z. Zellforsch. Mikrosk. Anat.* 72, 516–524.
Sterba, G., and Wolf, G. (1969). *Histochemie* 17, 57–63.
Sterba, G., and Wolf, H. (1970). *Acta Zool.* 51, 141–147.
Sterba, G., Pfister, C., and Naumann, W. (1965). *Z. Mikrosk.-Anat. Forsch.* 74, 33–38.
Sterba, G., Müller, H., and Naumann, W. (1967a). *Z. Zellforsch. Mikrosk. Anat.* 76, 355–376.
Sterba, G., Ermisch, A., Frezer, K., and Hartmann, G. (1967b). *Nature (London)* 216, 504.
Sterba, G., Wolf, G., and Scheuner, G. (1967c). *Naturwissenschaften* 54(18), 495.
Sterba, G., Hess, J., and Ermisch, A. (1969). *Pfluegers Arch.* 310, 277–280.
Sterba, G., Kleim, I., Naumann, W., and Petter, H. (1981). *Cell Tissue Res.* 218, 659–662.
Sterba, G., Kiessig, C., Naumann, W., Petter, H., and Kleim, I. (1982). *Cell Tissue Res.* 226, 427–439.
Studnička, F. K. (1899). *Sitzungsber. Böhm. Ges. Wiss., Math. Naturwiss. Kl.* 36, 1–10.
Studnička, F. K. (1900). *Anat. Hefte, Abt. 2* 15, 303–430.
Sturrock, R. (1984). *Anat. Anz.* 156, 21–30.
Stutinsky, F. (1950). *C.R. Seances Soc. Biol. Ses Fil.* 144, 1357–1360.
Takeuchi, I. K., and Takeuchi, Y. K. (1986). *Neurobehav. Toxicol. Teratol.* 8, 143–150.
Takeuchi, I. K., Kimura, R., Matsuda, M., and Shoji, R. (1987). *Acta Neuropathol.* 73, 320–322.
Takeuchi, I. K., Kimura, R., and Shoji, R. (1988). *Experientia* 44, 338–340.
Takeuchi, Y. K., and Sano, Y. (1983). *Anat. Embryol.* 167, 311–319.
Talanti, S. (1958). *Ann. Med. Exp. Biol. Fenn.* 36, Suppl. 9, 1–97.
Talanti, S. (1959). *Anat. Rec.* 134, 473–490.
Talanti, S. (1967). *Anat. Rec.* 159, 379–386.
Talanti, S. (1969). *Experientia* 25, 963–964.
Talanti, S. (1971). *Experientia* 27, 833.
Talanti, S., and Pasanen, V. (1968). *Life Sci.* 7, Part 1, 1245–1250.
Taniguchi, K., Taniguchi, K., and Mochizuki, K. (1985). *Jpn. J. Vet. Sci.* 47, 385–395.
Tramu, G., Pillez, A., and Leonardelli, J. (1983). *Cell Tissue Res.* 228, 297–311.
Tulsi, R. S. (1982). *J. Comp. Neurol.* 211, 11–20.
Tulsi, R. S. (1983). *Cell Tissue Res.* 222, 637–649.
Ueda, S., Ihara, N., Tanabe, T., and Sano, Y. (1988). *Brain Res.* 444, 361–365.
Upton, P. D., Dunihue, F. W., and Chambers, W. F. (1961). *Am. J. Physiol.* 201, 711–713.
Varano, L., Laforgia, V., D'Uva, V., Ciarcia, G., and Ciarletta, A. (1978). *Cell Tissue Res.* 192, 53–65.
Vigh, B., and Vigh-Teichmann, I. (1971). *Acta Biol. Acad. Sci. Hung.* 22(2), 227–243.
Vigh, B., and Vigh-Teichmann, I. (1973). *Int. Rev. Cytol.* 35, 189–251.
Vigh, B., Vigh-Teichmann, I., Koritsánsky, S., and Aros, B. (1970). *Z. Zellforsch. Mikrosk. Anat.* 109, 180–194.
Vigh, B., Vigh-Teichmann, I., and Aros, B. (1977). *Cell Tissue Res.* 183, 541–552.
Vigh-Teichmann, I., Vigh, B., Koritsánszky, S., and Aros, B. (1970). *Z. Zellforsch. Mikrosk. Anat.* 108, 17–34.
von Bomhard, K., Köhl, W., Schinko, I., and Wetzstein, R. (1974). *Z. Anat. Entwicklungsgesch.* 144, 101–122.

Vullings, H. G. B., and Diederen, J. H. B. (1985). *Cell Tissue Res.* **241**, 663–670.
Vullings, H. G. B., Diederen, J. H. B., and Smeets, A. J. M. (1983). *Comp. Biochem. Physiol. A* **74A**(2), 455–458.
Wakahara, M. (1974). *Cell Tissue Res.* **152**, 239–252.
Wake, K., Ueck, M., and Oksche, A. (1974). *Cell Tissue Res.* **154**, 423–442.
Weindl, A., and Joynt, R. J. (1973). *Arch. Neurol. (Chicago)* **29**, 16–22.
Weindl, A., and Schinko, I. (1975). *Brain Res.* **88**, 319–324.
Weissmann-Nanopoulos, D., Belin, M. F., Didier, M., Aguera, M., Partisani, M., Maitre, M., and Pujol, J. F. (1983). *Neurochem. Int.* **5**, 785–791.
Wenger, T., Klein, M. J., Stoeckel, M. E., Porte, A., and Stutinsky, F. (1969). *C. R. Seances Soc. Biol. Ses Fil.* **163**, 2436–2441.
Wenzel, J., Kunde, D., David, E., and Hecht, A. (1970). *Z. Mikrosk.-Anat. Forsch.* **82**, 243–263.
Wetzstein, R., Schwink, A., and Stanka, P. (1963). *Z. Zellforsch. Mikrosk. Anat.* **61**, 493–523.
Wiklund, L. (1974). *Cell Tissue Res.* **155**, 231–243.
Wiklund, L., and Møllgard, K. (1979). *J. Neurocytol.* **8**, 469–480.
Wiklund, L., Lundberg, J. J., and Møllgard, K. (1977). *Acta Physiol. Scand., Suppl.* **452**, 27–30.
Wingstrand, K. G. (1953). *Ark. Zool. (Stockholm)* **6**, 41–67.
Winkelmann, E. (1960). *Z. Mikrosk.-Anat. Forsch.* **66**, 147–176.
Wislocki, G. B., and Leduc, E. H. (1952). *J. Comp. Neurol.* **97**, 515–544.
Wislocki, G. B., and Leduc, F. H. (1954). *J. Comp. Neurol.* **101**, 283–310.
Wislocki, G. B., and Roth, W. D. (1958). *Anat. Rec.* **130**, 125–130.
Wislocki, G. B., Leduc, E. H., and Mitchell, A. J. (1956). *J. Comp. Neurol.* **104**, 493–517.
Wolf, G., and Sterba, G. (1972). *Acta Zool. (Stockholm)* **53**, 147–154.
Wolff, M. (1907). *Biol. Zentralbl.* **27**, 186–192.
Wood, J. H. (1983). *In* "Neurobiology of Cerebrospinal Fluid" (J. H. Wood, ed.), Vol. II, pp. 43–65. Plenum, New York.
Woollam, D. H. M., and Collins, P. (1980). *J. Anat.* **131**, 135–143.
Yamada, T. (1961). *Endocrinology (Baltimore)* **69**(4), 706–711.
Yulis, C. R., and Lederis, K. (1986). *Proc. Natl. Acad. Sci. U.S.A.* **83**, 7079–7083.
Yulis, C. R., and Lederis, K. (1988a). *Gen. Comp. Endocrinol.* **79**, 301–311.
Yulis, C. R., and Lederis, K. (1988b). *Cell Tissue Res.* **254**, 539–542.
Yulis, C. R., García, M. E., and Rodríguez, E. M. (1990). *Cell Tissue Res.* **259**, 543–550.
Zamir, N., Palkovits, M., and Brownstein, M. J. (1983). *Brain Res.* **280**, 81–93.
Ziegels, J. (1976). *Arch. Biol.* **87**, 429–476.
Ziegels, J. (1977). *C. R. Seances Soc. Biol. Ses Fil.* **171**, 1306–1308.
Ziegels, J. (1979). *J. Neural Transm.* **44**, 317–326.
Ziegels, J., and Devecerski, V. (1973). *Arch. Biol.* **84**, 231–241.
Ziegels, J., and Devecerski, V. (1976). *Arch. Biol.* **87**, 129–136.

Cerebellar Lectins

Jean-Pierre Zanetta, Sabine Kuchler, Sylvain Lehmann, Ali Badache,
Susanna Maschke, Philippe Marschal, Pascale Dufourcq, and Guy Vincendon
Laboratoire de Neurobiologie Moléculaire des Interactions Cellulaires, Centre
de Neurochimie du CNRS, 67000 Strasbourg, France

I. Introduction

The role of lectins in the nervous system was initially investigated at the same time as that of the cell adhesion molecules (CAMs) (Edelman, 1985, 1986) but did not attract much attention for many years. This was mainly due to the lack of identification of the specific carbohydrate-binding proteins involved in cell adhesion mechanisms. Nervous tissue lectins were essentially detected as agglutinating activities associated, probably (indirect evidence), with proteins. Since the first demonstration of a lectin activity in the electric organ of *Electrophorus electricus* (Teichberg *et al.*, 1975), relatively little attention was paid to lectins in the nervous tissue or its targets (muscle, for example). Developmentally regulated agglutinating activity in muscle was the first suggestion that a lectin could be associated with synapse formation in mammalian nervous tissue. Antibodies against muscle lectin (Gremo *et al.*, 1978) detected a related antigen in brain, but this antigen was still not identified as an endogenous brain component having lectin activity. In 1981, hyaluronectin, a protein antigen binding to immobilized hyaluronic acid columns, was discovered (Delpech and Halavent, 1981). Employing affinity histochemical techniques (with transient concanavalin A binding cerebellar glycoproteins as probes) binding sites due to endogenous lectins were observed (Reeber *et al.*, 1981; Zanetta *et al.*, 1984). Several protein subunits with the property of specific solubilization by mannose from young rat cerebella were detected. Two of them corresponded to a membrane-bound lectin called R1 (for receptor 1; Zanetta *et al.*, 1985a). Immunocytochemical localization was at the origin of its involvement in interneuronal recognition occurring as a first step of synaptogenesis. One year later, a β-galactoside-binding soluble protein was isolated from mammalian brain (Caron *et al.*, 1987) and a soluble mannose-binding lectin was isolated from rat cerebellum and characterized

123

chemically and immunochemically (Zanetta *et al.,* 1987c). The use of neoglycoproteins for affinity histochemistry and sequential fractionation of nervous tissue proteins on specially designed affinity columns for the purification of endogenous lectins allowed the detection of several of these compounds in nervous tissue (Gabius *et al.,* 1988; Gabius and Bardosi, 1990). Nevertheless, the field of lectins, in nervous as in other domains, remained of minor interest until very recently. The discovery of the essential role of carbohydrate-binding proteins in lymphocyte or leukocyte homing (Geoffroy and Rosen, 1989; Springer, 1990; Rosen *et al.,* 1990) revealed the potential role of glycobiological systems in adhesion or recognition phenomena. However, surprisingly enough, the cerebellum (despite its complexity) was the tissue where the fundamental role of these endogenous lectins was documented. This review summarizes major contributions in the field of cerebellar endogenous lectins, with special emphasis on their characterization, interactions with endogenous ligands, changes in localization during development, and roles. Special attention is directed to two molecules, the cerebellar lectins cerebellar soluble lectin (CSL) and R1, which are the only lectins so far isolated from the cerebellum, but as with most other lectins, they do not display organ specificity for the cerebellum nor for the nervous tissue. For more general information on the diversity of structures, functions, and uses of lectins, numerous reviews are available (Barondes, 1981, 1982, 1984; Barondes and Haywood-Reid, 1981; Barondes and Rosen, 1976; Barondes *et al.,* 1981; Beyer and Barondes, 1982; Childs and Feizi, 1980; Cook, 1986; Drickamer *et al.,* 1986; Gabius, 1987; Gabius *et al.,* 1986; Geoffroy and Rosen, 1989; Gitt and Barondes, 1986; Harris and Zalik, 1985; Hirabayashi *et al.,* 1987, 1989; Leffler *et al.,* 1989; Lis and Sharon, 1986; MacBride and Przybylski, 1986; Milos *et al.,* 1989, 1990; Monsigny *et al.,* 1983; Olden *et al.,* 1985; Quicho, 1986; Rauvala, 1983; Sharon, 1984; Sharon and Lis, 1989; Vraz *et al.,* 1986; Zalik and Milos, 1986; Zalik *et al.,* 1983, 1987).

II. Detection of Carbohydrate-Binding Sites in the Cerebellum

Only a few studies have documented the presence of carbohydrate-binding (endogenous lectins) sites in the cerebellum. The techniques employed the binding of glycoconjugates to cerebellar sections which had been fixed with aldehyde mixtures. Although these techniques are not without experimental pitfalls (for discussion, see Kuchler *et al.,* 1990a), nonetheless, they have proved their efficiency in demonstrating carbohydrate-binding sites, which have been clearly identified using immunocytochemical techniques.

A. Mannose-Binding Sites

The first report of the histochemical detection of lectins in cerebellar tissue was published in 1984 (Zanetta *et al.*, 1984). Two types of glycoconjugates were employed for detection of the potential endogenous lectins: (1) co-valent complexes with horseradish peroxidase (HRP) of endogenous cerebellar glycoproteins, called B1 and B2 (Reeber *et al.*, 1981) and isolated from young rat cerebella; (2) fluorescent cerebellar glycopeptides or complexes of cerebellar glycopeptides with HRP. After binding to cerebellar sections, the bound molecules were revealed with diaminobenzidine and subsequently observed either by light microscopy or directly with a fluorescence microscope. This method revealed a strong and transiently expressed binding of a special type of mannose-rich glycans to large neurons. This binding was essentially mannose dependent since it disappeared after treatments of the ligands with α-D-mannosidase, and was displaceable with mannose. By contrast, complex N-glycans did not display strong binding to cerebellar sections. These findings were at the origin of the discovery of the endogenous lectin R1 (Zanetta *et al.*, 1985a) and sustained the basis for the isolation of R1 and CSL as mannose-binding lectins (Zanetta *et al.*, 1985a, 1987c).

B. Other Carbohydrate-Binding Sites

Recently, the carbohydrate-binding sites present in sections of fixed tissue of the developing rat cerebellum were analyzed using another method which allowed the detection of endogenous lectins. It uses synthetic tools, biotinylated neoglycoproteins, in conjunction with subsequent staining with avidin–peroxidase. Neoglycoproteins are constructed by chemically coupling carbohydrate moieties (monosaccharides or oligosaccharides) to an inert carrier protein (generally bovine serum albumin, BSA). The sugar part of the neoglycoproteins included common constituents of the carbohydrate part of cellular glycoconjugates, namely, mannose, galactose, fucose, N-acetylglucosamine, N-acetylgalactosamine, and N-acetylneuraminic acid, to probe for the presence of respective endogenous receptors. Heparin can be also biotinylated. These molecules display a strong avidity for endogenous carbohydrate-binding proteins. Using this technique and appropriate blanks, specific positive reactions of different intensities were obtained for all neoglycoproteins and heparin. In two instances, the binding of neoglycoproteins could be compared to endogenous lectin-specific antibodies. Despite significant similarities, such a comparison revealed notable differences (attributed primarily to fixation and the presence of physiological ligands that can mask the active endogenous carbohydrate-binding proteins). The staining pattern with the individual probes disclosed specificity of localization with variable developmental regulation. Consequently, these results suggest that recognition processes during cerebellar development may include several types of carbohydrate determinants binding to their respective receptors.

III. The β-Galactoside-Binding Lectins

A. Evidence for Cerebellar β-Galactoside-Binding Lectins

In nervous tissue, most studies were concerned with lectins having agglutinating activity inhibited by β-galactosides. The first molecule of this type was discovered in *Electrophorus electricus* electric organ (Teichberg *et al.*, 1975), followed by a similar agglutinating activity in chick muscle and in a myogenic cell line (Nowak *et al.*, 1976). Lactose-inhibitable calcium-independent agglutinating activities were first identified in embryonic and in adult chicken brain (Kobiler and Barondes, 1977; Nowak *et al.*, 1977) and also in chicken retina and spinal cord (Eisenbarth *et al.*, 1978). Immunological studies (Kobiler *et al.*, 1978) demonstrated structural analogies between lectins found in various tissues. These findings were confirmed thereafter in the central nervous systems of birds (Beyer and Barondes, 1982) and mammals (Gremo *et al.*, 1978; Joubert *et al.*, 1987a,b). Purification of lactose-inhibitable lectin from mammalian brain was achieved only recently (Caron *et al.*, 1987), using similar techniques to those used for β-galactoside-binding lectins in other tissues. Rat and bovine brain β-galactoside-binding lectins display similar M_r for the native lectins (30,000 for the bovine and 33,000 for the rat lectins) and both are dimers, probably constituted of two identical subunits (Caron *et al.*, 1987). Recently (Bladier *et al.*, 1989), a similar lectin has been isolated from human brain, called HBL-14, which has immunological cross-reactivity with the RBL-16 molecule and is immunologically related to lectins isolated from human lung (Cerra *et al.*, 1984).

B. Roles of the β-Galactoside-Binding Lectins in Cerebellar Tissue

The roles of β-galactoside-specific lectins are still poorly documented in the nervous system. However, these lectins appear to be developmentally regulated and thus specific ontogenetic roles have been proposed. This seems to be the case in retina (Gremo *et al.*, 1978), in dorsal root ganglia

(Regan *et al.*, 1986), and in brain and cerebellum (Joubert *et al.*, 1989; Kuchler *et al.*, 1989b). In dorsal root ganglia, antigens reacting with antibodies to the lung lectins RL-14.5 and RL-29 are developmentally regulated and belong to overlapping but not coincident subsets of neurons (Regan *et al.*, 1986). Endogenous ligands recognized by specific antibodies are present in or on the same cells. This established for the first time a positive correlation between localization of β-galactoside-binding lectins on the one hand and their ligands on the other hand. But, since the ultrastructural localization of the two types of compounds is not known, it poses difficulty to discern their function. Implication of these molecules in ontogeny is expected, since these molecules display developmental changes in their levels and localization (Dodd and Jessell, 1986; Dodd *et al.*, 1988).

Immunocytochemical localization of antigens revealed by anti-HBL-14 antibodies has been performed during cerebellar development at both optical and ultrastructural levels in the cerebellum (Kuchler *et al.*, 1989b). In adult tissue, the lectin was present intracellularly in astrocytes and neurons of brain and cerebellum, whereas oligodendrocytes were free of immunolabeling. In agreement with biochemical studies (Joubert *et al.*, 1987a), β-galactoside-binding lectin displays a developmentally regulated pattern in neurons. The lectin is transiently present at the surface of young parallel fibers in the upper part of the cerebellar molecular layer. Since the lectin is virtually absent from the cell bodies from which the axons originate, it has been hypothesized that the lectin was produced and secreted by astrocytes, where it is very much concentrated (Kuchler *et al.*, 1989b). From these developmental studies, it was suggested that the lectin may play a role in *neurite fasciculation*. A similar hypothesis has been proposed from developmental studies in the forebrain (Joubert *et al.*, 1989).

During later development (between the 10th and 18th postnatal days in the rat), the antigen is developmentally concentrated in large neurons of the cerebellum, particularly in Purkinje cells (Kuchler *et al.*, 1989b). These changes in intensity of the immunocytochemical staining correspond to a transient increase in lectin level in cell bodies and dendrites, occurring at the period of maximal synaptogenesis. Although accumulation takes place in the dendritic spines (the structures where synapses are formed), this spurt of the lectin does not correspond to a cell recognition mechanism since the lectin is never externalized. Thus, a role in *intracellular transport* of an as yet unidentified β-galactoside ligand from the cell body to the postsynaptic density is expected (Kuchler *et al.*, 1989b).

Evidence was forwarded that the lectin was also present in the nuclei, in nuclear pores, as well as in nuclear regions not necessarily in association with chromatin (Kuchler *et al.*, 1989b). The lectin was also found in the nuclei of forebrain cells (Joubert *et al.*, 1989). The role of the lectin in nuclei remains obscure but considerable interest surrounds this localization in brain as well as in other tissues (Hubert *et al.*, 1989; Moutsatsos *et al.*, 1986).

Although similar results can be obtained using antibodies to human lung lectin (S. H. Barondes, personal communication), it is not known whether the different developmental patterns observed in different areas of the nervous tissue are due to the same molecule or correspond to differential expression of cell-specific isolectins as observed in other tissues (Leffler *et al.*, 1989).

Surprisingly, another type of β-galactoside-binding activity in the nervous tissue (and therefore found in the cerebellar white matter) is due to the myelin basic protein (Ikeda and Yamamoto, 1985). The property of this molecule (termed as MBP by neurochemists, and evidently different from the mannose-binding proteins) to bind β-galactosides was ignored by most neurochemists in the field, although the first evidence was obtained 10 years ago (Mullin *et al.*, 1981). Hemagglutination invoked by MBP can be inhibited by β-galactosides at the same concentration as that needed for inhibition of β-galactoside-binding lectins from other tissues. But, the binding of MBP to ganglioside GM_4 (Mullin *et al.*, 1981) and to glycophorin suggests that it recognizes more complex glycans. Recent immunological and sequence studies (Abbott *et al.*, 1989; Abbott and Feizi, 1991) and a comparison of the available data from literature reveal that short sequences found in most of the carbohydrate-recognition domains of the S-type β-galactoside-binding lectins are present in the MBP. Thus, the initial possibility that MBP is a lectin is compatible with sequence analysis. It is also an interesting observation for the role of MBP in myelin compaction: MBP is considered (Matthieu *et al.*, 1986) as the molecule responsible for binding between the cytoplasmic surface of the oligodendrocyte (the myelinating cell in the central nervous system). The finding that the MBP sequence responsible for experimental allergic encephalomyelitis (EAE) contains a tetrapeptide found in all β-galactoside-binding lectins (Abbott *et al.*, 1989; Abbott and Feizi, 1991) suggests that demyelination observed in EAE could be a consequence of the inhibition by antibodies of the carbohydrate-binding properties of MBP.

IV. Glycosaminoglycan-Binding Proteins

A. Hyaluronectin

Hyaluronectin has been isolated by Delpech and Halavent (1981) as a soluble brain glycoprotein binding to columns of immobilized hyaluronic acid. It was characterized as a diffused band of M_r ranging from 40,000 to

100,000 (Delpech and Halavent, 1981) with maximum material around 68,000. This binding is very specific for hyaluronic acid and for products of digestion of hyaluronic acid by bovine testis hyaluronidase. By immuno-cytochemical studies (Delpech *et al.*, 1987), its presence in cerebellar white matter and in the premigratory zone of the young rat cerebella was revealed. This specific localization was in agreement with the distribution of hyaluronic acid obtained by others using affinity histochemical techniques with the biotinylated hyaluronic acid-binding region of a chondroitin sulfate proteoglycan (Ripellino *et al.*, 1988). Carbohydrate specificity seems to be very narrow for hyaluronic acid. In the adult white matter, the hyaluronectin is especially concentrated in the nodes of Ranvier (Delpech, unpublished).

The role of hyaluronectin in white matter is still entirely speculative. The localization of hyaluronectin in the cerebellar premigratory zone, where hyaluronic acid has been transiently detected, strengthens the general idea that hyaluronic acid is involved in cell migration. What precisely are the mechanisms of this involvement remains to be clarified. Particularly, the positive implication of complexes formed between hyaluronic acid and the apparently monovalent hyaluronectin, especially in cell interaction mechanisms, is far from clear.

B. Heparin-Binding Proteins

A relatively high number of components have been isolated from the nervous tissue that have the ability to bind to columns made of immobilized heparin. Some of these compounds belong to the category of growth factors, and some of them have been identified to the previously isolated acidic and basic fibroblast growth factors (FGF). These molecules, as isolated, usually do not show adhesive properties, agglutinating activities, and polyvalency as expected for lectins. Thus, they are separate from the family of lectins. Nevertheless, a heparin-binding domain made of about 30 amino acids has been demonstrated for the basic FGF (Baird *et al.*, 1988). Yet other heparin-binding proteins have been isolated from brain after solubilization in octylglucoside (Mähönen and Rauvala, 1985; Rauvala and Pihlaskari, 1987). The 43,000 and the 52,000 heparin-binding molecules (Mähönen and Rauvala, 1985; Rauvala and Pihlaskari, 1987) are probably identical to an inhibitor of plasminogen activator isolated by Guenther *et al.* (1985) from glioma conditioned medium. They display strong adhesive properties for neuroblasts or neuroblastoma cells but appear to fail in promoting neurite outgrowth. Another molecule, called P30, has been isolated from rat brain as a heparin-binding protein. It probably exists as a dimer *in vivo* and displays strong adhesive properties for neuroblasts. This type of adhesion is associated with a considerable activity as neurite outgrowth promoting factor (Rauvala and Pihlaskari, 1987). However, recent identification of this molecule with amphoterin indicates that this molecule is probably not a lectin specific for heparin-related substances.

Heparin binding has been recently involved in a great variety of cell adhesion mechanisms. For instance, cell adhesion invoked by N-CAM can be inhibited by heparin (Cole *et al.*, 1986). Identical observations have been made for laminin, fibronectin, and other not yet identified molecules involved in adhesion. It should be stressed, however, that inhibitions occur at a very high heparin concentration (more than 100 μg/ml for 50% inhibition). Similarly, inhibition using Fab fragments of antibodies obtained against the "heparin binding site" is poor but in the same order of magnitude as that obtained with antibodies to the "homophilic binding site" of N-CAM (Cole *et al.*, 1986).

For all these molecules it is not clear if they have a carbohydrate-binding site as lectins or if they have affinities for macromolecules possessing repetitive disposition of negative charges. This is a major point to be clarified in future studies in this field as the evidence that interactions between these molecules and endogenous ligands actually occur *in vivo* is mounting. But, the role of heparin in inhibiting cell adhesion or glial cell proliferation (Robertson and Goldstein, 1988) indicates an important role for a heparin-binding mechanism in the nervous tissue.

V. Laminin and CAMs

Laminin has been described as an agglutinin (Ozawa *et al.*, 1983) and, since this component is present in the basal membrane of brain tissue, it should be expected that it invokes agglutination. But, since laminin has heparin-binding properties (Edgar *et al.*, 1984), it is possible that agglutination may result from such a binding. Whether laminin is a lectin or not remains to be elucidated. In the cerebellum, a role of laminin has been postulated in contact guidance of cell migration. The major argument for such a role is the finding that laminin is present *in vivo* at the surface of astrocytes (Lieisi, 1985a,b) as it is *in vitro*. However, its endogenous role as a substratum for neuron migration as well as for neurite outgrowth has been challenged (David, 1988; Sanes, 1985).

Finally, it has been observed that L1 (Ng-CAM) interaction with N-CAM enhances the adhesive properties of the latter (Kadmon *et al.*, 1990). The interaction between the two CAMs is mainly dependent on the presence of mannose-rich glycans (Kadmon *et al.*, 1990). Similarly, it should be remembered that recent experiments of transfection of the P0 gene (a CAM) in transformed cells induced the hypothesis of lectin-like activities for molecules up to now considered as CAMs (Schneider-Schaulies *et al.*, 1990). These hypotheses provide a new insight into the field of cell adhesion.

VI. Mannose-Binding Lectins

Two mannose-binding lectins have been isolated from the rat cerebellum. The first lectin, called R1 (for receptor 1), is a membrane-bound lectin insoluble in the absence of detergent (Zanetta *et al.*, 1985a). The second is a soluble lectin, called CSL (cerebellar soluble lectin), and is solubilized in the absence of detergents (Zanetta *et al.*, 1987c). The initial idea of the presence of mannose-binding lectins in rat cerebellum emanated from the search for a role of a specific class of cerebellar glycoproteins, transiently expressed during development (Zanetta *et al.*, 1978). These glycoproteins, potential endogenous ligands of lectins, have a very interesting property: they are highly insoluble glycoproteins, and, in particular, are poorly soluble in neutral detergents. Consequently, it was hypothesized that, during homogenization of the tissue, the hypothetical receptors of the glycans of these glycoproteins may bind to their ligands, thus rendering them insoluble. Addition of mannose would be the way to solubilize these molecules specifically. The first paper based on this method (the solubilization in the presence of the carbohydrate inhibitor is classical for isolation of lectins, but the sequential extraction procedure is original) reported a simplified sequential extraction procedure which allowed the solubilization (Zanetta *et al.*, 1984) and the separation of several protein subunits (called R1–R5). They were isolated by preparative gel electrophoresis, injected into rabbits for production of antibodies. The R1 antigen was the only one that gave a positive response to immunization. It was identified as the dimer of R2 and was unrelated to the R4 and R5 subunits. The other subunits were found, thereafter, to be soluble in the absence of detergent and thus separated from R1 subunits according to their solubility properties (Zanetta *et al.*, 1987c). They were isolated by preparative gel electrophoresis, injected into rabbits for production of antibodies. Antibodies, in turn, were employed to isolate the antigens by immunoaffinity chromatography and this procedure was used successfully for both CSL and R1 purification (Zanetta *et al.*, 1987c; Marschal *et al.*, 1989). Such a purification protocol allowed the isolation of molecules having preserved their carbohydrate-binding properties, the characteristic of lectins. The denominations of the two molecules are thus operational and certainly poorly mediatic (as were the majority of the lectins isolated before 1990). This prosaic terminology is certainly not the result of an undefined biological role. The change in definition of the term R1 from *receptor 1* to *recognin 1* is not out of date, and the change in definition of CSL from *cerebellar soluble lectin* to *cell-sealing lectin* is also entirely justified by the demonstrated role of this molecule.

A. Structure and Properties of R1

Isolation of Lectin R1 and Carbohydrate Specificity of R1

Using several immobilized monoclonal or polyclonal antibodies, active lectin R1 was obtained as two bands of 45,000 and 65,000. The 45,000 band was derived from a unique proteolytic cleavage during the purification of the initial lectin of M_r 65,000. Both molecules display carbohydrate-binding properties. *In vivo,* lectin R1 behaves as a dimer of M_r 130,000. Lectin R1 is totally unrelated to other mannose-binding proteins, including CSL (see below), and it displays different carbohydrate-binding properties with other mannose-binding lectins, with the exception of CSL, which shows similar, although not identical, carbohydrate specificity (Marschal *et al.,* 1989). Mannose is a relatively poor inhibitor for R1 and shows inhibition at 75 mM, whereas the other monosaccharides do not inhibit at all. Glycosaminoglycans are poorly inhibitory, the best being hyaluronic acid. Horseradish peroxidase isoenzyme 6 and especially isoenzyme 8 (11 μg/ml) were good inhibitors of R1, as ovalbumin or its glycopeptides. Endogenous glycoproteins B1 and B2 transiently expressed in the cerebellum (Zanetta *et al.,* 1978; Reeber *et al.,* 1981) were very good inhibitors (2–3 μg/ml). The glycopeptides isolated from these molecules inhibited at a concentration of 1 μg/ml and corresponded to the best inhibitors yet identified. Studies with glycopeptides of known structures (Marschal *et al.,* 1989) suggested that the first *N*-acetylglucosamine residue of *N*-glycans is part of the binding site of the lectins. In the cerebellum (Marschal *et al.,* 1989), R1 displayed a specificity for a relatively small number of glycoprotein subunits.

B. Role of Lectin R1 in the Nervous Tissue

In the nervous tissue, R1 is specifically localized in neurons throughout the different neural cells; it is absent from oligodendrocytes and astrocytes. However, it is also present in nonneural cell types: some but not all endothelial cells and in macrophages. The localization of R1 in the two types of nonspecific nervous cells is not surprising since R1 is probably identical to one of the receptors for circulating mannosyl glycoproteins found in liver sinusoidal cells (endothelial and Kupffer cells; Zanetta *et al.,* 1987a). A molecule binding to immobilized HRP and specifically eluted with mannose can be isolated from rat liver that reacts with anti-R1 antibodies. However, there are no data concerning the role of R1 in the cerebellar endothelial cells and in macrophages. But, a role in receptor-mediated endocytosis of glycoconjugates or in homing of cells of the immune system constitutes the most probable involvement of R1. By contrast, as shown below, its role in neurons has been documented.

1. Role of R1 in Normal Cerebellar Synaptogenesis

In the cerebellum (see Fig. 3 for a simplified illustration of the cerebellar structure), lectin R1 is highly concentrated in the cell body and dendrites of neurons, especially large neurons. There is very little lectin in Purkinje cells at birth and in the adult animal. However, the lectin level increases strongly in the cell body and dendrites at the period corresponding to synaptogenesis (Zanetta *et al.,* 1985a; Dontenwill *et al.,* 1985). This localization suggests that R1 may play a role in neuronal recognition during synaptogenesis. At the level of electron microscopy, it appears that the lectin is transiently externalized at the surface of the dendrites of Purkinje cells at the period of synaptogenesis (Zanetta *et al.,* 1985a; Dontenwill *et al.,* 1985). The lectin participates in endocytotic activity since it is found in special types of coated pits and coated vesicles. These structures are peculiar since they correspond to phagocytosis of portions of the membranes of parallel fibers by Purkinje cells and to double-walled coated vesicles. The fine localization of R1 in the Purkinje cell dendrites shows that only the dendrites where R1 is expressed in large amounts are the sites of this endocytosis (Figs. 1 and 2).

Since this endocytosis takes place simultaneously with the disappearance of glycoproteins of the parallel fibers, the hypothesis was made that R1 binds to specific glycoproteins at the surface of parallel fibers and participates in their receptor-mediated internalization in Purkinje cells. This was supported by ultrastructural studies (Eckenhoff and Pysh, 1979; Palacios-Prü *et al.,* 1981). This was also compatible with the data concerning the degradation of the glycans of glycoproteins of parallel fibers. These glycans are actually degraded in large amounts between the 13th and the 21st postnatal days (Reeber *et al.,* 1980). However, the study of the localization of α-D-mannosidase indicated that this enzyme, supposed to be involved in glycan degradation (and showing a peak of activity at this period), was uniquely localized in the Purkinje cells and not in or between parallel fibers (Zanetta *et al.,* 1983).

The direct evidence for this internalization of glycoprotein components

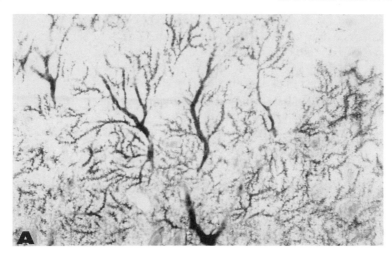

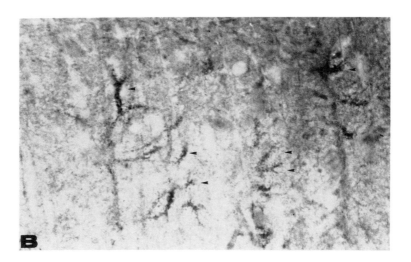

FIG. 1   Immunoperoxidase localization of the Purkinje cell specific glycoprotein subunit (PSG; Reeber *et al.,* 1981) and lectin R1 in the molecular layer of the rat cerebellum at the period of synaptogenesis. The immunostaining for PSG (A) reveals all the Purkinje cell dendrites, whereas that for R1 (B) reveals only a small proportion of them. The Purkinje cell dendrites showing high concentration of lectin R1 are those in which endocytosis of the membrane of parallel fibers by Purkinje cell occurs (see Fig. 2). Magnification: ×325.

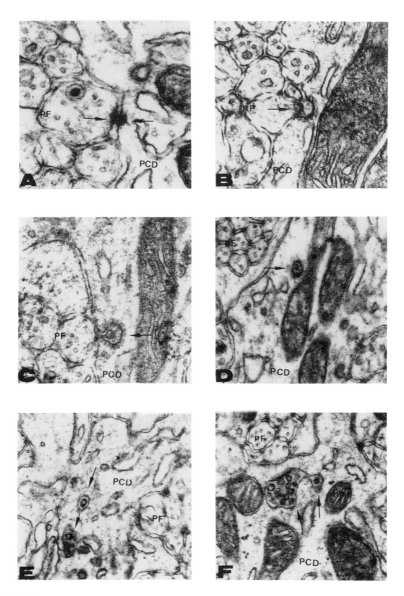

FIG. 2 Ultrastructural illustrations of the interactions between Purkinje cell dendrites and parallel fibers at the early stages of synaptogenesis [the sequence of events illustrated here is mostly speculative, but the evidence for such a mechanism was provided by Palacios-Prü *et al.* (1981)]. (A) First contact between a parallel fiber (PF) and a Purkinje cell dendrite (PCD), showing intracellular membrane specializations and extracellular junctional material. (B) The formation of a coated pit in the PCD is accompanied by the formation of a protuberance in the parallel fiber (arrow). (C) A portion of the PF membrane is internalized into the PCD. (D) Double-walled coated vesicles (arrow) are found close to the PCD surface. (E and F) Progressive transition between double-walled coated vesicles and multivesicular bodies (arrows) as an intermediate step to lysosomes which are localized in larger dendrites (proximal to the cell body). Magnification: ×47,000 for A–D and ×35,000 for E and F.

of the membranes of parallel fibers by Purkinje cells, followed by their degradation into these cells, was obtained by using chloroquine (Dontenwill *et al.*, 1983). Chloroquine is known to neutralize the endosomal and lysosomal compartments and thus block dissociation of receptor–ligand complexes and lysosomal enzymes. Under these conditions, the accumulation of glycoprotein components of the parallel fibers into nonfunctional lysosomes of Purkinje cells is observed (Dontenwill *et al.*, 1983).

This recognition of axonal glycoprotein by a dendritic lectin, which belongs to two cells which thereafter establish synapses, probably constitutes the first stage of synapse formation. If this recognition does not occur (due to transient action of tunicamycin during synthesis of the axonal glycoproteins), synapses cannot be formed (Zanetta *et al.*, 1987b). If the level of the glycoproteins is reduced (under the effect of thyroid hormone deficiency; Zanetta *et al.*, 1985b) the number of synapses is reduced (Nicholson and Altman, 1972a,b). Thus, a clear correlation between the presence of these glycoproteins (initially identified as the transiently expressed, concanavalin A-binding glycoproteins B1, B2, B4, and B5) and synaptogenesis has been documented.

The nature of the endogenous glycoprotein ligands of R1 is particularly interesting. The glycans of these transiently expressed glycoproteins also bind to concanavalin A (Zanetta *et al.*, 1978; Reeber *et al.*, 1981), to CSL (Marschal *et al.*, 1989), and to Elec-39 monoclonal antibody, an antibody of the HNK-1 family revealing oncofetal glycan antigen (Bon *et al.*, 1987; Mailly *et al.*, 1989; Musset *et al.*, 1987; Kuchler *et al.*, 1991a). The B4 and B5 bands previously identified have recently been characterized as a doublet glycoprotein band of M_r 28,000 and 31,000 called the "31-kDa glycoprotein" (Kuchler *et al.*, 1989d). It is a glypiated glycoprotein transiently present at the surface of the membranes of young neurons. The specific antibody has been raised against an antigen of the membrane of T lymphocytes solubilized using the phosphatidylinositol-specific phospholipase C (Pierres *et al.*, 1987).

2. Role of R1 in Repair after Cerebellar Lesions

In the adult, R1 is located exclusively intracellularly and the axonal glycoproteins with their special type of glycans are no longer present on the axonal surface. But, both partners of the glycobiological recognition system can reappear (Lehmann *et al.*, 1991a) at the surface of dendrites (for R1) and on axons (for the glycoproteins) during a neosynaptogenesis in the adult induced by production of lesions of these axons (Chen and Hillman, 1982). The same mechanism of endocytosis of glycoprotein components of the axonal membrane in the target cell can be observed morphologically (Lehmann *et al.*, 1991a). In particular, the lectin R1 and the 31-kDa glycoprotein can be found in the endocytotic micrographs depicting internalization of portions of the membrane of parallel fibers by Purkinje cells. This observation constitutes the first demonstration that reexpression of recognition molecules, as a basis for a repair mechanism in the adult, can occur, analogous to that occurring during normal construction of the CNS. It may be recalled that, under the same conditions, polysialylated sequences characteristic of embryonic N-CAM, and suspected to play a role in neuronal repair, are not reexpressed on the neuronal partners. In contrast, these polysialylated glycans are expressed, in the area of the lesion only, in reactive astrocytes (Lehmann *et al.,* 1991a). Such an astrocytic expression of embryonic N-CAM sequence in reactive astrocytes has also been observed by others (Le Gal la Salle and Valin, 1991).

3. Role of R1 in Other Parts of the Nervous Tissue

Whether such a mechanism of cell recognition is specific for the cerebellum or whether a similar mechanism operates in other parts of the central and peripheral nervous system (PNS) cannot be answered completely at the present time. However, it is documented that lectin R1 is not restricted to the cerebellum and is present (and transiently expressed) in other parts of the CNS and in the PNS (sensitive targets and muscles; Thomas *et al.,* in preparation). Similarly, the same types of glycoprotein ligands are present at the surface of the afferent (incoming) fibers. Thus, similar types of mechanism involving the same lectin and, potentially, the same glycans shared by the same or by different glycosylated polypeptide chains are indicated. Knowledge regarding these molecules during normal synaptogenesis and repair may provide clues to understanding the mechanism of neuronal recognition.

C. Isolation and Carbohydrate Specificity of Lectin CSL

Lectin CSL was isolated from young rat cerebella by sequential extractions of the tissue. CSL was concentrated in a fraction specifically solubilized in the presence of 0.5 M mannose in the absence of detergent (Zanetta *et al.,* 1987c). The purified protein bands isolated first by preparative gel electrophoresis (Zanetta *et al.,* 1987c) were used to produce antibodies in rabbits. Antibodies, in turn, were employed to isolate the antigens by immunoaffinity chromatography (Zanetta *et al.,* 1987c) from the material solubilized in the presence of mannose, as indicated above. Active CSL lectin (possessing agglutinating activity) was obtained as a major doublet band of M_r 33,000 and 31,500 with a minor component of M_r 45,000. Antibodies raised in rabbits against individual protein bands reacted with all three components, as revealed by immunoaffinity chromatography. This demonstrated a clear immunochemical relationship between the three components. Amino acid composition showed an homology between the 33,000 and 31,500 components (Zanetta et al., 1987c) and suggested a basic isoelectric point.

Based on a different experimental approach, a mannose-binding protein has been detected in pig and human white matter (Bardosi et al., 1988; Gabius et al., 1988) by affinity chromatography on immobilized mannose columns. In human sciatic nerve, the molecular weight of this molecule is identical to that of CSL revealed by its specific antibodies. Similarly, a partially purified protein fraction solubilized from axons by heparin (DeCoster and DeVries, 1989) probably contains a high proportion of CSL. The arguments for this analogy are (1) identical M_r, affinity for heparin, and basic isoelectric point; (2) similar localization and interactions with axons; (3) similar proliferating activity for Schwann cells. By contrast, the CSL differs from other mannose-binding proteins isolated from the liver or blood in several properties, although their M_r are in close vicinity (but significantly different). The major difference is that MBPs possess calcium-dependent carbohydrate-binding activities, whereas the activity of CSL is not calcium dependent.

Lectin CSL displays similar, although not identical, carbohydrate specificity with lectin R1 (Marschal et al., 1989). Mannose is a relatively poor inhibitor and shows inhibition at 37.5 mM, whereas the other monosaccharides did not inhibit. Glycosaminoglycans were poorly inhibitory, the best being heparin (acting at the concentration of 185 μg/ml, a concentration range at which heparin-binding proteins are inhibited). Horseradish peroxidase isoenzyme 6 and especially isoenzyme 8 (11 μg/ml) are good ligands of CSL. In contrast, ovalbumin, a good inhibitor of R1, requires a much higher concentration (2.5 mg/ml) for inhibition of agglutination induced by CSL. Endogenous glycoproteins B1 and B2, transiently expressed in the cerebellum (Zanetta et al., 1978; Reeber et al., 1981), are very good inhibitors (2–3 μg/ml). The glycopeptides isolated from these molecules inhibit at a concentration of 1 μg/ml and correspond to the best inhibitors yet identified. Studies with glycopeptides of known structures (Marschal et al., 1989) suggest that the first N-acetylglucosamine residue of the interacting N-glycans is part of the binding site of the lectins. In the cerebellum (Marschal et al., 1989) and in other parts of the nervous tissue (Kuchler et al., 1988, 1989a,c,d, 1990a,b; Lehmann et al., 1990), CSL displays specificity for a small number of glycoprotein subunits. This indicates that, in most nontransformed cells, the endogenous lectin CSL has fewer endogenous ligands. This may be the basis for specific cell adhesion. For instance, the myelin-associated glycoprotein MAG and the major glycoprotein of the PNS, P0 are important identified ligands of CSL

(Kuchler *et al.*, 1989a). Only a few of the polypeptide chains of these molecules bind to CSL, although they are all concanavalin A-binding (Kuchler *et al.*, 1989a; Badache *et al.*, 1991). Several of these chains have glycans recognized by HNK-1 or Elec-39 antibodies. Experiments of competition of binding to MAG of CSL and Elec-39 antibody indicate that CSL binds to some Elec-39-reacting glycans, although some Elec-39-binding glycans are not binding to CSL.

D. Role of Endogenous Lectin CSL in the Nervous Tissue

1. CSL and Contact Guidance of Neuron Migration

The mechanism of contact guidance of neuron migration is a general phenomenon occurring during nervous tissue ontogenesis. This concept emerged from the morphological studies of normal or mutant cerebella deficient in cell migration (Ramón y Cajál, 1911; Rakic, 1971; Rakic and Sidman, 1973; Sidman and Rakic, 1973; Sotelo and Changeux, 1974a,b; Altman, 1972a,b,c; Palay and Chan-Palay, 1974). As shown in Fig. 3, in the cerebellum, granule cells proliferate in the external zone in the proximity of the pia mater, i.e., the external germinal layer (egl). Subsequently, the cells are ready to migrate in the premigratory zone situated at the transition of the molecular layer (ml) and egl. Then, cells cross the ml and the Purkinje cell layer (PC), finally reaching the internal granular layer (igl). Such a migration occurs along preexisting radial astrocytic processes (Bergmann fibers). The cultures of cerebellar explants provided meaningful tools to understand the molecular basis of contact guidance. It has been observed that protease inhibitors (Moonen *et al.*, 1982; Lindner *et al.*, 1983, 1986) or Fab fragments derived from antibodies specific for some glycoprotein components of the family of cell adhesion molecules (namely, L1/Ng-CAM molecules; Lindner *et al.*, 1983, 1986; Hemmendinger and Caviness, 1988) inhibited neuron migration in cultured cerebellar explants.

The initial idea that CSL may be involved in this phenomenon emanated from the localization of CSL in the cerebellar premigratory zone (Kuchler *et al.*, 1987), its accumulation in Bergmann fibers, and its externalization (Kuchler *et al.*, 1989c). Studies of the role of CSL in the adhesion mechanism in astrocyte cell cultures showed that CSL and its glycoprotein ligands were responsible for the formation of contacts between the astrocytes (Kuchler *et al.*, 1989c). This conclusion was based on several observations: (1) the lectin is synthesized by astrocytes and externalized (immunocytochemistry at the ultrastructural level); (2) CSL ligands are present in astrocytes (detection on blots with labeled CSL) and at their surface (adhesion on CSL coated layer); (3) the contacts between astrocytes in

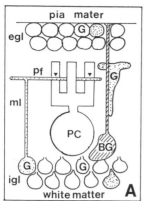 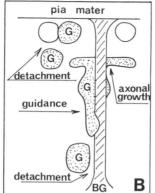 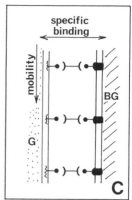

FIG. 3   (A) Simplified scheme of the cerebellar cortex illustrating the steps of maturation of the granule cells. Only three types of cells were taken into accounts: granule cells (G), Purkinje cells (PC), and Bergmann glial cells (BG). Arrowheads indicate the position of the synapses formed by parallel fibers (pf) on the Purkinje cell dendrites. See text for further explanation. (B) Steps involved in granule cell migration. The cells of the egl are highly compacted and migration first results from the separation of these cells in an area called the premigratory zone. This step is followed by the attachment to Bergmann glial fibers and guided migration. The final step is the detachment of the granule cell from the astrocytic guide. (C) Scheme of the postulated mechanism of attachment of granule cells to Bergmann glial fibers. It involves the formation of CSL ()—() bridges between the glycans (●—) of glycoproteins found at the surface of the astrocyte and of the neuron. This model ensures a strong and specific binding between the two membranes and a relatively free mobility of the neuron because the neuronal ligand is a glypiated glycoprotein. It shows comprehensively why anti-CSL and anti-31-kDa antibodies efficiently inhibit migration.

cultures can be disrupted by addition of very low amounts of anti-CSL Fab fragments (Kuchler *et al.*, 1989c).

The second observation was the localization of the 31-kDa glycoprotein ligand of CSL at the surface of premigrating neurons of the egl and in the areas of contacts between the migrating neurons and the Bergmann fibers (Kuchler *et al.*, 1989d). The potential role of the 31-kDa glycoprotein in such a migration phenomenon was substantiated by biochemical analysis of a cerebellar-mutant strain of mice deficient in neuron migration, the weaver mice. As observed by Mallet *et al.* (1974, 1976), this mutant showed an anomaly of the protein composition of its immature neurons: the absence of a doublet band at M_r 28,000 and 31,000. Although these molecules were suggested to be histone H-1, they had very similar properties to the 31-kDa glycoprotein. The histone hypothesis has not been sustained. But, from biological studies (Hatten *et al.*, 1984, 1986), it is clear that the anomaly of migration observed in the *weaver* mutant results from a deficient adhesion property of the *weaver* granule cells.

The final demonstration of the role of CSL in the migration of neurons was obtained using the system of cerebellar explants in cultures. Fab fragments obtained from the antibodies raised against CSL are much more powerful inhibitors (20 to 200-fold) of neuron migration in cerebellar explants than the Fab fragments obtained from other antibodies against potential CAMs (Lehmann et al., 1990). A similar powerful inhibition was obtained using antibodies to the 31-kDa neuronal glycoprotein, a ligand of CSL which is concentrated at the surface of migrating neurons and localized as CSL (Kuchler et al., 1989d) in the areas of contact between migrating neuroblasts and astrocyte processes. The underlying mechanism (Lehmann et al., 1990) seems to be that CSL is synthesized and externalized by astrocytes. The polyvalent lectin allows formation of bridges between glycans of glycoproteins present on the surface of neuroblasts (Kuchler et al., 1989d) and similar glycans shared by different glycoproteins of the astrocyte membrane (Kuchler et al., 1989c). Thus, CSL plays the role of an adhesion molecule, ensuring the tight contact between the migrating cell and its guide. The 31-kDa glycoprotein is anchored in the membrane through a phosphatidylinositol glycan (Pierres et al., 1987). Under such an arrangement, a free relative mobility of the surfaces in the plane of the membrane may be envisaged, whereas binding of the glycan through CSL to the astrocyte surface will ensure a strong and specific transverse interaction. In this system, proteases (Moonen et al., 1982; Lindner et al., 1983, 1986) and, predictably, phospholipases, glycanases, endo-N-acetylglucosaminidases, or mannosidases may inhibit migration as efficiently as the antibodies against CSL and the 31-kDa glycoprotein.

2. Endogenous Lectin CSL and Myelination

Myelin is a specialized compact membrane assembly around axons composed of the processes of the oligodendrocytes in the CNS and of the myelinating Schwann cells in the PNS. The studies of the mechanisms involved in myelination and myelin compaction constitute a broad field of neurobiology, since myelin plays an important role in the conductance of nerve impulses both in the CNS and PNS. The dramatic consequences of demyelination are illustrated by multiple sclerosis. Several hypothesis have been presented for explaining myelin compaction. From the studies of the mld dysmyelinating mutant (Doolittle and Schweikart, 1977), it has been proposed (Matthieu et al., 1986) that myelin basic protein (MBP) is involved in the mechanism of myelin compaction at the cytoplasmic surface of the oligodendrocyte membrane. The adhesion could be due to interaction of the positively charged MBP with negatively charged sulfogalactocerebrosides, but it could be due to an interaction of MBP with other galactolipids since a β-galactoside-binding activity has been associated with it (Ikeda and Yamamoto, 1985). Similarly, the proteolipid protein (PLP) has been suggested to be involved in myelin compaction at the level of the extracellular face of the oligodendrocyte (Dautigny *et al.*, 1986; Duncan *et al.*, 1987). This assumption was based on the observation that myelin of the *jimpy* mutant, having a specific anomaly of PLP protein (Nave *et al.*, 1986), is not compacted. Furthermore, PLP molecules aggregate very easily. However, so far as the interactions between the extracellular surface of myelinating cells are concerned, several hypotheses assumed that myelin glycoproteins were involved in myelin compaction, particularly the MAG (Matthieu, 1981). This hypothesis could not be sustained because MAG is not a constituent of compact myelin. In contrast, it may participate in the junctions between axons and myelinating cells. In the PNS tissue, a major glycoprotein constituent of myelin is glycoprotein P0 (Kitamura *et al.*, 1976, 1981; Lemke and Axel, 1985; Sakamoto *et al.*, 1987). P0 is a 29,000 glycoprotein of the family of CAMs (Salzer and Colman, 1989) found in compact myelin (Trapp *et al.*, 1984) and possessing cell adhesion properties. Transfection of the P0 gene in the Chinese hamster ovary cells (CHO cells) clearly increases the adhesiveness of these cells (Filbin *et al.*, 1990). P0 glycoprotein has been involved in myelin compaction through four different kinds of interactions: (1) hydrophobic interactions between external surfaces of Schwann cells (Lemke and Axel, 1985), (2) a mechanism involving the glycan (Lemke and Axel, 1985), or (3) a mechanism involving the adhesion specificity of CAMs (Salzer and Colman, 1989), and (4) glycan lectin interaction (Kuchler *et al.*, 1989a; Zanetta *et al.*, 1990a; Badache *et al.*, 1991). A number of molecules in the family of the CAMs have been found in myelin of the CNS or PNS (Martini and Schachner, 1986; Poltorak *et al.*, 1987). Most of the myelin glycoproteins so far identified (MAG and P0) share a common L2/HNK-1 epitope (McGarry *et al.*, 1983; Inuzuka *et al.*, 1984; O'Shannessy *et al.*, 1985), also detected on a glycolipid-containing sulfated glucuronic acid residue (Inuzuka *et al.*, 1984).

a. Presence of CSL in Myelin   In the CNS, CSL is present in oligodendrocytes and in myelin of the central nervous tissue (Kuchler *et al.*, 1987, 1988), as observed by immunocytochemical techniques at the level of light and electron microscopy. In cultures of oligodendrocytes (Espinosa de Los Monteros *et al.*, 1986; McCarthy and DeVellis, 1980), CSL lectin is present intracellularly, on the external face of the oligodendrocyte membrane, in the areas of contact between oligodendrocytes and in compact myelin produced by the cultured oligodendrocytes (Kuchler *et al.*, 1988). CSL is directly involved in myelin compaction since CSL glycoprotein ligands are present at the oligodendrocyte surface and since very small amounts (4 μg/ml) of anti-CSL Fab fragments are able to dissociate compact myelin (Kuchler *et al.*, 1988). The role of CSL in myelin compaction was further documented by studying dysmyelinating mutants (Kuchler *et al.*, 1990b, 1991b): CSL displays a nonhomogeneous distribution and is present only in the areas where myelin is normally compacted and is totally absent from the areas where myelin is fully disorganized (Kuchler *et al.*, 1990b, 1991b).

In the PNS, immunocytochemical detection at the ultrastructural level reveals the presence of CSL in compact myelin of the adult rat sciatic nerve (Kuchler *et al.*, 1989a). The lectin is very concentrated intracellularly in myelinating Schwann cells and can be detected in low amounts in compact myelin. CSL immunoreactivity is detected very early in rat sciatic nerve (postnatal day 1). It is present in oblong cells which become more and more elongated and then produce processes surrounding axons at the period of myelination (Kuchler *et al.*, 1989a). Recent studies on Schwann cell cultures demonstrated that CSL is present in Schwann cells in cultures, although the levels seem to be less than in oligodendrocytes.

b. Ligands of CSL in Myelinating Cells Two major ConA-binding ligands of CSL were found in cultured oligodendrocytes. One, with a M_r in the range of 100,000, is probably MAG, whereas the second, with a M_r of 16,000, remains unidentified (Kuchler *et al.*, 1988). These ligands are in part at the surface (Fressinaud *et al.*, 1988), since oligodendrocytes adhere rapidly on CSL coat with an immediate and significant increase in proliferation.

Studies performed on purified myelin fractions from normal animals indicate that the population of CSL glycoprotein ligands found in myelin preparations was much more complex than in oligodendrocyte cultures (Kuchler *et al.*, 1990b, 1991b). A doublet glycoprotein band in the region of 100,000 was seen, along with several glycoprotein bands with M_r of 50,000, 31,000, 29,000, 24,000, 19,000, and 16,000. Glycoproteins with similar M_r had been detected previously in purified myelin fractions (Linington *et al.*, 1984; Matthieu *et al.*, 1974; Neskovic *et al.*, 1986; Poduslo, 1981; Poduslo *et al.*, 1977, 1980; Quarles *et al.*, 1979; Schluesener *et al.*, 1987; Zanetta *et al.*, 1977). The 16,000 protein corresponds to that found in oligodendrocyte cultures. In *mld, quaking,* and *jimpy* dysmyelinating mutants, the bands at 24,000 and 19,000 were absent (Kuchler *et al.*, 1990b, 1991b). In contrast, the 16,000 constituent was unchanged, suggesting that it does not constitute a structural component of compact myelin. The 31,000 glycoprotein corresponds to the 31-kDa axonal glycoprotein ligand of CSL (Kuchler *et al.*, 1989a,d).

The study of CSL glycoprotein ligands during sciatic nerve development (Kuchler *et al.*, 1989a) indicated considerable quantitative variations.

Four major components are present at day 1, with M_r of 100,000, 50,000, 31,000, and 16,000. The 16,000 glycoprotein decreases slowly between days 1 and 15, disappearing almost completely in the adult. It corresponds to the same molecule that is found in oligodendrocyte cultures, and is considerably increased in myelin of *mld, jimpy,* and *quaking* mutants. The 31,000 band decreases progressively until, in the adult, it is present in very small amounts. It corresponds to the 31-kDa axonal glycoprotein previously identified in CNS axons (Kuchler *et al.,* 1989a,d). The 50,000 component decreases progressively with age and does not seem to be specific to the PNS, since it corresponds to a similar component in the CNS. The high M_r molecule (100,000) decreases considerably with maturation, but it is still present in low amounts in the adult sciatic nerve; it is MAG. Major modifications were observed for a compound having M_r at 29,000 which has all the characteristics of P0 glycoprotein. The 31,000 glycoprotein, definitely different from the P0, corresponds to the 31-kDa axonal glycoprotein (Kuchler *et al.,* 1989a,d).

All the components binding to CSL are also ConA-binding. However, the quantity of CSL-binding material is not proportional to that having ConA-binding properties. In other words, it is evident that, for one given molecular glycoprotein entity, not all the glycans are CSL binding. In contrast, it is probable that changes in proportion of polypeptide chains binding to CSL occur during development, which are different from the changes in proportion of polypeptide chains binding to ConA. Recently (Badache *et al.,* 1991), we have been successful in isolating, from CNS rat myelin, CSL-binding MAG (5% of the chains) from non-CSL-binding MAG (around 95% of the chains), based on their solubility characteristics.

c. Adhesive and Signaling Properties of CSL during Myelination   So far as myelin compaction is concerned, bridging of the external surface of the myelinating cell membrane by interactions of the polyvalent lectin CSL with glycans of surface glycoproteins appears to be a possibility. The basis for this is the localization of CSL and the effect of anti-CSL Fab fragments. The ligands involved in this compaction *in vitro* could be only MAG and the 16,000 glycoprotein in oligodendrocyte cultures. Other low-M_r glycoproteins in rat CNS myelin and P0 in PNS myelin have also been suggested. The localization of both CSL and 31-kDa axonal glycoprotein in the areas of contact provides support for their involvement in the initial contact between axons and myelinating cells during the initial period of myelination and the maintenance of this contact. The postulated glycoproteins of the myelinating cell membrane have still not been identified and could correspond to MAG, or the 50,000 or the 16,000 glycoprotein, both in the CNS and PNS. The observation that CSL is absent from nonmyelinating Schwann cells whereas the axons are still immunoreactive for

31-kDa glycoprotein is in agreement with the postulated role of CSL. But, in certain instances, CSL could be more than an adhesion molecule. It can induce signals for proliferation of myelinating cells (Fressinaud *et al.*, 1988) or differentiation of neurons. It can be assumed that initial contact between the neuroblast and oligodendrocyte surface *in vivo* could be the signal for a symbiotic development of neurons and oligodendrocytes. This is the salient property of CSL as a cell adhesion recognition lectin (CARL).

d. CSL and Myelin Cell Adhesion Molecules   The postulated role of CSL in adhesion mechanisms occurring during myelination is conflicting with the homophilic mechanisms involving CAMs. MAG and P0 are molecules related to CAMs of the supergene family of immunoglobulins. MAG–MAG and P0–P0 interactions have been postulated for explaining surface adhesion in myelin. Transfections in CHO cells of the genes of MAG and P0 induce an increased adhesion of CHO cells. These results were considered as compatible with a homophilic mechanism of adhesion involving the homophilic interaction domain of CAMs (Filbin *et al.*, 1990; D'Urso *et al.*, 1990). But, in these experiments, the possibility that carbohydrate-binding protein might be present was not evaluated. Similar experiments performed on other CAMs [carcinoembryonic antigen (CEA) and nonspecific carcinoembryonic antigen (NCA); Oikawa *et al.* (1989)] suggested that heterophilic mechanisms could be responsible for increased adhesivity of cells. In order to investigate this important point, we have analyzed (Lehmann *et al.*, 1991b) the possibility that CSL was present in CHO cells and could be (at least in part) responsible for adhesion in these cells. This is actually the case in nontransfected CHO cells, where CSL seems to be a determinant molecule for adhesion mechanisms. Thus, today, whether adhesion takes place through homophilic interactions or through a glyco-biological system of recognition involving lectins and glycans is still under discussion. But, the pilot experiment in this field is clearly depicted: Do the anti-CSL Fab fragments inhibit the adhesion invoked by transfection of the genes of CAMs in CHO cells?

e. Role of CSL in Multiple Sclerosis   The strong effect of anti-CSL antibodies on destruction of myelin (Kuchler *et al.*, 1988) was a good indication that CSL could be a key molecule in multiple sclerosis (MS). MS is considered as an autoimmune disease in which most of the clinical features can be associated with the destruction of only one structure of the nervous tissue: the myelin of the CNS (for review, see Adams, 1989). In fact, several types of anomaly can also be observed, including vascular destruction in the areas of plaques and partial destruction of the ependymal cell layer (ependymal cells are lining the cerebral ventricles). Considering the CSL as an immunological target in MS one could explain (1) myelin destruction [low amounts of anti-CSL Fab fragments dissociated myelin produced by cultured oligodendrocytes (Kuchler *et al.,* 1988)], and (2) destruction of the ependymal cell layer [CSL is concentrated in the tight junctions between these cells (Perraud *et al.,* 1988)].

In order to test the potential presence of anti-CSL antibodies in MS patients, we set up an immunoblotting test, involving partially purified CSL as the antigen and cerebrospinal fluid of patients as the source of antibody (Zanetta *et al.,* 1990b,c). The results obtained on more than 500 patients showed that more than 85% of the MS patients had substantial levels of anti-CSL antibodies in their cerebrospinal fluid. In reverse, 85% of the non-MS patients did not have anti-CSL antibodies. The presence of anti-CSL antibodies is certainly an early event in the progress of the disease, since the test revealed positive results in 17 patients at a time when the diagnosis of MS had not been confirmed.

The implication of the systematic presence of anti-CSL antibodies in the cerebrospinal fluid of MS patients, as well as their presence in the plasma of MS patients (Zanetta *et al.,* 1990c), suggests that CSL is a key molecule in the etiopathology of MS. Recent studies (Zanetta *et al.,* 1991b) suggest that CSL is the major (if not exclusive) immunological target in MS.

3. CSL in Other Systems

CSL has been shown to be an important molecule for adhesion mechanisms in other systems involving both nervous and nonnervous cells. One interesting model concerns aggregating cell cultures (Guenthert-Lauber and Honegger, 1985). In this system, it has been shown that early EGF treatment of aggregating cells causes a superaggregation phenomenon. It has been shown (Ténot *et al.,* 1989), that this phenomenon is associated with a relatively specific and strong synthesis and accumulation (about 10-fold the control value) of CSL molecules.

An important observation was reported recently (Zanetta *et al.,* 1991a), demonstrating the possible role of CSL in malignant transformation. The CSL is not a molecule specific to the nervous system, rather it can be found in variable amounts in several other tissues. Frequently, it is present in the tight junctions between cells, suggesting a role as an adhesion molecule. This is the case in hepatocytes, where levels of CSL and its glycoprotein ligands are relatively low. A glycoprotein with a M_r of 110,000 that is present in low amounts constitutes the major ligand of CSL. In cultured hepatoma cells, the levels of CSL are very similar to those found in normal cultured cells or *in vivo* in the liver. But, we have found that all hepatoma cells contained dramatically increased levels of CSL glycoprotein ligands (Zanetta *et al.,* 1991a). Furthermore, the pattern was similar for the different hepatoma cells. Such an increase in levels between transformed and control cells and similarity between transformed cells were not observed when glycoproteins were revealed with ConA. We assume that the change observed with CSL is specific for this lectin and is concerned with very minor glycoprotein ligands specifically revealed by the lectin.

The significance of this change in CSL ligands has been interpreted in terms of a new hypothesis of cell transformation based on the essential role of CSL as the adhesion molecule, called the hypothesis of the "non-sense signal" (Zanetta *et al., 1991a*). It is assumed that contacts between normal cells (adhering through CSL) occur through CSL and a small number of its glycoprotein ligands, these ligands being able to transduct a signal to the cells in contact. In malignant cells, the various numerous ligands of CSL transduce different signals, the resultant of which is a *non-sense* for the cells. Thus, contact can take place, but in the absence of the expected response occurring in normal cells. This hypothesis is based on the reported results on hepatoma cells and on a variety of systems. For instance, oligodendrocytes seeded on a CSL coated layer can proliferate. They cannot proliferate on ConA layers. Since ConA recognizes CSL-binding ligands, the variable reaction in oligodendrocytes is only that ConA binds a large variety of cell surface ligands. The unique, meaningful signal invoked by adhesion on CSL is not operating. A similar observation was made for axonal growth where neurons grow neurites when seeded on CSL but do not when seeded on ConA. Further studies are necessary to assess if this hypothesis can be generalized. The general observation that CSL specifically binds glycans reacting with antibodies binding to oncofetal glycan antigens is of special interest in this field.

VII. Conclusions and Perspectives

In sharp contrast to the intense development of the study of lectins in other tissues, the field of endogenous nervous lectins remained quite stagnant for several years, until the isolation of molecules associated with lectin activities provided stimulation. Several such molecules have been isolated, and it is likely that a larger number of lectins may be isolated in the near future from the nervous tissue. In sharp distinction to CAMs (possible ligands for endogenous lectins), there is essential evidence that lectins possess binding properties for endogenous ligands, and thus seem to be endowed with recognition properties. The main unresolved problem is that of colocalization (locus and time, certified by ultrastructural studies) of lectins and their ligands and the functional significance of such an interaction. Considering what is known about lectins in various organisms from virus to human, it is clear that they do have a role in cell recognition, cell adhesion, intracellular traffic, internalization of external molecules, and transmembrane signaling. These properties may be fundamental for nervous system ontogenesis, function, and pathology. However, clearly, various endogenous brain lectins display specific roles, as cell recognition lectins or as cell adhesion–recognition lectins (CARL), or as cell transit lectins. Their roles could change during ontogenesis and this could be a major argument for their involvement in nervous tissue plasticity. Considering the adhesion mechanisms, the observation that some lectins (Kuchler *et al.,* 1991a) can bind the carbohydrate moiety of CAMs considered as participating in cell adhesion (Schwarting *et al.,* 1987) opens new concepts for the role of lectins in the brain. The binding of CSL (and to a lesser extent of R1) to its endogenous ligands occurs through special types of glycans. Particularly, HNK-1- or Elec-39-reactive glycan epitopes seem to play an important role in this phenomenon. Thus, it is tempting to hypothesize that CSL is actually a cell adhesion lectin having specific binding for carbohydrate epitopes implicated in cell adhesion. In conclusion, we may say that CSL (and to a lesser extent R1) is possibly one of the "missing links" in the mechanism of cell adhesion. Furthermore, it has been demonstrated that CSL displays its adhesive properties in various cell types: hepatocytes, kidney, intestine, etc. (unpublished results). In normal cells, the density and number of CSL glycoprotein ligands are very small. In contrast, in transformed cells, the density and number of CSL glycoprotein ligands are increased dramatically (Zanetta *et al.,* 1991a). This increase in CSL ligands correlates (but not superimposes) with the observed increased level of oncofetal antigens detected by antibodies of the HNK-1 family in transformed cells (Abot and Balch, 1981; Baba *et al.,* 1986; Ilyas *et al.,* 1990). These observations prompted us to hypothesize that cell transformation is caused by the generation of *polysemic* signals to cells which are maintained in contact through CSL and its numerous membrane-bound glycoprotein ligands. The net result is a non-sense signal causing transformation.

Acknowledgments

The authors thank Prof. A. N. Malviya for reviewing the manuscript and A. Meyer and M. Zaepfel for technical assistance. Recent works of the research group in Strasbourg have been supported by grants from the Association Française contre les Myopathies, Ligue française contre la Sclérose en Plaques, and the National Multiple Sclerosis Society.

References

Abbott, W. M., and Feizi, T. (1991). *J. Neuroimmunol.* **31**, 179–179.

Abbott, W. M., Mellor, A., Edwards, Y., and Feizi, T. (1989). *Biochem. J.* **259**, 283–290.

Abot, T., and Balch, C. M. (1981). *J. Immunol.* **127**, 1024–1029.

Adams, C. W. M. (1989). "A Color Atlas of Multiple Sclerosis." Wolfe Medical Publications Ltd., London.

Altman, J. (1972a). *J. Comp. Neurol.* **145**, 353–398.

Altman, J. (1972b). *J. Comp. Neurol.* **145**, 399–464.

Altman, J. (1972c). *J. Comp. Neurol.* **145**, 465–514.

Baba, H., Sato, S., Inuzuka, T., Nishisawa, M., Tanaka, M., and Miyatake, T. (1986). *J. Neuroimmunol.* **13**, 89–97.

Badache, A., Burger, D., Villarroya, H., Robert, Y., Kuchler, S., Steck, A. J., and Zanetta, J.-P. (1991). Submitted for publication.

Baird, A., Schubert, D., Ling, N., and Guillemin, R. (1988). *Proc. Natl. Acad. Sci. U.S.A.* **85**, 2324–2328.

Bardosi, A., Dimitri, T., and Gabius, H.-J. (1988). *Acta Neuropathol.* **446**, 1–7.

Barondes, S. H. (1981). *Annu. Rev. Biochem.* **50**, 207–231.

Barondes, S. H. (1982). *Trends Neurochem. Sci.* **378**, 280–287.

Barondes, S. H. (1984). *Science* **223**, 1259–1264.

Barondes, S. H., and Haywood-Reid, P. L. (1981). *J. Cell Biol.* **91**, 568–572.

Barondes, S. H., and Rosen, S. D. (1976). *In* "Neuronal Recognition" (S. H. Barondes, ed.), pp. 332–356. Plenum, New York.

Barondes, S. H., Beyer, E. C., Springer, W. R., and Cooper, N. C. (1981). *J. Supramol. Struct. Cell. Biochem.* **16**, 233–242.

Beyer, E. C., and Barondes, S. H. (1982). *J. Cell Biol.* **92**, 23–27.

Bladier, D., Joubert, R., Avellana-Adalid, V., Kemeny, J. L., Doinel, C., Amouroux, J., and Caron, M. (1989). *Arch. Biochem. Biophys.* **269**, 433–439.

Bon, S., Meflah, K., Musset, F., Grassi, J., and Massoulié, J. (1987). *J. Neurochem.* **49**, 1720–1731.

Caron, M., Joubert, R., and Bladier, D. (1987). *Biochim. Biophys. Acta* **925**, 290–296.

Cerra, R. F., Haywood-Reid, P. L., and Barondes, S. H. (1984). *J. Cell Biol.* **98**, 1580–1589.

Chen, S., and Hillman, D. E. (1982). *Brain Res.* **240**, 205–220.

Childs, R. A., and Feizi, T. (1980). *Cell Biol. Int. Rep.* **4**, 755–000.

Cole, G. J., Loewy, A., and Glaser, L. (1986). *Nature (London)* **320**, 445–447.

Cook, G. M. W. (1986). *J. Cell Sci., Suppl.* **4**, 265–272.

Dautigny, A., Mattei, M.-G., Morello, D., Alliel, P. M., Pham-dinh, D., Amar, L., Arnaud, D., Simon, D., Mattei, J.-F., Guenet, J.-L., Jollès, P., and Avner, P. (1986). *Nature (London)* **321**, 867–869.

David, S. (1988). *J. Neurocytol.* **17**, 131–144.

DeCoster, M. A., and DeVries, G. H. (1989). *J. Neurosci. Res.* **22**, 283–289.

Delpech, A., Delpech, B., Girard, N., and Chauzy, C. (1987). *NATO Adv. Study Inst. Ser., Ser. H* **5**, 334–341.

Delpech, B., and Halavent, C. (1981). *J. Neurochem.* **36**, 855–859.

Dodd, J., and Jessell, T. M. (1986). *Exp. Biol.* **124**, 225–238.

Dodd, J., Morton, S. B., Karagogeos, D., Yamamoto, M., and Jessell, T. M. (1988). *Neuron* **1**, 105–116.

Dontenwill, M., Devilliers, G., Langley, O. K., Roussel, G., Hubert, P., Reeber, A., Vincendon, G., and Zanetta, J.-P. (1983). *Dev. Brain Res.* **10**, 287–299.

Dontenwill, M., Roussel, G., and Zanetta, J.-P. (1985). *Dev. Brain Res.* **17**, 245–252.

Doolittle, D. P., and Schweikart, K. M. (1977). *J. Hered.* **68,** 331–332.

Drickamer, K., Kordal, M. S., and Reynolds, L. (1986). *J. Biol. Chem.* **261,** 6878–6887.

Duncan, I. D., Haammang, J. P., and Trapp B. D. (1987). *Proc. Natl. Acad. Sci. U.S.A.* **84,** 6287–6291.

D'Urso, D., Brophy, P. J., Staugaitis, S. M., Gillespie, C. S., Frey, A. B., Stempak, J. G., and Colman, D. R. (1990). *Neuron* **4,** 449–460.

Eckenhoff, M. F., and Pysh, J. J. (1979). *J. Neurocytol.* **8,** 623–638.

Edelman, G. M. (1985). *Annu. Rev. Biochem.* **54,** 135–169.

Edelman, G. M. (1986). *Annu. Rev. Cell Biol.* **2,** 81–116.

Edgar, D., Timpl, R., and Thoenen, H. (1984). *EMBO J.* **3,** 1463–1468.

Eisenbarth, G. S., Ruffolo, R. R., Jr., and Walsh, F. S. (1978). *Biochem. Biophys. Res. Commun.* **83,** 1246–1252.

Espinosa de Los Monteros, A., Roussel, G., and Nussbaum, J.-L. (1986). *Dev. Brain Res.* **24,** 117–125.

Filbin, M. T., Walsh, F. S., Trapp, B. D., Pizzey, J. A., and Tennekoon, X. (1990). *Nature (London)* **344,** 871–872.

Fressinaud, C., Kuchler, S., Sarlière, L. L., Vincendon, G., and Zanetta, J.-P. (1988). *C. R. Seances Acad. Sci., Sér. 3* **307,** 863–868.

Gabius, H.-J. (1987). *In Vivo* **1,** 75–84.

Gabius, H. J., and Bardosi, A. (1990). *Histochemistry* **93,** 581–592.

Gabius, H.-J., Engelhardt, R., Rehm, S., Barondes, S. H., and Cramer, F. (1986). *Cancer J.* **1,** 19–22.

Gabius, H.-J., Kohnke, B., Hellmann, T., Dimitri, T., and Bardosi, A. (1988). *J. Neurochem.* **51,** 756–763.

Geoffroy, J. S., and Rosen, S. D. (1989). *J. Cell Biol.* **109,** 2463–2469.

Gitt, A., and Barondes, S. H. (1986). *Proc. Natl. Acad. Sci. U.S.A.* **83,** 7603–7607.

Gremo, F., Kobiler, D., and Barondes, S. H. (1978). *J. Cell Biol.* **79,** 491–499.

Guenther, J., Nick, H.-P., and Monard, D. (1985). *EMBO J.* **4,** 1963–1966.

Guenthert-Lauber, B., and Honegger, P. (1985). *Dev. Neurosci.* **7,** 286–295.

Harris, H., and Zalik, S. E. (1985). *J. Cell Sci.* **79,** 105–117.

Hatten, M. E., Liem, R. K. H., and Mason, C. A. (1984). *J. Neurosci.* **4,** 1163–1172.

Hatten, M. E., Liem, R. K. H. and Mason, C. A. (1986). *J. Neurosci.* **6,** 2676–2683.

Hemmendinger, L. M., and Caviness, V. S. Jr. (1988). *Dev. Brain Res.* **38,** 290–295.

Hirabayashi, J., Oda, Y., Oohara, T., Yamagata, T., and Kasai, K. (1987). *Biochim. Biophys. Acta* **916,** 321–327.

Hirabayashi, J., Ayaki, H., Soma, G. I., and Kasai, K. I. (1989). *FEBS Lett.* **250,** 161–165.

Hubert, J., Seve, A. P., Facy, P., and Monsigny, M. (1989). *Cell Differ. Dev.* **27,** 69–81.

Ikeda, K., and Yamamoto, T. (1985). *Brain Res.* **329,** 105–108.

Ilyas, A. A., Chou, D. K. H., Jungalwala, F. B., Costello, C., and Quarles, R. H. (1990). *J. Neurochem.* **55,** 594–601.

Inuzuka, T., Quarles, R. H., Noronha, A. B., Dobersen, M. J., and Brady, R. O. (1984). *Neurosci. Lett.* **51,** 105–111.

Joubert, R., Caron, M., and Bladier, D. (1987a). *Brain Res.* **36,** 146–150.

Joubert, R., Caron, M., Deugnier, M. A., Rioux, F., Sensenbrenner, M., and Bisconte, J. C. (1987b). *Cell. Mol. Biol.* **31,** 131–138.

Joubert, R., Kuchler, S., Zanetta, J.-P., Bladier, D., Avellana-Adalid, V., Caron, M., Doinel, C., and Vincendon, G. (1989). *Dev. Neurosci.* **11,** 397–413.

Kadmon, G., Kowitz, A., Altevogt, P., and Schachner, M. (1990). *J. Cell Biol.* **110,** 209–218.

Kitamura, K., Suzuki, M., and Uyemura, K. (1976). *Biochim. Biophys. Acta* **455,** 806–816.

Kitamura, K., Sakamoto, Y., Suzuki, M., and Uyemura, K. (1981). *In* "Glycoconjugates" (T. Yamatawa, T. Osawa, and S. Handa, eds.), pp. 273–274. Jpn. Sci. Soc. Press, Tokyo.

Kobiler, D., and Barondes, S. H. (1977). *Dev. Biol.* **60**, 326–330.

Kobiler, D., Beyer, E. C., and Barondes, S. H. (1978). *Dev. Biol.* **64**, 265–272.

Kuchler, S., Vincendon, G., and Zanetta, J.-P. (1987). *C. R. Seances Acad. Sci., Sér. 3* **305**, 317–320.

Kuchler, S., Fressinaud, C., Sarlieve, L. L., Vincendon, G., and Zanetta, J.-P. (1988). *Dev. Neurosci.* **10**, 199–212.

Kuchler, S., Herbein, G., Sarliève, L. L., Vincendon, G., and Zanetta, J.-P. (1989a). *Cell. Mol. Biol.* **35**, 581–596.

Kuchler, S., Rougon, G., Marschal, P., Lehmann, S., Reeber, A., Vincendon, G., and Zanetta, J.-P. (1989b). *Dev. Neurosci.* **11**, 414–427.

Kuchler, S., Perraud, F., Sensenbrenner, M., Vincendon, G., and Zanetta, J.-P. (1989c). *Glia* **2**, 437–445.

Kuchler, S., Rougon, G., Marschal, P., Lehmann, S., Reeber, A., Vincendon, G., and Zanetta, J.-P. (1989d). *Neuroscience* **33**, 111–124.

Kuchler, S., Zanetta, J.-P., Vincendon, G., and Gabius, H.-J. (1990a). *Eur. J. Cell Biol.* **52**, 87–97.

Kuchler, S., Zanetta, J.-P., Zaepfel, M., Badache, A., Sarliève, L. L., Gumpel, M., Baumann, N., and Vincendon, G. (1990b). *Dev. Neurosci.* **12**, 382–397.

Kuchler, S., Zanctta, J.-P., Bon, S., Zaepfel, M., Massoulié, J., and Vincendon, G. (1991a). *Neuroscience* **41**, 551–562.

Kuchler, S., Zanetta, J.-P., Zaepfel, M., Badache, A., Sarliève, L. L., Vincendon, G., and Matthieu, J.-M. (1991b). *J. Neurochem.* **56**, 436–445.

Leffler, H., Masiarz, F. R., and Barondes, S. H. (1989). *Biochemistry* **28**, 9222–9229.

Le Gal la Salle, G., and Valin, A. (1991). *C. R. Seances Acad. Sci., Ser. 3* **312**, 43–47.

Lehmann, S., Kuchler, S., Théveniau, M., Vincendon, G., and Zanetta, J.-P. (1990). *Proc. Natl. Acad. Sci. U.S.A.* **87**, 6455–6459.

Lehmann, S., Kuchler, S., Gobaille, S., Marschal, P., Badache, A., Reeber, A., Vincendon, G., and Zanetta J.-P. (1991a). *Brain Res. Bull.* (in press).

Lehmann, S., Kuchler, S., Badache, A., Zacpfel, M., Meyer, A., Vincendon, G., and Zanetta J.-P. (1991b). *Eur. J. Cell Biol.* **66**, in press.

Lemke, G., and Axel, R. (1985). *Cell (Cambridge, Mass.)* **40**, 501–508.

Liesi, P. (1985a). *EMBO J.* **4**, 1163 1170.

Liesi, P. (1985b). *EMBO J.* **4**, 2505–2511.

Lindner, J., Rathjen, F. G., and Schachner, M. (1983). *Nature (London)* **305**, 427–430.

Lindner, J., Guenther, J., Nick, H., Zinser, G., Antonicek, H., Schachner, M., and Monard, D. (1986). *Proc. Natl. Acad. Sci. U.S.A.* **83**, 4568–4571.

Linington, C., Webb, M., and Woodhams, P. L. (1984). *J. Neuroimmunol.* **6**, 387–396.

Lis, H., and Sharon, R. (1986). *Annu. Rev. Biochem.* **55**, 35–67.

MacBride, R. G., and Przybylski, R. J. (1986). *In Vitro* **22**, 568–574.

Mähönen, Y., and Rauvala, H. (1985). *Eur. J. Cell Biol.* **36**, 91–97.

Mailly, P., Ben Younes-Chennoufi, A., and Bon, S. (1989). *Neurochem. Int.* **15**, 517–530.

Mallet, J., Huchet, M., Shelanski, M., and Changeux, J.-P. (1974). *FEBS Lett.* **46**, 243–246.

Mallet, J., Huchet, M., Pougeois, R., and Changeux, J.-P. (1976). *Brain Res.* **103**, 291–312.

Marschal, P., Reeber, A., Neeser, J. R., Vincendon, G., and Zanetta, J.-P. (1989). *Biochimie* **71**, 645–653.

Martini, R., and Schachner, M. (1986). *J. Cell Biol.* **103**, 2439–2448.

Matthieu, J.-M. (1981). *Neurochem. Int.* **3**, 355–363.

Matthieu, J.-M., Daniel, A., Quarles, R. H., and Brady, R. O. (1974). *Brain Res.* **81**, 348–353.

Matthieu, J.-M., Roch, J.-M., Omlin, F. X., Rambaldi, I., Almazan, G., and Braun, P. E. (1986). *J. Cell Biol.* **103**, 2673–2682.

McCarthy, K. D., and DeVellis, J. (1980). *J. Cell Biol.* **85,** 890–902.
McGarry, R. C., Helfand, S. L., Quarles, R. H., and Roder, J. C. (1983). *Nature (London)* **306,** 376–378.
Milos, N. C., Ma, Y. L., and Frunchak, Y. N. (1989). *Cell Differ. Dev.* **28,** 203–209.
Milos, N. C., Ma, Y. L., Varma, P. V., Bering, M. P., Mohamed, Z., Pilarski, L. M., and Frunchak, Y. N. (1990). *Anat. Embryol.* **182,** 319–327.
Monsigny, M., Kieda, C., and Roche, A. C. (1983). *Biol. Cell.* **47,** 95–110.
Moonen, G., Grau-Wagemans, M. P., and Selak, I. (1982). *Nature (London)* **298,** 753–755.
Moutsatsos, I. K., Davis, J. M., and Wang, J. L. (1986). *J. Cell Biol.* **102,** 477–483.
Mullin, B. R., Decandis, F. X., Montanaro, A. J., and Reid, J. D. (1981). *Brain Res.* **222,** 218–221.
Musset, F., Frobert, Y., Grassi, J., Vigny, M., Boulla, G., Bon, S., and Massoulié, J. (1987). *Biochimie* **69,** 147–156.
Nave, K.-A., Lai, C., Bloom, F. E., and Milner, R. J. (1986). *Proc. Natl. Acad. Sci. U.S.A.* **83,** 9264–9268.
Neskovic, N. M., Roussel, G., and Nussbaum, J.-L. (1986). *J. Neurochem.* **47,** 1412–1418.
Nicholson, J. L., and Altman, J. (1972a). *Brain Res.* **44,** 13–24.
Nicholson, J. L., and Altman, J. (1972b). *Brain Res.* **44,** 25–32.
Nowak, T. P., Haywood, P. L., and Barondes, S. H. (1976). *Biochem. Biophys. Res. Commun.* **68,** 650–657.
Nowak, T. P., Kobiler, D., Roel, L. E., and Barondes, S. H. (1977). *J. Biol. Chem.* **252,** 6026–6030.
Oikawa, S., Inuzuka, C., Kuroki, M., Matsuoka, Y., Kosaki, G., and Nakazato, H. (1989). *Biochem. Biophys. Res. Commun.* **164,** 39–45.
Olden, K., Bernard, B. A., Humphries, M. J., Yeo, T. K., White, S. L., Newton, S. A., Bauer, H. C., and Parent, J. B. (1985). *Trends Biochem. Soc.* **10,** 78–82.
O'Shannessy, D. J., Willison, H. J., Inuzuka, T., and Quarles, R. H. (1985). *J. Neuroimmunol.* **9,** 255–268.
Ozawa, M., Sato, M., and Muramatsu, T. (1983). *J. Biochem. (Tokyo)* **94,** 479–485.
Palacios-Prü, E. L., Palacios, L., and Mendoza, R. V. (1981). *J. Submicrosc. Cytol.* **13,** 145–167.
Palay, S. L., and Chan-Palay, V. (1974). *In* "Cerebellar Cortex" (S. L. Palay and V. Chan-Palay, eds.). Springer-Verlag, Berlin.
Perraud, F., Kuchler, S., Gobaille, S., Labourdette, G., Vincendon, G., and Zanetta, J.-P. (1988). *J. Neurocytol.* **17,** 745–752.
Pierres, M., Barbet, J., Naquet, P., Pont, S., Régnier-Vigouroux, A., Bard, M., Devaux, C., Marchetto, S., and Rougon, G. (1987). *UCLA Symp. Mol. Cell. Biol.* [N. S.] **73,** 293–300.
Poduslo, J. F. (1981). *J. Neurochem.* **36,** 1924–1931.
Poduslo, J. F., Everly, J. L., and Quarles, R. H. (1977). *J. Neurochem.* **28,** 977–986.
Poduslo, J. F., Harman, J. L., and McFarlin, D. E. (1980). *J. Neurochem.* **34,** 1733–1744.
Poltorak, M., Sadoul, R., Keilhauer, G., Landa, C., Fahrig, T., and Schachner, M. (1987). *J. Cell Biol.* **105,** 1893–1899.
Quarles, R. H., McIntyre, L. J., and Pasnak, C. F. (1979). *Biochem. J.* **183,** 213–221.
Quicho, F. A. (1986). *Annu. Rev. Biochem.* **55,** 287–315.
Rakic, P. (1971). *J. Comp. Neurol.* **141,** 283–312.
Rakic, P., and Sidman, R. L. (1973). *J. Comp. Neurol.* **152,** 103–132.
Ramón y Cajál, S. (1911). "Histologie du système nerveux de l'homme et des vertébrés," Vols. I and II. Maloine, Paris.
Rauvala, H. (1983). *Trends Biochem. Sci.* **376,** 323–325.
Rauvala, H., and Pihlaskari, R. (1987). *J. Biol. Chem.* **262,** 16625–16635.

Reeber, A., Vincendon, G., and Zanetta, J.-P. (1980). *J. Neurochem.* **35,** 1273–1277.

Reeber, A., Vincendon, G., and Zanetta, J.-P. (1981). *Brain Res.* **229,** 53–65.

Regan, L. J., Dodd, J., Barondes, S. H., and Jessell, T. M. (1986). *Proc. Natl. Acad. Sci. U.S.A.* **83,** 2248–2252.

Ripellino, J. A., Bailo, M., Margolis, R. U., and Margolis, R. K. (1988). *J. Cell Biol.* **106,** 845–855.

Robertson, P. L., and Goldstein, G. W. (1988). *Brain Res.* **447,** 341–345.

Rosen, S. D., Geoffroy, J. S., Lasky, L. A., Singer, M. S., Stachel, S., Stoolman, L. M., True, D. D., and Yednock, T. A. (1990). *Glycobiology* **111,** 69–90.

Sakamoto, Y., Kitamura, K., Yoshimura, K., Nishijima, T., and Uyemura, K. (1987). *J. Biol. Chem.* **262,** 4208–4214.

Salzer, J. L., and Colman, D. R. (1989). *Dev. Neurosci.* **11,** 377–390.

Sanes, J. R. (1985). *Nature (London)* **315,** 714–715.

Schluesener, H. J., Sobel, R. A., Linington, C., and Weiner, H. L. (1987). *J. Neuroimmunol.* **12,** 4016–4021.

Schneider-Schaulies, J., von Brunn, A., and Schachner, M. (1990). *J. Neurosci. Res.* **27,** 286–297.

Schwarting, G. A., Jungalwala, F. B., Chou, D. K. H., Boyer, A. M., and Yamamoto, M. (1987). *Dev. Biol.* **120,** 65–76.

Sharon, N. (1984). *Biol. Cell.* **51,** 239–246.

Sharon, N., and Lis, H. (1989). *Science* **246,** 227–234

Sidman, R. L., and Rakic, P. (1973). *Brain Res.* **62,** 1–35.

Sotelo, C., and Changeux, J.-P. (1974a). *J. Cell Biol.* **53,** 271–289.

Sotelo, C., and Changeux, J.-P. (1974b). *Brain Res.* **77,** 484–491.

Springer, T. A. (1990). *Nature (London)* **346,** 425–434.

Teichberg, V. I., Silman, I., Beitsch, D., and Resheff, D. (1975). *Proc. Natl. Acad. Sci. U.S.A.* **72,** 1383–1387.

Ténot, M., Kuchler, S., Zanetta, J.-P., Vincendon, G., and Honegger, P. (1989). *J. Neurochem.* **53,** 1435–1441.

Trapp, B. D., Quarles, R. H., and Suzuki, K. (1984). *J. Cell Biol.* **99,** 595–606.

Vraz, A., Meromsky, L., and Lotan, R. (1986). *Cancer Res.* **46,** 3667–3672.

Zalik, S. E., and Milos, N. (1986). *In* "Developmental Biology: A Comprehensive Synthesis" (L. M. Browder, ed.), Vol 2, pp. 145–194. Plenum, New York.

Zalik, S. E., Milos, N., and Ledsham, I. (1983). *Cell Differ.* **12,** 121–127.

Zalik, S. E., Thomson, L. W., and Ledsham, I. M. (1987). *Proc. Natl. Acad. Sci. U.S.A.* **84,** 6345–6348.

Zanetta, J.-P., Sarliève, L. L., Mandel, P., Vincendon, G., and Gombos, G. (1977). *J. Neurochem.* **29,** 827–838.

Zanetta, J.-P., Roussel, G., Ghandour, M. S., Vincendon, G., and Gombos, G. (1978). *Brain Res.* **142,** 301–319.

Zanetta, J.-P., Meyer, A., Dontenwill, M., Basset, P., and Vincendon, G. (1982). *J. Neurochem.* **39,** 1601–1606.

Zanetta, J.-P., Roussel, G., Dontenwill, M., and Vincendon, G. (1983). *J. Neurochem.* **40,** 202–208.

Zanetta, J.-P., Reeber, A., Dontenwill, M., and Vincendon, G. (1984). *J. Neurochem.* **42,** 334–339.

Zanetta, J.-P., Dontenwill, M., Meyer, A., and Roussel, G. (1985a). *Dev. Brain Res.* **17,** 233–243.

Zanetta, J.-P., Dontenwill, M., Reeber, A., Vincendon, G., Legrand, C., Clos, J., and Legrand, J. (1985b). *Dev. Brain Res.* **21,** 1–6.

Zanetta, J.-P., Bingen, A., Dontenwill-Kiefer, M., Reeber, A., Meyer, A., and Vincendon, G. (1987a). *Cell. Mol. Biol.* **33,** 423–434.

Zanetta, J.-P., Dontenwill, M., Reeber, A., and Vincendon, G. (1987b). *NATO Adv. Study Inst. Ser., Ser. H* **2,** 92–104.

Zanetta, J.-P., Meyer, A., Kuchler, S., and Vincendon, G. (1987c). *J. Neurochem.* **49,** 1250–1257.

Zanetta, J.-P., Kuchler, S., Marschal, P., Zaepfel, M., Meyer, A., Badache, A., Reeber, A., Lehmann, S., and Vincendon, G. (1990a). *NATO Adv. Study Inst. Ser., Ser. H* **43,** 433–450.

Zanetta, J.-P., Warter, J.-M., Kuchler, S., Marschal, P., Rumbach, L., Lehmann, S., Tranchant, C., Reeber, A., and Vincendon, G. (1990b). *Lancet* **335,** 1482–1484.

Zanetta, J.-P., Warter, J.-M., Lehmann, S., Tranchant, C., Kuchler, S., and Vincendon, G. (1990c). *C. R. Seances Acad. Sci., Ser. 3* **311,** 327–331.

Zanetta, J.-P., Staedel, C., Kuchler, S., Zaepfel, M., Meyer, A., and Vincendon, G. (1991a). *Carbohydr. Res.* **213,** 117–126.

Zanetta J.-P., Kuchler, S., Lehmann, S., Dufourcq, P., Badache, A., Meyer, A., Zaepfel, M., Warter, J.-M., and Vincendon, G. (1991b). Submitted for publication.

The Role of Jasmonic Acid and Related Compounds in the Regulation of Plant Development

Yasunori Koda

Department of Botany, Faculty of Agriculture, Hokkaido University,
Sapporo 060, Japan

I. Introduction

We have been investigating the mechanism of potato tuberization from hormonal point of view for more than 10 years. We have isolated a specific tuber-inducing substance from the leaves of the potato plant (Koda *et al.*, 1988) and identified it as 3-oxo-2-(5'-β-D-glucopyranosyloxy-2-*cis*-pentenyl)cyclopentane-1-acetic acid (Yoshihara *et al.*, 1989). The substance is a glucoside (Fig. 1A) and we named the aglycone of the substance "tuberonic acid" (Fig. 1B). The chemical structure of tuberonic acid (TA) is closely related to that of jasmonic acid (JA; Fig. 1C). JA also has strong potato tuber-inducing activity (Koda *et al.*, 1991a). Recently, we found that tuberization in yam plants (*Dioscorea batatas* Decne.) seems to be controlled by JA (Koda and Kikuta, 1991). We also found that JA is capable of inducing *in vitro* tuberization of Jerusalem artichoke plants (*Helianthus tuberosus* L.). These observations suggest the involvement of JA in tuberization in various plants.

Methyl jasmonate (JA-Me, jasmonic acid methyl ester; Fig. 1D) was first isolated as a major component of the aroma of the essential oils of *Jasminum grandiflorum* L. and *Rosemarinus officinalis* L. (Demole *et al.*, 1962; Crabalona, 1967) and now it is recognized as an important ingredient in the perfume industry.

JA was first isolated as a plant growth inhibitor from the medium in which *Lasiodiplodia theobromae* Griff. and Maubl. (synonym *Botryodiplodia theobromae* Pat.), a common tropical fungus, was cultured (Aldridge *et al.*, 1971). This was the first report indicating that JA has a regulatory effect on plant growth. Cucurbic acid (CA; Fig. 1E), a compound that is structurally related to JA, was isolated as a growth inhibitor from seeds of

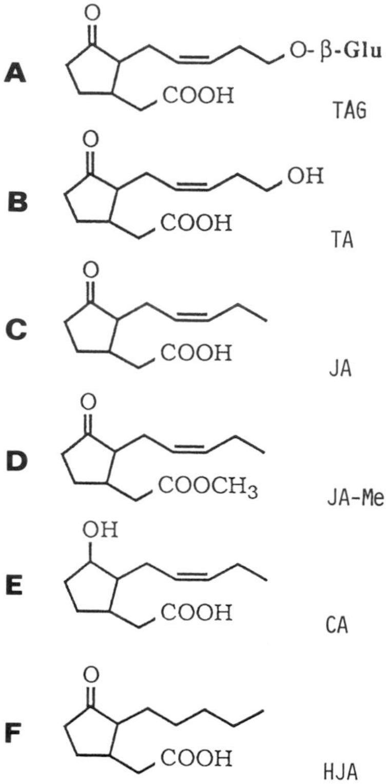

FIG. 1   Chemical structures of jasmonic acid-related compounds present in plants. (A) Glucosyl tuberonic acid (TAG), (B) tuberonic acid (TA), (C) jasmonic acid (JA), (D) methyl jasmonate (JA-Me), (E) cucurbic acid (CA), (F) dihydrojasmonic acid (HJA).

Cucurbita pepo (Fukui *et al.,* 1977a,b). After 1980, several reports were published that showed that JA and JA-Me are present in plant tissues and that these compounds exert some inhibitory effects on plant growth, such as the inhibition of the growth of seedlings (Yamane *et al.,* 1980, 1981; Dathe *et al.,* 1981) and the promotion of leaf senescence (Ueda and Kato, 1980).

Besides being present in plants and fungus, JA-Me was also found in hairpencils of oriental fruit moths (*Grapholitha molesta*), and a mixture including JA-Me was capable of attracting pheromone-releasing females from a distance of several centimeters (Baker *et al.,* 1981). However, a subsequent study suggested that JA-Me was not synthesized in the moth tissues, but originated from their feed, green apples (Lofstedt *et al.,* 1989).

JA and related compounds, henceforth designated as JA-related compounds, are now known to be distributed widely among higher plants and play important roles in the regulation of plant development. Recently, these compounds have been reported to be potent inducers of expression of several genes, such as the gene of proteinase inhibitor in tomato plants (Farmer and Ryan, 1990) and the genes of vegetative storage proteins in soybean plants (Anderson *et al.*, 1989). Various growth-inhibiting activities of JA and JA-Me were sometimes compared to those of abscisic acid (ABA), but recent information suggests that the mechanisms of inhibition of JA and JA-Me are different from those of ABA.

The purpose of this article is to present a review of our current knowledge about the role of JA and related compounds in regulation of plant development, and to give a more personal account of results obtained recently in our laboratory.

II. Ubiquitous Presence of Jasmonic Acid-Related Compounds in Higher Plants

A. Jasmonic Acid and Its Derivatives

1. Jasmonic Acid and Methyl Jasmonate

During the course of investigation into gibberellins in immature seeds of *Phaseolus vulgaris*, Yamane and co-workers noticed the presence of some compounds which inhibit the growth of rice seedlings and isolated the compounds from the material, one being ABA and the other being JA. The final yield of JA from 1 kg of immature seeds was 0.7 mg. They also confirmed the occurrence of JA in immature seeds of *Dolichos lablab* L., and in fresh leaves and insect galls of *Castanea crenata* Sieb. et Zucc. (Yamane *et al.*, 1981). Because of the inhibitory effect of JA on germination of pollen of tea plant (*Camellia sinensis* L.) at a concentration above 20 mg/liters ($\sim 10^{-4}$ M), they assumed that JA is an endogenous regulator of pollen germination, and isolated JA and JA-Me from the pollens and anthers of three *Camellia* species (Yamane *et al.*, 1982). The amount of JA in them was between 7.8 and 34.3 μg/kg fresh weight. In *Camellia japonica* and *Camellia sasanqua*, JA-Me was also found, the amounts being 11.9 and 16.5 μg, respectively. However, JA-Me showed no inhibitory effect on the pollen germination.

As it is well known that the development of the pericarp and the seeds are dependent on each other, Dathe *et al.* (1981) thought that studies on the endogenous hormones in the pericarp should contribute to an understanding of the hormonal regulation of fruit development. They surveyed endogenous growth inhibitors in the developing pericarp of broad bean (*Vicia faba* L.) by measuring the inhibition of growth of wheat seedlings. In addition to ABA, they isolated JA from the pericarp. They examined the change in the level of JA during the growth of the pericarp by a bioassay, the highest level of as much as 3 mg/kg fresh weight being attained prior to the termination of pod elongation.

Ueda and Kato (1980) found that an extract from wormwood shoots (*Artemisia absinthium* L.) contained a compound which could promote the senescence of leaf segments of oat, and identified the compound as JA-Me. The final yield of the substance from 1 kg of fresh material was 0.1 mg. The authors also isolated two substances which promoted senescence of oat leaf segments from immature leaves of *Cleyera ochnacea,* one being ABA and another being JA (Ueda and Kato, 1982a).

Tsurumi *et al.* (1985) found that *Mimosa pudica* pulvinules excised in the morning were opened by treatment with indoleacetic acid (IAA). Pulvinules excised in the evening, however, showed a poor response to IAA. Because the susceptibility of the material to IAA was recovered by washing, the authors supposed an occurrence of an inhibitor in the material harvested in the evening. Tsurumi and Asahi (1985) purified the inhibitor from the plants and identified it as JA. The final yield of JA from 1 kg of fresh *Mimosa* plants was 40 μg.

Meyer *et al.* (1984) investigated the occurrence of JA and JA-Me in various parts of many seed plants using different methods such as gas chromatography (GC), GC-mass spectrometry (GC-MS), high-performance liquid chromatography (HPLC), radioimmunoassay, and bioassay. By means of GC-MS, JA was identified in tissues and organs of nine species of Leguminosae (*Vicia faba, V. narbonensis, Pisum sativum, Phaseolus vulgaris, P. coccineus, Glycine max, Calliandra haematocephala, Dolichos lablab, Lupinus albus*) and of eight angiosperms other than Leguminosae (*Quercus robur, Fagus sylvatica, Cucurbita maxima, Malus sylvestris, Citrus aurantifolia, C. sinensis, Solanum tuberosum, Helianthus annuus*). Young apple fruits (*Malus sylvestris*) contained both JA and JA-Me. The highest amount of JA was found in the fruit parts, the amounts determined by GC in immature pericarp of soybean (*Glycine max*) and young fruits of kidney bean (*Phaseolus vulgaris*) being 1.26 mg and 1.14 mg/kg fresh weight, respectively. Although an identification was not carried out, the authors found JA-Me in Douglas fir plants (*Pseudotsuga menziesii*), a gymnosperm.

Lopez *et al.* (1987) studied the changes in the level of JA in various parts of soybean fruits during the growth of the plants using radioimmunoassay. The highest levels were found in pericarp tissues containing vascular bundles at all stages investigated. The levels were between 2.6 and 4.5 mg/kg fresh weight. Among the seed parts, hilum and testa contained more JA than embryos. The levels in testa and hilum were between 1.3 and

2.3 mg/kg fresh weight and that in embryo was less than 0.4 mg/kg fresh weight.

2. Dihydrojasmonic Acid

Saturation of the double bond in the pentenyl side chain at the C-2 position of JA gives dihydrojasmonic acid (HJA) (Fig. 1F). Miersch *et al.* (1989) isolated HJA from immature fruits of the broad bean in addition to JA. The final yield of HJA was 0.18 mg/kg fresh weight.

3. Amino Acid Conjugates of Jasmonic Acid

Bohlmann *et al.* (1984) isolated *N*-(acetoxy)jasmonoyl-phenylalanine from *Praxelis clematidea*. Brückner *et al.* found *N*-[(−)-jasmonoyl]-*S*-tyrosine (1986) and *N*-[(−)-jasmonoyl]-*S*-tryptophan (1988) in flowers of broad bean using radioimmunoassay. Although the roles of these conjugates are unknown at present, they seem to suggest an important route in the metabolism of JA.

B. Cucurbic Acid

The elongation of a young intact plant is usually induced by application of gibberellins and not by auxins. Fukui *et al.* expected the occurrence of new plant growth regulators in cucumber seedlings because the hypocotyl is elongated by both IAA and gibberellin, and they surveyed the growth regulators in cucumber seeds with the guidance of the gibberellin bioassay using dwarf rice seedlings. Finally, they isolated three gibberellins as growth promoters and several growth inhibitors, including cucurbic acid (CA), cucurbic acid glucoside, and methyl cucurbate glucoside (Fukui *et al.*, 1977a), and determined the chemical structure of CA and its derivatives (Fukui *et al.*, 1977b). These were the first reports that showed the presence of JA-related compounds in higher plants and the inhibitory effect of the compounds on plant growth.

C. Tuberonic Acid and Glucosyl Tuberonic Acid

Tuberization in potato plants is controlled predominantly by photoperiod. Short days promote tuberization whereas long days inhibit the process. The response to the photoperiod interacts with many other factors, including genotype, temperature, nitrogen level, and age of mother tubers (see Koda *et al.*, 1988). With grafting experiments, Gregory (1956) and

Chapman (1958) demonstrated the occurrence of a tuberization stimulus which is formed in the leaves under short days and transmitted to underground parts to induce tuberization. Although many studies were done to identify the stimulus (see reviews by Melis and Van Staden, 1984; Ewing, 1985), most were concentrating on the effects of known plant hormones on the tuberization. Only a few attempts have been made to isolate the stimulus directly from the leaves. Madec (1963) reported that the sap expressed from potato plants which had been grown under short days was capable of inducing tuberization when injected into a plant under long days. However, Simonds (1965) could not confirm Madec's result. This discrepancy seem to be due to the absence of an accurate assay method for the activity. Using cultures of single-node segments of potato stem *in vitro* (see Section V,B,1), Koda and Okazawa (1988) demonstrated the occurrence of two acidic substances which have tuber-inducing activity in the leaves and in physiologically old tubers. The tuber-inducing activity in the leaves increased under short days and remained constant under long days, suggesting that these substances are the tuberization stimulus postulated by Gregory (1956).

One of the two substances was isolated from the leaves (Koda *et al.*, 1988) and identified as glucosyl tuberonic acid [3-oxo-2-(5'-β-D-glucopyranosyl-2'-*cis*-pentenyl)cyclopentane-1-acetic acid] (Yoshihara *et al.*, 1989). TA was also found in the potato leaves as well as in beet (*Beta vulgaris*) leaves (Y. Koda, K. Ohkawa, and Y. Kikuta, unpublished data).

III. Biosynthesis and Metabolism of Jasmonic Acid-Related Compounds

A. Biosynthesis

1. Biosynthetic Pathway

Investigations into the biosynthetic pathway of JA started independently of JA. Lipoxygenase is a common enzyme in plant tissues. The enzyme catalyzes the incorporation of molecular oxygen into certain polyunsaturated fatty acids having a *cis,cis*-1,4-pentadiene system to form a fatty acid hydroperoxide. Since the enzyme deteriorates the quality of seeds of soybean and some cereals as food, the enzyme in these seeds has been studied intensively (Eskin *et al.*, 1977). To plant biochemists, the enzymatic capability to produce fatty acid hydroperoxide has been a puzzling phenomenon, because the highly reactive hydroperoxide could damage cellular components. Zimmerman and Feng (1978) demonstrated that the hydroperoxide of linolenic acid could be converted to an 18-carbon cyclic fatty acid, 12-oxo-*cis,cis*-10,15-phytodienoic acid (12-oxo-PDA).

After the work by Dathe *et al.* (1981) which showed the presence of JA in the pericarp of broad bean at a high concentration, Vick and Zimmerman (1983) noticed the similarity in the structure between 12-oxo-PDA and JA. They prepared [U-^{14}C]linolenic acid by feeding flax embryos with sodium [^{14}C]acetate, and then prepared [^{14}C]12-oxo-PDA from the labeled linolenic acid by an acetone powder of flax seed extract. When thin transverse sections of pericarp of broad bean were incubated with [U-^{14}C]12-oxo-PDA, four radioactive metabolites could be obtained. They identified three of them as 3-oxo-2-(2'-pentenyl)cyclopentaneoctanoic acid (OPC-8:0), 3-oxo-2-(2'-pentenyl)cyclopentanebutanoic acid (OPC-4:0) (see Fig. 2), and JA. Their subsequent experiment showed that six plant species can metabolize 12-oxo-PDA to JA (Vick and Zimmerman, 1984). The plant species are eggplant (*Solanum melongena* L.), flax (*Linum usitatissium* L.), sunflower (*Helianthus annuus* L.), oat (*Avena sativa* L.), wheat (*Triticum aestivum* L.), and corn (*Zea mays* L.). The presence of enzymes which convert 12-oxo-PDA to JA in several plant species, including both dicots and monocots, suggests the ubiquitous presence of the pathway in higher plants.

Hamberg (1988) showed that 13-hydroperoxylinolenic acid was metabolized to a short-lived allene oxide, 12,13-epoxy-9,11,15-octadecatrienoic acid (12,13-EOT), by hydroperoxide dehydrase obtained from defatted meal of corn. There were two different pathways to form 12-oxo-PDA from the allene oxide. One was a nonenzymatic pathway via α-ketol (12-oxo-13-hydroxy-9,15-octadecadienoic acid) leading to a racemic mixture of 12-oxo-PDA. This nonenzymatic reaction of allene oxide produced 12-oxo-PDA and α-ketol in a ratio of about 0.14 : 1.00. The other pathway was enzyme-catalyzed cyclyzation leading to optically pure 12-oxo-PDA. The enzyme was named allene oxide cyclase.

The complete reaction sequence from linolenic acid to JA is shown in Fig. 2. At first, lipoxygenase catalyzes O_2 incorporation at C-13 of linolenic acid. Then, 13-hydroperoxylinolenic acid is transformed into a short-lived allene oxide derivative (12,13-EOT) by hydroperoxide dehydrase. The allene oxide is converted to 12-oxo-PDA by allene oxide cyclase. The ring double bond of 12-oxo-PDA is saturated by reductase (Vick and Zimmerman, 1986) and the reaction produces OPC-8:0. Next, six carbons are removed from the carboxyl side chain by β-oxidation occurring three times successively, and finally, JA is formed. At present, there is no direct information about biosyntheses of TA and CA. TA and CA seem to be easily formed from JA by a hydroxylation at C-5 of the pentenyl side chain, and by reduction of the keto group at the C-3 position of the cyclopentane to the hydroxy group, respectively. However, a possibility still exists that

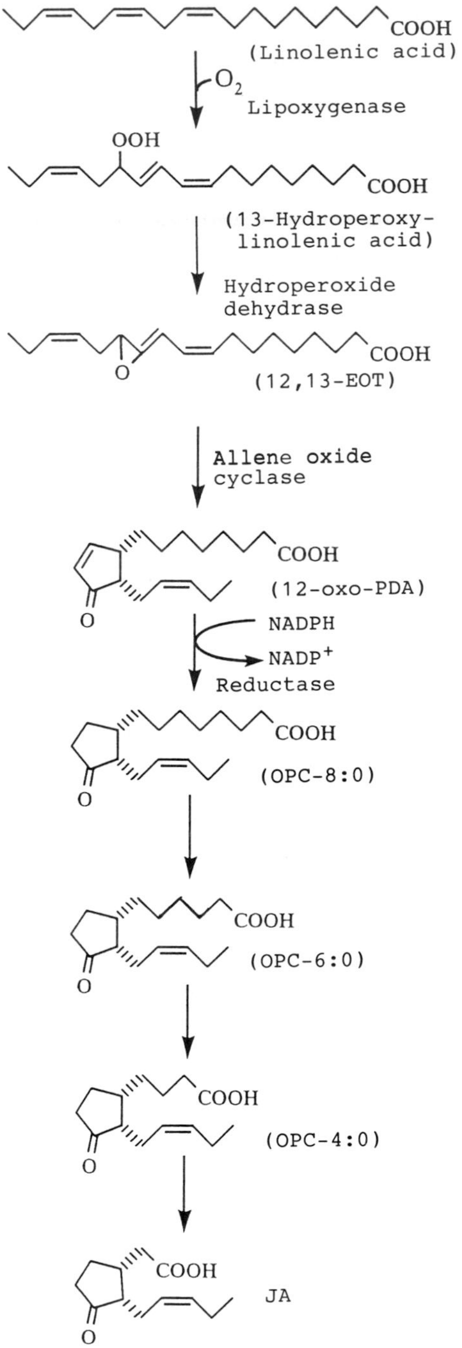

FIG. 2   Biosynthetic pathway of jasmonic acid (Vick and Zimmerman, 1984, 1986).

these compounds are not formed directly from JA but formed from pre-cursors of JA.

Physiological roles of metabolites of linolenic acid other than JA-related compounds remain to be elucidated. Vick and Zimmerman (1984) noticed the similarity in structure of many of these metabolites to eicosanoids in mammalian tissues and speculated that some of the metabolites may behave as potent metabolic regulators in plant tissues.

Since the highest activities of the enzymes that are involved in biosynthesis of JA were found in young, actively growing plant tissues (Vick and Zimmerman, 1981, 1982, 1986), Vick and Zimmerman (1986) speculated that JA has another, more subtle role in metabolism which is not related to growth inhibition and senescence. However, a high enzyme activity does not always mean a high amount of product of the enzyme. It is possible that one of the rate-limiting factors of JA biosynthesis is the level of free linolenic acid, not the activities of the enzymes. The levels of free fatty acids in plant tissues are known to be very low. In phospholipids of membrane of higher plants, much linolenic acid is contained at the sn-2 site (Ansell and Spanner, 1982). Although a basic question, whether lipoxygenase is able to oxidize polyunsaturated fatty acids that are still in the intact phospholipids remains open, it seems likely that a release of the fatty acids from phospholipids by a phospholipase is necessary for the action of lipoxygenase (Kockritz et al., 1985). It is also possible that the activity of phospholipase A_2 (EC 3.1.1.4), which catalyzes the liberation of fatty acids from the sn-2 site of phospholipids (Van den Bosch, 1982), is one of the rate-limiting factors of biosynthesis of JA-related compounds.

2. Enzymes Involved in Biosynthesis of Jasmonic Acid

a. Lipoxygenase From the above mentioned studies, it is well documented that JA is synthesized from linolenic acid by a series of enzymes. The initial reaction in JA synthesis is catalyzed by lipoxygenase (EC 1.13.11.12). The presence of multiple forms of lipoxygenase was reported in many plants (Eskin et al., 1977), and they have been classified into three types according to the optimal pH and the positional specificities of the hydroperoxides produced (Ohta et al., 1986). The enzyme catalyzes O_2 incorporation either at C-9 or at C-13 of linolenic acid. JA is a product only when oxygenation occurs at C-13. However, many plant lipoxygenases catalyze O_2 incorporation predominantly at C-9. For example, the proportion of 9-hydroperoxide formed by the enzyme isolated from potato tubers was 95% (Galliard and Phillips, 1971). The 9-hydroperoxide may follow a different course of metabolism. Furthermore, not all the 13-hydroperoxide is metabolized to JA, part is converted to α- and γ-ketols by hydroperoxide isomerase (dehydrase) (Zimmerman and Vick, 1970) and part is cleaved to oxoacid and aldehydes by hydroperoxide lyase (Vick and Zimmerman, 1976; Sekiya *et al.,* 1979; Kim and Grosch, 1981). These observations suggest that only a part of free linolenic acid is metabolized to JA.

b. Hydroperoxide Dehydrase and Allene Oxide Cyclase   Vick and Zimmerman (1987) examined the metabolism of 13-hydroperoxylinolenic acid in leaf protoplast of spinach. Enzyme activities were assayed by measuring the amount of product by GC after methylation. When the ruptured protoplasts were divided into cytoplasm and chloroplast fractions by centrifugation, the activity of hydroperoxide dehydrase was predominantly found in the chloroplast fraction. The enzyme activity was not found in mitochondria and peroxisomes. The enzyme was associated with membranes of chloroplast. The molecular weight of the enzyme was 220,000, as estimated by gel filtration. The enzyme seemed to be highly hydrophobic and, in the absence of a detergent, aggregated to form high-molecular-weight complexes. The enzyme had its optimum activity in the range of pH 5 to 7.

Hamberg (1988) studied the biosynthesis of 12-oxo-PDA in enzyme preparations obtained from defatted meal of corn. He fractionated corn homogenates by differential centrifugation and found that hydroperoxide dehydrase activity was associated with the membrane fraction (105,000 *g* precipitate). On the other hand, allene oxide cyclase was found in the soluble fraction (105,000 *g* supernatant). The molecular weight of allene oxide cyclase was estimated to be ~45,000 by gel filtration.

c. 12-oxo-PDA Reductase   Vick and Zimmerman (1986) characterized 12-oxo-PDA reductase from seeds and seedlings of corn. The enzyme activity was assayed by measuring the amount of reduced product (OPC-8:0) by GC after methylation. The side chains of naturally occurring 12-oxo-PDA are in the cis conformation with respect to the plane of the cyclopentane ring. Vick and Zimmerman, however, used the trans isomer of the compound as a substrate for the enzyme, because the cis isomer is unstable and easily epimerized to the trans isomer (see Section IV). The molecular weight of 12-oxo-PDA reductase, estimated by gel filtration, was 54,000. The enzyme had a broad pH optimum from pH 6.8 to 8.0. The value of K_m for 12-oxo-PDA was 190 μM. The enzyme preferred NADPH to NADH as a proton donor, the K_m values for NADPH and NADH being 13 μM and 4.2 mM, respectively. They also examined the changes in the reductase activity during the germination of corn seeds. There was a moderate amount of the activity in the ungerminated seeds, but by the fifth day of germination the activity had increased 5-fold. Some of the increased reductase activity seemed to be due to newly synthesized enzyme, because newly formed shoots of corn had the enzyme activity.

B. Metabolism

Hitherto, only two studies dealing with metabolism of JA-related compounds have been published. Meyer *et al.* (1989) synthesized (±)-dihydro[2-^{14}C]jasmonic acid (HJA) and fed the compound to derooted 6-day-old barley seedlings from the bases of the shoots. After a feeding period of 72 hr, ~90% of the compound was taken up by the shoots. The methanol extract of the shoots was divided into ethyl acetate-soluble and water-soluble fractions, and then each fraction was subjected to thin-layer chromatography (TLC). About 60% of the radioactivity was found in the ethyl acetate fraction and the rest in the aqueous fraction. TLC analysis revealed that 24% of the radioactivity in the ethyl acetate fraction was HJA and there were six metabolites in the fraction. The aqueous fraction contained two metabolites. The major metabolites in the ethyl acetate fraction and aqueous fraction were identified as 3-oxo-2-(4'-hydroxypentyl)-cyclopentane-1-acetic acid and its β-ᴅ-glucoside, respectively. A small amount of 3-oxo-2-(5'-hydroxypentyl)cyclopentane-1-acetic acid was also found in the ethyl acetate fraction. In this report, some preliminary results obtained in experiments with [2-^{14}C]JA were also presented. JA was metabolized faster than HJA and disappeared in the shoots after 72 hr. In their subsequent work, Meyer *et al.* (1991) reported structural elucidation of eight minor metabolites of HJA. The metabolites were conjugates of leucine, isoleucine, and valine.

Although the pathway of degradation of JA remains unknown, these reports suggest the possibility that a hydroxylation is the first reaction of the pathway just as it is of gibberellic acid. TA is one of the hydroxylation products of JA. As shown in Figs. 9, 10, and 11, TA has no inhibitory effects on plant growth but has strong tuber-inducing activity (Fig. 5). It seems likely that the hydroxylation at C-5 of the pentenyl side chain of JA, which forms TA, occurs to get rid of the inhibitory effects of JA, not to be a first step of breakdown of JA.

IV. Stereochemistry of Jasmonic Acid-Related Compounds

A. Jasmonic Acid

JA has four stereoisomers because of the presence of two chiral carbons at the C-1 and C-2 positions of the cyclopentane ring (Fig. 3). Since the

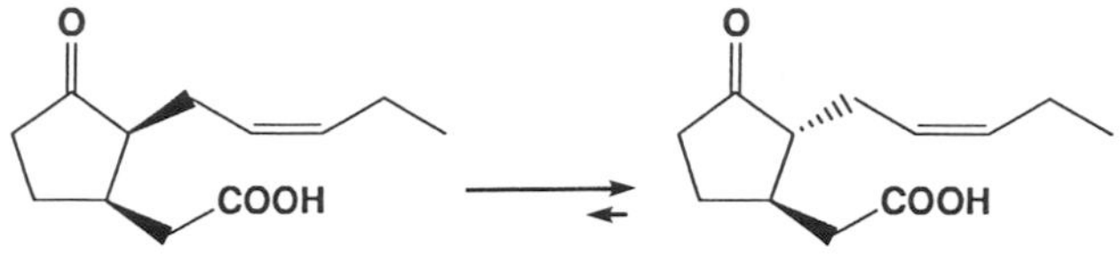

FIG. 3   Absolute chemical structures of stereoisomers of JA. (1*R*,2*S*)- and (1*S*,2*R*)-JA, which have cis conformation of the side chains with respect to the plane of the cyclopentane ring, are easily epimerized to the trans isomers (1*R*,2*R* and 1*S*,2*S*).

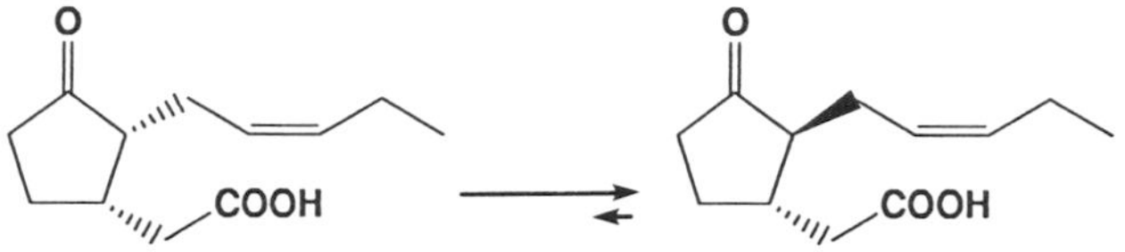

nomenclature of the isomers has been confused, *RS* notation is used in this review to avoid such confusion. Isomers of 1*R*,2*S* and 1*S*,2*R* have cis conformation of the side chains with respect to the plane of the cyclopentane ring, and 1*R*,2*R* and 1*S*,2*S* have trans conformation. The cis isomers are easily epimerized to the trans isomers, presumably through enol intermediates. Usually, in an equilibrium state, the proportion of the cis isomer to the trans isomer is about 8 : 92. This kind of epimerization is known to be stimulated under basic and acidic conditions, and at high temperature (Vick *et al.*, 1979).

   Vick and Zimmerman (1984) suggested that the absolute stereochemistry of naturally occurring JA is 1*R*,2*S* and this compound is epimerized to 1*R*,2*R* during the purification procedure. Miersch *et al.* (1986) isolated both (1*R*,2*S*)-JA [referred to as (+)-7-iso-JA by the authors] and (1*R*,2*R*)-JA [(−)-JA] from immature fruit of broad bean. The authors estimated, by the addition of ¹⁴C-labeled cis isomer, the ratio of the naturally occurring isomers (1*R*,2*S* : 1*R*,2*R*) to be 65 : 35 in immature fruit. They also compared the biological activities of both isomers. Growth of wheat and GA₃-stimulated dwarf rice seedlings was inhibited more effectively by (1*R*,2*S*)-JA than by (1*R*,2*R*)-JA. Senescence-promoting activity of (1*R*,2*S*)-JA was stronger than that of the (1*R*,2*R*)- isomer. The results suggest that the absolute stereochemistry of JA and JA-Me affects their various biological activities. Hamberg *et al.* (1988) examined the stereochemistry of 12-oxo-PDA and deduced that the configurations of the carbons which bear the side chains of 12-oxo-PDA are identical to the configurations of the corresponding carbons of (1*R*,2*S*)-JA.

Synthetic JA was separated into the trans and cis isomers by HPLC according to the method described in Section V,B,3 and their potato tuber-inducing activities were compared (Y. Koda and Y. Kikuta, unpublished data). The cis isomer of JA had stronger tuber-inducing activity than did the trans isomer (Fig. 4).

B. Tuberonic Acid

TA also has four stereoisomers. Although we have not yet established the absolute stereochemistry of naturally occurring TA, it seems to have the 1*R*,2*S* configuration, just as JA. We reported that the side chains of TA seem to be in the cis conformation because isolated TAG readily changes to another inactive compound (Yoshihara *et al.*, 1989). This observation suggests that the trans isomer has no or very little tuber-inducing activity.

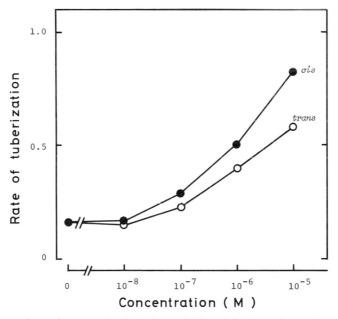

FIG. 4 Comparison of potato tuber-inducing activities of the cis and trans isomers of JA. Synthetic JA was separated into the cis and trans isomers by HPLC and assayed for tuber-inducing activity.

Synthetic TA was separated into the cis and trans isomers by HPLC as described in Section V,B,3 and their tuber-inducing activities were compared (Y. Koda and Y. Kikuta, unpublished data). As was expected, the cis isomer exhibited much stronger activity than did the trans isomer (Fig. 5). The activity of the trans isomer was very low. The *cis*-TA and TAG have no inhibitory effects on plant growth, such as promotion of leaf senescence of oat leaves, inhibition of soybean callus growth, and inhibition of seedling growth of lettuce (Y. Koda and Y. Kikuta, unpublished data).

Epimerization of *cis*-TAG to the trans isomer was observed when a solution of the cis isomer in 10% methanol was kept at room temperature for 2 days (Fig. 6)

C. Cucurbic Acid

CA has eight stereoisomers because the carbon at the C-3 position is chiral. The presence of the hydroxyl group at the C-3 position of the ring precluded the epimerization, and CA can be isolated in its natural stereo configuration. Fukui *et al.* (1977b) demonstrated that the stereo configuration of CA is (1*R*,2*S*)-3*S*-hydroxy-2-(2′-pentenyl)cyclopentane-1-acetic acid.

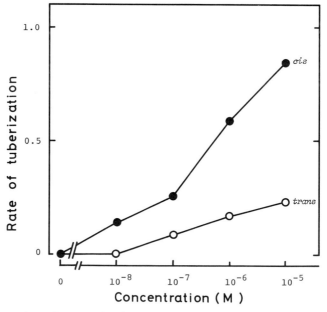

FIG. 5   Comparison of potato tuber-inducing activities of the cis and trans isomers of TA.

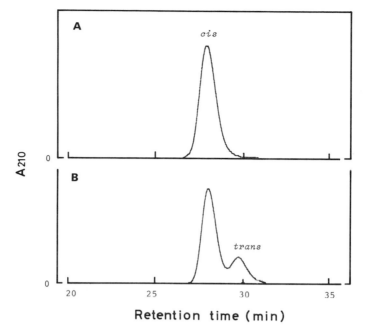

FIG. 6   Profiles of elution of TAG after HPLC, demonstrating the epimerization from the cis to the trans isomer. Purified *cis*-TAG (A) and the purified acid stored for 2 days at room temperature in 10% methanol (B) were chromatographed on a Novapak C_{18} column in 8% acetonitrile that contained 0.1% acetic acid.

V. Isolation and Quantitative Analysis

A. Extraction and Purification

1. Extraction and Fractionation by Solvent Partitioning

JA-related compounds are easily extracted from plant materials by methanol, ethanol, or acetone. The material should be homogenized immediately after harvest with sufficient solvent to prevent enzymatic changes of the compounds. Although the stability of the compounds under light is unknown, it is preferable to avoid strong light exposure during the extraction procedure.

After removal of the plant residue by filtration, the solvent is removed by evaporation, and the resultant aqueous residue is acidified to pH 3.0 with 1 M HCl and partitioned three times with ethyl acetate. Partitioning with diethyl ether has also been employed (Dathe *et al.*, 1981; Tsurumi and Asahi, 1985). Partition coefficients (concentration in organic solvent/

concentration in aqueous phase) for JA and TA between aqueous buffers (50 mM phosphate, pH 3.0 and 7.0) and ethyl acetate are given in Table I. Solutions of TA and JA at a concentration of 10^{-4} M in the buffer were shaken with an equal volume of water-saturated ethyl acetate, and the amount of the compounds in each phase was determined by HPLC. The solubility of TA in water is considerably higher than that of JA. Recoveries of TA and JA by three-time-successive extraction with equal volumes of ethyl acetate from the buffer at pH 7.0 are 50.9 and 87.5%, respectively, while those at pH 3.0 are 97.5 and 100%, respectively. This observation indicates that partitioning with organic solvent should be carried out under acidic conditions.

The ethyl acetate fraction is divided into acidic and neutral fractions by extraction of basic buffer (usually 5% sodium bicarbonate) in the usual way. The fractions are dried over anhydrous sodium sulfate and evaporated to dryness. It is important to note that epimerization of JA-related compounds can easily occur under acidic or basic conditions (see Section IV,A). Therefore, partitioning with basic buffer and acidification of the buffer solution should be carried out as quickly as possible. In the acidic ethyl acetate fraction, free acids of JA-related compounds, such as JA, CA, and TA, are found. In the neutral fraction, methyl esters of them are present. On the other hand, water-soluble derivatives of JA-related compounds such as glucosides and glucosyl esters are expected to be present in the aqueous fraction that remained after the first ethyl acetate extraction.

2. Purification of Free Acids and Methyl Esters

Usually, further purification is carried out by chromatographies on columns of charcoal, silica gel, and silica gel octadecyl silyl (ODS). During the purification procedure of potato tuber-inducing substances, we tested the availabilities of silica gel column chromatography and TLC, and found that these chromatographies caused a considerable loss of the tuber-inducing activity in a fraction. This observation appears to indicate that silica gel

TABLE I

Partition Coefficients for JA and TA between Aqueous Buffers and Ethyl Acetate

pH of aqueous phase (50 mM phosphate)	Partition coefficients	
	JA	TA
3.0	177.5	2.29
7.0	0.996	0.268

stimulates epimerization of TA and TAG, resulting in a considerable loss of their activity (Fig. 5). Therefore, silica gel chromatographies are not an adequate method for purification of TA and its derivatives.

Column chromatography using granular charcoal seems to be the most suitable method for purification of JA-related compounds. Dried samples are usually dissolved in 10–20% acetone and loaded onto the column. Elution is carried out with increasing acetone content in water. JA was eluted with 35–65% acetone (Yamane *et al.*, 1981), and JA-Me was eluted with 70–80% acetone (Ueda and Kato, 1980). This elution pattern is affected by various factors, such as the column size, volume of eluant, and amount of sample.

Although various procedures for purification of JA have been reported, one example of the procedure is described as follows (Koda and Kikuta, 1991). The acidic ethyl acetate fraction obtained from 400 g of fresh leaves was dissolved in 100 ml of 30% ethanol, sonicated for a few minutes, and filtered. The filtrate was passed through a 100-ml column of charcoal which had previously been washed with ethanol and then with water, with final equilibration in 30% ethanol. The loaded column was washed with 400 ml of 30% ethanol and eluted with 400 ml of ethanol. The eluate was evaporated to dryness and the resultant residue was fractionated on a column of silica gel ODS (24 mm i.d. × 360 mm; RQ-2, Fuji Gel) in 60% methanol that contained 0.1% acetic acid at a flow rate of 3 ml/min. Elution volumes of TA and JA from the column were approximately 100–120 ml and 220–250 ml, respectively. The fraction of the eluate corresponding to JA was collected and evaporated to dryness. The residue was dissolved in chloroform and insoluble substances were removed by filtration. Further purification was carried out by HPLC. The chloroform-soluble fraction was evaporated to dryness and fractionated on a Novapak C_{18} column (8 mm i.d. × 100 mm; Radialpak cartridge, Waters) in 60% methanol that contained 0.1% acetic acid at a flow rate of 1 ml/min. The retention time, under these conditions, of authentic JA was ~8.4 min, and the fraction of the eluate corresponding to JA was collected. The fraction was rechromatographed on the same column in 30% acetonitrile that contained 0.1% acetic acid. The elution profile, monitored by absorbance at 210 nm, showed a sharp peak of compound that eluted with the same retention time as authentic JA (15.1 min). The recovery of JA by this purification procedure, which was calculated by the method of standard addition, was 65%.

TA can be purified in a similar manner. However, the TA fraction usually contains many impurities which interfer with its purification. For isolation of TA, therefore, repeated HPLCs using different columns and solvents are necessary.

Ueda and Kato (1980) purified JA-Me from the neutral ethyl acetate

fraction obtained from shoots of wormwood by charcoal column chromatography, silica gel column chromatography, and silica gel TLC.

3. Purification of Water-Soluble Derivatives

The water-soluble derivatives of JA-related compounds can be purified by
a similar procedure (Koda *et al.*, 1988). At first, much impurity which is
strongly adsorbed on the charcoal column is removed by ethanol precipitation. The aqueous fraction that remained after the first ethyl acetate
extraction is neutralized with 1 *M* NaOH and evaporated to reduce its
volume as much as possible. To prevent bumping of the solution, a small
amount of *n*-butanol is added to the solution. Then, four volumes of
ethanol is added and the mixture is kept at −20°C for 1 hr. The precipitate
is removed by filtration, and ethanol is evaporated from the filtrate. The
resultant aqueous residue is loaded onto a charcoal column, and the column is washed with 30% methanol and eluted with ethanol. Further purification is carried out by various chromatographies, guided by bioassay
for JA-related compounds. The purification procedure for TAG was considerably complicated by the presence of many impurities in the fraction
(Koda *et al.*, 1988).

B. Quantitative Analysis

1. Bioassay

Bioassays for JA are usually designed to use its inhibitory activity, such as
senescence promotion and inhibition of the seedling growth. As plant
extracts are full of toxic compounds, it is difficult to estimate JA level in a
crude extract by a bioassay. Therefore, considerable purification of a
fraction is necessary prior to bioassay.

a. Oat Leaf Assay Using senescence-promoting activity of JA and
JA-Me, these compounds can be detected by an oat leaf assay which was
originally described by Shibaoka and Thimann (1970). Ueda and Kato
(1980) used the assay with a slight modification. Oat seeds (*Avena sativa* L.
cv. Victory) are germinated in vermiculite under continuous white fluorescent light (~20 $W \cdot m^{-2}$) at 25°C for 7 days. The upper 3-cm leaf segments of the first leaves are excised from the seedlings, and 10 of these
segments are placed on a filter paper moistened with 5 ml of test solution in
a petri dish. Test compounds are dissolved in McIlvain citrate phosphate
buffer solution, diluted 1 : 10 (pH 4.7), that contains 10^{-6} *M* benzyladenine
and 0.2% Tween 20. In our laboratory, 100 mg/liter of benzylpenicillin

potassium is added to the solution to prevent bacterial contamination. After incubation for 3 days at 25°C in the dark, the remaining chlorophyll in the segments is extracted twice with boiling 80% ethanol. The amount of chlorophyll is determined from the absorbance at 665 nm. Figure 7 shows a comparison of senescence-promoting activities of JA, JA-Me, and ABA. Of the compounds tested, JA-Me has the strongest activity. The threshold concentration for detection of JA and JA-Me is 10^{-6} M (~1 μg/5 ml test solution). TA cannot be detected by this assay method because TA has no effect on leaf senescence.

b. Seedling Assays Dathe *et al.* (1981) isolated JA as a plant growth inhibitor from the pericarp of broad bean. Inhibitor activity was determined by a wheat seedling assay which had been established as a bioassay for gibberellin (Dathe *et al.*, 1978). Wheat seeds (*Triticum aestivum* L. cv. Carola) are germinated for 3 days at 20°C in the dark. Five seedlings are transferred to a test tube that contains 1 ml of test solution and grown

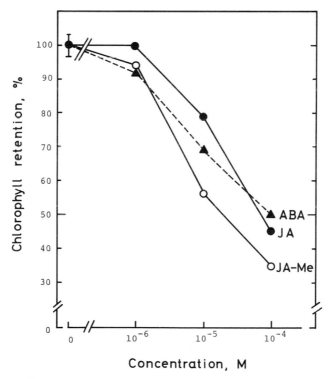

FIG. 7 Effects of JA, JA-Me, and ABA on senescence of oat leaf segments in the presence of 10^{-6} M benzyladenine. Value of the control represents a mean ± standard error ($n = 5$).

under a photoperiod of 16 hr of light and 8 hr of darkness. After 5 days, the length of the seedlings is measured. The authors did not describe the threshold concentration of JA for detection by this assay method.

A dwarf rice seedling assay has been used by some workers (Yamane *et al.*, 1980; Meyer *et al.*, 1984). Dwarf rice seeds (*Oryza sativa* L. cv. Tnaginbozu) are germinated on moistened filter paper at 30°C in the dark. After 3 days, seedlings are selected for uniformity (coleoptile length of 4–10 mm) and five seedlings are transferred to a test tube that contains 1 ml of water supplemented with 10^{-6} M GA_3 and the sample to be tested. Then, the seedlings are grown under a 16-hr photoperiod with a 25/20°C day/night temperature regime and at a high humidity. After 1 week, the length of the second leaf sheath is measured.

c. Potato Stem Segment Assay   Since many JA-related compounds have potato tuber-inducing activity (Koda *et al.*, 1991a; see Section VI,K,1), cultures of single-node segments of potato stem *in vitro* which have been used for the detection of tuber-inducing activity (Koda and Okazawa, 1988) are appropriate as an assay for JA-related compounds. A brief outline of the assay method is shown in Fig. 8. Etiolated potato shoots (*Solanum tuberosum* L. cv. Irish Cobbler) are obtained according to the method described by Koda and Okazawa (1983a). Single-node segments of stem are prepared from the etiolated shoots and sterilized with 1% sodium hypochlorite solution for 1 hr. Then, three single-node segments are planted horizontally in a 100-ml flask that contains 10 ml of basal medium (usually White's medium) supplemented with the sample to be tested. Five replicates are prepared. The concentration of sucrose in the medium is 2%. The medium is adjusted to pH 5.6 and solidified with 0.6% Bacto-agar before autoclaving for 7 min. The cultures are maintained at 25°C in the dark for 3 weeks. At the end of each culture, the rate of tuberization is calculated as the number of tuberized laterals divided by the number of emerged laterals. The threshold concentration for detection of TA, TAG, JA, and JA-Me is 10^{-7} M (~200 ng/10 ml of medium).

2. Radioimmunoassay

Knöfel *et al.* (1984) developed a radioimmunoassay (RIA) for JA according to Weiler's method (Weiler, 1982). JA was coupled by the carbodiimide method to bovine serum albumin (BSA) and the antiserum was raised in rabbits against JA–BSA–conjugate. Tritium-labeled JA was prepared by a metal-catalyzed hydrogen isotope exchange (Pleiss *et al.*, 1983) and used as a tracer. Since affinity of the antiserum to JA-Me was higher than that to JA, the authors used JA-Me as a standard and the samples were methylated by diazomethane prior to use. Cross-reactivity studies with compounds which are structurally related to JA demonstrated that the antiserum was specific to JA. The radioimmunoassay had a measuring range of

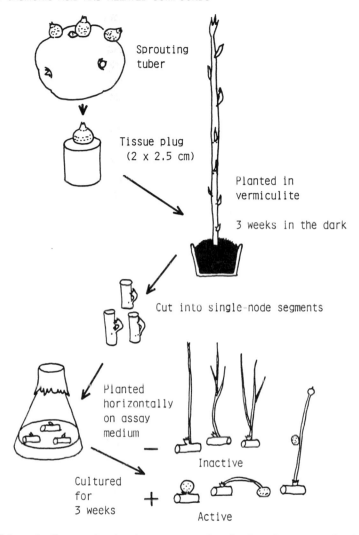

Sprouting
tuber

Tissue plug
(2 x 2.5 cm)

Planted in
vermiculite

3 weeks in the dark

Cut into single-node segments

Planted
horizontally
on assay
medium

—

Inactive

Cultured
for
3 weeks

+

Active

FIG. 8   Schematic diagram showing the assay procedure for detecting potato tuber-inducing activity.

2–200 ng for (−)-JA-Me [(1R,2R)-JA-Me]. The authors stated that extensive purification is not necessary prior to determination.

3. High-Performance Liquid Chromatography and Gas Chromatography

HPLC and GC can be used for quantitative analysis of JA-related compounds if authentic samples of them are available. Since the resolving power of HPLC is lower than that of GC, extensive purification is necessary before quantitative analysis by HPLC. As described in Section V,A, JA can be purified using a silica gel ODS column, such as Novapak C_{18} (Waters), with 60% methanol that contains 0.1% acetic acid or with 30% acetonitrile that contains 0.1% acetic acid. The elution profile is monitored by absorbance at 210 nm. However, this column cannot separate diastereomers of JA (cis and trans with respect to the plane of the cyclopentane ring; see Section IV,A). On the other hand, a silica gel column, such as μ-Porasil (Waters), can separate cis-JA ($1R,2S$ and $1S,2R$) from trans-JA ($1S,2S$ and $1R,2R$). The solvent system is 80% chloroform in n-hexane that contains 0.1% acetic acid, and the elution profile is monitored by absorbance at 280 nm. Retention times of trans and cis isomers are 14.8 and 18.2 min, respectively. Diastereo mixtures of JA-Me cannot be separated by this column. Diastereo mixtures of TA can be separated by a Novapak C_{18} column under the following conditions: solvent, 6% tetrahydrofuran that contains 0.1% acetic acid; flow rate, 1 ml/min; detected at 210 nm. Retention times of cis- and trans-TA are 15.9 and 18.0 min, respectively. If there is a sharp peak corresponding to a known JA-related compound on the chromatogram, the amount is calculated from a standard curve constructed from measurements of peak areas generated by known amounts of the compound.

Anderson (1985) described a HPLC assay for determination of ABA and JA in plant extracts, based on the coupling of the compounds with a fluorescent hydrazide to give stable fluorescent products. The method allowed the detection of both compounds at 5 pmol ($\sim$1 ng) and provided a linear response to 15 nmol ($\sim$30 μg).

Meyer et al. (1984) measured the amount of JA-Me by GC under the following conditions: steel column (3 m × 4 mm); 10% EG SS-X on Gas Chrom P (125–150 μm); carrier gas, N_2, 46 ml/min; isothermal column temperature, 180°C; hydrogen flame ionizing detector. Retention time of JA-Me was 12 min.

VI. Physiological Roles of Jasmonic Acid-Related Compounds

A. Senescence Promotion

1. Structure–Activity Relationships

After the isolation of JA-Me from wormwood plants as a senescence-promoting factor (Ueda and Kato, 1980), several studies have investigated the senescence-promoting activity of JA and JA-Me. Ueda et al. (1981)

synthesized various compounds which are structurally related to JA and compared their inhibitory effects on kinetin-induced retardation of senescence of oat leaves. Among the compounds tested, JA-Me was the most effective compound. HJA-Me also had strong senescence-promoting activity. JA had moderate activity and HJA showed no activity. Ueda *et al.* prepared analogs of HJA-Me which had different length of *n*-alkyl substituents at the C-2 position and compared their activity. The most active compound had an *n*-pentyl side chain (HJA-Me). Increasing or decreasing the length resulted in a decrease of the activity. Methyl 4-acetylnonanoate, which partially resembles HJA-Me but has no cyclopentane ring, had considerable activity. The authors suggested that those parts of structure that are indispensable for the strong senescence-promoting activity are the cyclopentanone, the methyl acetate substituent at the C-1 position, and the pentenyl or pentyl substituent at the C-2 position.

2. Degradation of Chlorophyll and Ribulose-1,5-Bisphosphate Carboxylase

It is well known that leaf senescence is inhibited by light and cytokinin, and promoted by abscisic acid and ethylene. Leaf senescence is the sum of various physiological processes that involve the loss of chlorophyll, proteins, and nucleic acids, disassembly of membrane, and alterations in the gene expression program and in enzyme activities (Stoddart and Thomas, 1982). Among them, the loss of chlorophyll has been used as a parameter of the senescence because it is the most prominent and it is easy to quantify.

Weidhase *et al.* (1987a) treated segments of barley leaves with JA-Me and benzyladenine and examined the effects of these compounds on the degradation of chlorophyll and ribulose-1,5-bisphosphate carboxylase (RuBPCase) under continuous light and in the dark. JA-Me promoted degradation of chlorophyll as well as the rapid decrease in the level of RuBPCase, while total protein content in the leaf segments was not greatly affected. The senescence-promoting activity of JA-Me differed in the light and in the dark. For example, the initial rates of chlorophyll and RuBP-Case breakdown were markedly higher in the light than in the dark in the presence of $4.5 \times 10^{-5} M$ JA-Me, suggesting an involvement of JA-Me in some light-dependent processes. Although an application of $4.5 \times 10^{-5} M$ benzyladenine inhibited the breakdown of chlorophyll and RuBPCase in barley leaves, the substance could not prevent a rapid decrease in the levels of chlorophyll and RuBPCase in the early phase of senescence induced by JA-Me. On the other hand, benzyladenine added 24 hr after JA-Me application resulted in a recovery of chlorophyll and RuBPCase at the later stages, suggesting a rapid inactivation of JA-Me in the leaf segments. Since JA-Me did not enhance the breakdown of chlorophyll in the

isolated chloroplasts, the authors speculated that JA-Me enhances senescence of chloroplast by promoting cytoplasmic events, such as promotion
of proteolytic and lipolytic activities, which eventually could result in the
degradation of constituents of chloroplast. The involvement of proteases
in the senescence of barley and wheat leaves was suggested by some
workers (Miller and Huffaker, 1982; Wittenbach *et al.*, 1982). Grossman
and Leshem (1978) showed that senescence of leaves of pea was accompanied by a significant increase in lipoxygenase activity and free
radical production.

Ethylene enhances the senescence of cucumber cotyledons. Abeles *et
al.* (1988) treated the hypocotyls obtained from 2-week-old seedlings with
ethylene and found that two kinds of peroxidases were synthesized *de
novo* during the senescence induced by ethylene, and the enzymes were
capable of degrading isolated chlorophyll. To test the hypothesis that
peroxidase plays a role in the degradation of chlorophyll (Matile, 1980;
Yamaguchi and Minamide, 1985), the authors compared the levels of
chlorophyll and peroxidase in cucumber cotyledons treated with ethylene,
JA-Me, and various plant hormones (Abeles *et al.*, 1989). No correlation
was found between the level of chlorophyll and that of peroxidase. In this
study, they found that JA-Me was as effective as ethylene in inducing the
senescence. The action of JA-Me and ethylene together was greater than
that of each alone. In addition to this observation, silver thiosulfate (STS),
which inhibits the ability of ethylene to induce chlorosis, had no effect on
the ability of JA-Me to promote chlorophyll degradation. Based on these
results, the authors suggested that ethylene was not involved in the ability
of JA-Me to induce senescence and that ethylene and JA-Me regulate
different parts of the senescence processes.

Cuello *et al.* (1990) reported that the age of barley leaves affected the
senescing response of the leaves to the senescence-promoting compounds
ABA, ethylene, and JA-Me. ABA and ethylene induced stronger senescence in old leaves than in young ones, while the senescence-promoting
activity of JA-Me was not largely affected by the age of the leaves. This
result suggests the presence of diverse mechanisms in leaf senescence.

3. Jasmonate-Induced Proteins

Weidhase *et al.* (1987b) treated first leaves of barley with JA or JA-Me by
floating them on a $4.5 \times 10^{-5} M$ aqueous solution for 4 days and examined
the changes in polypeptides in the leaves by SDS-polyacrylamide gel
electrophoresis. They found an accumulation of abundant proteins in
senescing leaves treated with JA or JA-Me, and named the proteins
jasmonate-induced proteins (JIPs). JIPs were not found in leaves treated
with water. Labeling and inhibitor experiments indicated *de novo* synthe-

sis of JIPs. ABA induced polypeptides which were immunologically and electrophoretically identical with JIPs, but ethylene (ethephon) failed to do so.

A subsequent experiment (Mueller-Uri *et al.*, 1988) demonstrated that specific mRNAs for JIPs and the JIPs labeled *in vivo* can be detected within 5 hr after addition of JA-Me. Since JIPs were detectable much earlier than any other senescence symptoms, such as degradation of chlorophyll and RuBPCase, the authors mentioned that formation of JIPs is a cause rather than a consequence of senescence. JA-Me also caused the cessation of synthesis of normal proteins but did not affect their respective mRNAs, suggesting that JA-Me affected gene expression not only at the transcriptional level but also at the translational level. Herrmann *et al.* (1989) found that leaves of a large number of plant species, including both monocots and dicots, formed JIPs in response to JA or JA-Me. There was a great variability in the number and relative molecular masses of JIPs in various plant species, and the JIPs were immunologically different from each other. The variability was found even in different cultivars of barley.

The results of Herrmann *et al.* clearly indicated that JA and JA-Me are capable of affecting gene expression in senescing leaves mainly at the transcriptional level. However, one of the puzzling facts that needs to be explained in the future is the enormous complexity of JIPs. It is also possible that JIPs are involved only in "artificial" and not in "natural" senescence, because the proteins were not found in barley leaves floated on water for 4 days in the dark (Weidhase *et al.*, 1987b). Previous work (Weidhase *et al.*, 1987a) indicated that, in the leaves under the same conditions, considerable senescence was observed and the amounts of chlorophyll and RuBPCase were reduced by nearly 50%.

Mueller-Uri *et al.* (1988) noticed that the massive alteration of gene expression induced by JA-Me is similar to that induced by various stresses, and compared JIPs to proteins which were induced by heat shock, but no similarity was found between them. The possibility that JA-Me is an inducer of stress was discussed in detail by Parthier (1990).

B. Effects on Greening of Etiolated Leaves and Oxygen-Evolving Activity of Chloroplasts

The physiological processes of greening of etiolated leaves which involve the synthesis of chlorophyll appear to be opposite to that of senescence. Fletcher *et al.* (1983) treated etiolated cotyledons of cucumber with ABA and JA-Me, and then placed them under light to see the effects of these compounds on greening of the cotyledon. While ABA inhibited the greening at concentrations above $10^{-6} M$, JA-Me did not affect it at any concentrations tested (10^{-6}–$10^{-3} M$). The results indicate that JA-Me does not prevent synthesis of chlorophyll. As stated above, senescence of cucumber cotyledon was promoted by JA-Me and ethylene, and not by ABA (Abeles *et al.*, 1989). These observations suggest that a inhibition of chlorophyll accumulation is not involved in the senescence induced by JA-Me.

Maslenkova *et al.* (1990) examined the effect of JA on oxygen-evolving activity of chloroplasts isolated from barley leaves. They cultivated 3-day-old barley seedlings in JA solution (10^{-6}, 10^{-5}, and 10^{-4} M) for 7 days and then isolated chloroplasts from the leaves. JA inhibited the growth of the seedlings. The Hill activity of the chloroplasts isolated from the plants treated with 10^{-4} M JA was reduced by 46%, and the oxygen flash yields were reduced by ~70%, compared with those of chloroplasts obtained from control plants. The kinetics characteristics of oxygen evolution indicated that JA decreased the value of the total number of oxygen-evolving centers. JA seems to affect the degree of structuring of the granal region or the structural integrity of the electron transport chain itself in barley chloroplasts.

C. Death Hormone

The term "death hormone" has been used (Nooden and Leopord, 1978) when discussing senescence in monocarpic plants. Monocarpic senescence, which occurs near the maturation of fruit, is a dramatic phenomenon and visible in most fields before harvest. The senescence favors movement of nutrients that remain in the vegetative parts to the fruit or seeds. As the removal of fruit prevents senescence, one hypothesis to account for this phenomenon is that developing fruit produce a senescence factor (death hormone) which is exported from the fruit to the rest of the plant, resulting in its senescence and death. Engvild (1989) reviewed the death hormone hypothesis and described the possibility that JA and JA-Me may be the death hormones. Nooden *et al.* (1990) applied various compounds, including ABA, ACC, and JA-Me, to soybean cuttings and tested for their senescence-promoting activity. Among the compounds tested, only JA-Me showed a slight senescence-promoting activity. However, the authors paid no attention to the result. Since soybean fruit contains a high level of JA (Lopez *et al.*, 1987), it is possible that JA, JA-Me, or some unknown compounds related to JA are death hormones.

D. Effects on Plant Hormones

1. Effect on Ethylene Production

A Polish group published several reports which described stimulation of ethylene production by JA-Me in tomato and apple fruits (Saniewski and

Czapski, 1985; Saniewski *et al.*, 1987a). They applied JA-Me at a concentration of 22 or 44 mM in lanolin paste to the surface of the fruits and measured ethylene production. The stimulation of ethylene production by JA-Me in tomato fruits was affected by the developmental stages of the fruits, the production of ethylene being enhanced much more in red ripe fruits than in green ones (Saniewski and Czapski, 1985). However, the activity of ethylene-forming enzyme (EFE) in green fruits was greatly enhanced by JA-Me. The stimulation of ethylene production observed in red fruits seemed to be due to stimulations of both the activities of EFE and 1-aminocyclopropane-1-carboxylic acid (ACC) synthase (Saniewski *et al.*, 1987b). An application of JA-Me to preclimacteric apples resulted in stimulation of both ethylene production and ACC accumulation. But, the same treatment to postclimacteric apples induced an inhibition of ethylene production and a reduction of ACC level (Saniewski *et al.*, 1987a).

Although these results indicate that applications of JA-Me at extremely high concentrations are able to affect the ethylene production, the involvement of JA-Me in natural ripening of fruits remains unknown. As described above, ethylene did not appear to be involved in the senescence-promoting activity of JA-Me (Abeles *et al.*, 1989). The difference in the effect of JA-Me on ethylene production suggests that the effect is specific to species and tissues.

2. Effect on Abscisic Acid Carrier

Carrier-mediated uptake of ABA has been demonstrated in apical regions of root of runner bean (Astle and Rubery, 1983). The uptake was stimulated by acidic external pH, suggesting that the transmembrane pH gradient is a driving force of the uptake. Since, in various biological systems, activities of JA and JA-Me have been compared to that of ABA, Astle and Rubery (1985) investigated the effects of JA and JA-Me on the uptake of ABA by suspension-cultured runner bean cells. Increasing concentrations of JA-Me inhibited ABA uptake by the cultured cells. Half-maximum inhibition required ~2.5 $\times$ 10^{-5} M JA-Me. JA-Me did not affect the uptake of 5,5-dimethyloxazolidine-2,4-dione (DMO), which was used as a probe for intracellular acidification. This observation suggests that the inhibition of ABA uptake by JA-Me is brought about by an interaction of JA-Me with the ABA carrier, not by decreasing the transmembrane pH gradient. In contrast, JA decreased uptakes of ABA and DMO in parallel, and the relative decrease in ABA uptake was much less than that caused by JA-Me. The authors speculated that JA decreases ABA uptake by acidifying the cytoplasm and so lowering the driving force for uptake. The physiological role of the inhibition of ABA uptake by JA-Me, and the reasons why JA and JA-Me have different effects on ABA uptake and acidification of cytoplasm remain open. Astle and Rubery assumed that local esterification and deesterification in the cytoplasm would play an important role.

The authors also showed that cultured cells are freely permeable to external JA-Me over external pH values ranging from 3.0 to 7.0, whereas cells are relatively impermeable to external ABA at an external medium pH of 6.0 to 7.0. JA was assumed to be similar to ABA and other lipophilic weak acids with respect to diffusive uptake.

E. Inhibition of Growth of Seedlings and Coleoptiles

The ability of JA and its derivatives to inhibit growth of rice seedlings was reported by Yamane et al. (1980). They synthesized various compounds which are structurally related to JA and applied them to 2-day-old seedlings of dwarf rice. The seedlings were incubated under continuous light at 30°C for 7 days and then the length of the second leaf sheath was measured. JA-Me had the strongest growth-inhibiting activity among compounds tested, the length of the second leaf sheath treated with JA-Me at a concentration of 10^{-5} M being nearly one-half of that of the control. The activity of JA was slightly lower than that of JA-Me, and HJA was as active as JA. Increasing or decreasing the length of the n-alkyl substituent at the C-2 position of the cyclopentane ring resulted in a gradual decrease in the activity, and removal of the side chain caused a complete loss of the activity. CA and CA-Me have the activity and 3-deoxy-JA-Me was inactive. The authors concluded that those parts of the structure that are indispensable for the activity are CH_2COOH or CH_2COOCH_3 at the C-1 position, cis-pentenyl or n-pentyl group at the C-2 position, and keto or hydroxyl group at the C-3 position. The structural requirements for growth-inhibiting activity are almost equal to those for senescence-promoting activity (Ueda et al., 1981).

Figure 9 shows the effect of the cis isomers of JA, TA, and TAG on the growth of lettuce seedlings (Y. Koda and Y. Kikuta, unpublished data). Lettuce seeds (Lactuca sativa L. cv. Grand Rapids) were surface-sterilized with a 1% solution of sodium hypochlorite for 20 min and germinated on agar-solidified Murashige-Skoog medium with half-strength inorganic salts, at 25°C under 12 hr of light and 12 hr of darkness. After 5 days, seedlings were derooted and transferred to test medium that contained 10^{-6} M gibberellic acid (GA_3) and compounds to be tested. After culture under the same conditions for 12 days, the length of each plant was measured. JA strongly inhibited the growth induced by GA_3 at concentrations above 10^{-6} M, while TA had no inhibitory effects. TAG showed a slightly promotive effect on the growth.

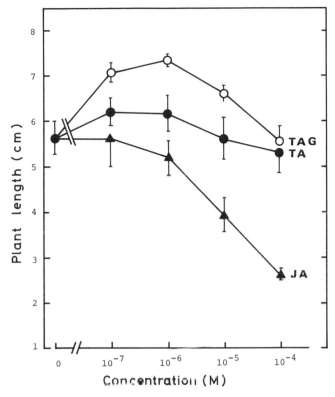

FIG. 9   Effects of the cis isomers of JA, TA, and TAG on growth of derooted lettuce seedlings in the presence of 10^{-6} M GA$_3$. Each value represents a mean ± standard error ($n = 5$).

The growth of coleoptiles of oat (*Avena sativa* L.) is stimulated largely by indole-3-acetic acid (IAA) and the stimulation is caused by cell elongation. Yamane *et al.* (1981) reported that JA had no effect on the growth of the coleoptiles induced by IAA. To confirm their result, we examined the effect of the cis isomers of JA, TA, and TAG on the growth. To evaluate the efficiency of our assay system, the effect of ABA, which is known to inhibit such growth (Rehm and Cline, 1973), was also examined. In our bioassay system, JA slightly but significantly inhibited the coleoptile growth induced by 3×10^{-7} M IAA at concentrations above 10^{-6} M, but the inhibiting activity was much lower than that of ABA (Fig. 10). By contrast, *cis*-TA and *cis*-TAG slightly promoted the growth at lower concentrations. The result seems to indicate that JA is capable of inhibiting the cell elongation induced by IAA.

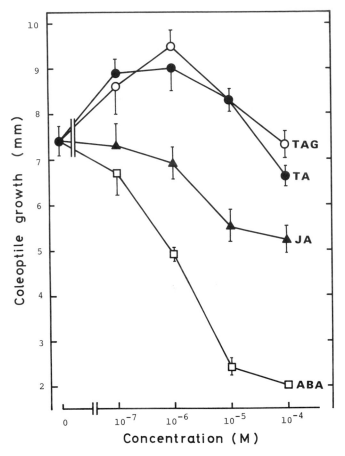

FIG. 10 Effects of the cis isomers of JA, TA, and TAG on straight growth of oat coleoptiles induced by $3 \times 10^{-7} M$ IAA.

F. Inhibition of Callus Growth

In the presence of IAA at an appropriate concentration (usually 2×10^{-5} M), the growth of soybean callus which is derived from cotyledons is fully dependent on exogenous cytokinin. The growth is brought about mainly by cell division, and is not affected by ABA (Blumenfeld and Gazit, 1970). Ueda and Kato (1982b) reported inhibitory effects of JA and JA-Me on the growth of the callus induced by cytokinin (kinetin). The inhibiting activity of JA was higher than that of JA-Me and an application of JA at a concentration of 10^{-6} M resulted in 60% inhibition. The results suggest that JA and JA-Me are capable of inhibiting cell division.

The authors also examined effects of these two compounds on the growth of radish cotyledons induced by cytokinin. Both JA and JA-Me slightly inhibited the growth, which is principally due to cell expansion (Letham, 1971).

Effects of the cis isomers of JA, TA, and TAG on the growth of soybean callus induced by 10^{-8} M zeatin riboside are shown in Fig. 11 (Y. Koda and Y. Kikuta, unpublished data). JA inhibited the callus growth at concentrations above 10^{-8} M, while TA and TAG had almost no inhibitory effects on it.

G. Inhibition of Opening of Stomata and Pulvinules

ABA has been reported to inhibit both stomatal opening and pulvinule opening (Zeiger, 1983; Satter and Galston, 1981). Opening of the pulvinules of *Mimosa pudica* is promoted either by light or by IAA in the dark. Tsurumi and Asahi (1985) compared effects of JA on movement of the pulvinule and on transpiration of the pinnae with those of ABA. An

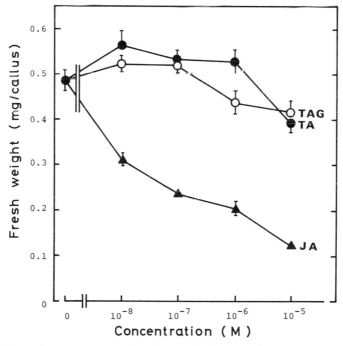

FIG. 11   Effects of the cis isomers of JA, TA, and TAG on growth of soybean callus induced by 10^{-8} M zeatin riboside. Fresh weight of callus was measured after 4 weeks. Each value represents a mean ± standard error ($n = 9$).

application of 10^{-4} M JA to excised pulvinules completely inhibited both IAA- and light-induced openings of the pulvinules, and the inhibitory activity was stronger than that of 10^{-4} M ABA. JA also caused a rapid closure of pulvinules which had previously been opened by IAA or light. The closure-inducing activity of JA was much higher than that of ABA. JA at a concentration of 10^{-4} M had no effect on transpiration of the pinnae, while ABA at the same concentration considerably reduced it. The authors speculated that JA and ABA affected separate sites, both of which were related to the movement of the pulvinules.

Satler and Thimann (1981) reported that JA is capable of inducing the closure of the stomata of *Avena* leaves. Since both stomatal opening and pulvinule opening are considered to be stimulated by the efflux of protons from the guard cells and the cells in the pulvinule (Zeiger, 1983; Satter and Galston, 1981), it seems probable that JA-Me suppresses the efflux of protons. Raghavendra and Reddy (1987) studied the effects of several compounds, including JA-Me and ABA, on stomatal movement in isolated epidermis of *Commelina benghalensis*. JA-Me inhibited the stomatal opening at concentrations above 10^{-9} M, and the opening was completely inhibited at 10^{-6} M. The activity of ABA was lower than that of JA-Me. The authors demonstrated that JA-Me suppressed the efflux of protons from the guard cells, just as ABA does, speculating that the action of JA-Me was similar to that of ABA.

H. Stimulation of Accumulation of Specific Proteins in Leaves

1. Vegetative Storage Proteins

JA and JA-Me have been reported to be capable of inducing some specific proteins, besides JIPs. Anderson (1988) found that JA at concentrations of 2×10^{-6}–3×10^{-5} M increased the level of a 30-kDa polypeptide in soybean callus cultured photomixotrophically. Furthermore, in leaves of soybean plants and *Glycine soja,* JA increased the 30-kDa peptide and another of 28 kDa (Anderson *et al.,* 1989). Both of these peptides were identified as vegetative storage protein (VSP), which is known to be a storage protein in soybean leaves (Wittenbach, 1983). The increase in VSP seemed to be specific for JA, because ABA, benzyladenine, or gibberellic acid did not alter the level of VSP. In photomixotrophic cells of soybean, the increase was observed at concentrations as low as 10^{-8} M (Anderson, 1991). VSP-inducing activity of JA was much higher than that of JA-Me.

In intact plants, VSP accumulates in vacuoles of leaf mesophyll cells and bundle sheath cells before flowering, declines during early pod-fill, and reaccumulates during late-fill. Removal of pods from the plants results in a large accumulation of VSP in the leaves (Wittenbach, 1983). VSP is considered to be particularly important for the temporal storage of nitrogen. Staswick (1990) reviewed the regulation of the expression of VSP genes and stated that VSP mRNA was increased in cultured soybean cells within 1 hr after the addition of $4 \times 10^{-6} M$ JA and continued to rise for at least 24 hr.

Mason and Mullet (1990) examined various factors which affect the expression of VSP genes, *VSP*A and *VSP*B encoding VSP-α (28-kDa) and VSP-β (31-kDa) polypeptides, respectively. When soybean seedlings were transferred to vermiculite that contained $5 \times 10^{-6} M$ JA-Me, the level of *VSP*B mRNA in the seedlings increased greatly after 12 hr, especially in mature root and stem tissues, whereas the level of *VSP*A mRNA increased a little. They also found that wounding the leaflet caused a threefold to fourfold increase in both *VSP*A and *VSP*B transcripts in the wounded tissue after 12 hr. The authors speculated that endogenous levels of JA and JA-Me modulate *VSP* mRNA both in intact plants and in wounded plants.

2. Proteinase Inhibitors

Activation of defensive genes in plants by pathogen and herbivore attacks, or by other mechanical wounding, which is usually caused by a variety of signaling molecules that are released following the initial wounding of the tissue, results in accumulation of defensive compounds in cells around the wounded area. Ethylene is a possible candidate for the signaling molecules. Among the defensive compounds that are synthesized in response to the wounding are proteinase inhibitor proteins.

Farmer and Ryan (1990) reported induction of the accumulation of defensive proteinase inhibitor proteins by JA-Me. Spray application of JA-Me to leaves of tomato plants induced the accumulation of proteinase inhibitor I to levels higher than could be induced by wounding. The presence of JA-Me in the atmosphere of chambers containing the plants also induced the accumulation of inhibitors I and II. The presence of JA-Me above 10 nl (JA-Me is oily) in a 1.25-liter chamber was enough to induce the accumulation. Tomato plants began to accumulate the proteinase inhibitors about 5 hr after the initial exposure to JA-Me. JA, which has a much lower vapor pressure than JA-Me, weakly induced the proteinase inhibitors in the leaves. Tobacco and alfalfa plants also responded to exposure to JA-Me and accumulated their respective trypsin inhibitor in the leaves.

The authors examined whether a species of plant that contained JA-Me in the leaves could induce expression of proteinase inhibitor genes in nearby tomato plants. Small tomato plants were incubated in an airtight chamber together with 5 g of fresh leafy branches of *Artemisia tridentata,* with no direct physical contact between the plants. Within 2 days, the leaves of tomato plants exhibited elevated levels of proteinase inhibitors I and II. The authors identified JA-Me in leaves of *A. tridentata* as the volatile signaling molecule for induction of accumulation of the proteinase inhibitors in tomato leaves. From these results, the authors speculated that JA-Me, or other volatile signals, released by such plants as *A. tridentata* have multiple effects on nearby plants, either by inducing the expression of defensive genes, or genes involved in other responses such as senescence.

Recently, we found that wounding potato leaves resulted in a rapid increase in the level of JA in the leaves (Y. Koda and Y. Kikuta, unpublished data). The evidence quoted above and this observation suggest the possibility that JA and JA-Me act as an alarm signal in a plant community. The signal which is formed in the wounded plant is transmitted to nearby plants through the air and induces activation of their defensive genes.

I. Stimulation of Growth of Meristem

Ravnikar and Gogala (1990) reported some stimulative effects of JA on the growth of meristems of potato plants. They added JA to a medium in which potato meristems were being cultured and found that the number of meristems which developed into shoots was doubled by addition of 10^{-6} M JA. The effect was found only in meristems obtained from the dark-grown plants and not in those from light-grown ones. They observed that plantlets grown on medium with JA were more vigorous than those on control medium. Although it is possible that the stimulation by JA is brought about by inhibition of formations of callus and root, it is more likely that JA has unknown stimulative effects on the development of the plantlets, because JA at concentrations as low as 10^{-7} M stimulated the growth of potato plantlets (M. Ravnikar, personal communications). They also reported that JA at the same concentration was effective for elimination of potato mosaic virus (PVM) from the meristems (Ravnikar and Gogala, 1989).

J. Determination of Stem Growth Habit of Soybean Plants

The possible involvement of JA-related compounds in the control of the stem growth habit of soybean plants was examined by Koda *et al.* (1991b). Soybean plants show diversity in stem growth habit which ranges from the determinate type to the indeterminate type. Stem growth of determinate plants abruptly terminates near the beginning of the flowering. Both the termination of stem growth and flowering in soybean plants are induced by short days. Woodworth (1933) showed that inheritance of the determinate and indeterminate phenotypes is under monogenic control (Woodworth, 1933), and Bernard (1972) developed isolines of determinate and indeterminate types. Their observations indicate that the endogenous factor responsible for the termination of stem growth is different from that responsible for flowering. Since the termination seems to be brought about by cessation of growth at the apical meristem, it is possible that some growth inhibitor(s) that forms in the leaves in response to short days causes the termination of stem growth. Since no differences were found in the levels of ABA in leaves of determinate and indeterminate isolines at any time during the growth of the plants (Koda et al., 1989), it seems likely that ABA is not a factor that directly controls the stem growth habit.

JA-like activities in leaves of both isolines, which were detected by a bioassay for tuber-inducing activity (see Section V,B,1,c), were very low 20 days before the commencement of flowering. JA-like activity increased rapidly thereafter and reached a maximum near the time of flowering. Although the activities in leaves of both isolines fluctuated in a similar manner, the activity in the determinate isoline was much higher than that in the indeterminate isoline after flowering. The presence of JA in the leaves of the determinate isoline was confirmed by HPLC and mass spectrometry. The level of JA in the leaves of the determinate plants was $\sim 10^{-6}$ mol/kg, on an average of four samples that were harvested at different times. Exogenous application of JA to cultured shoot apices of the indeterminate isoline strongly inhibited the growth. Above a concentration of 10^{-6} M, some shoots without any growth at all were observed. These results suggest the involvement of JA in the control of the growth habit of soybean plants.

It is also possible that JA-related compounds are involved in the control of the various growth habits of different types of plants, such as the maintenance of rosette type and the inhibition of bolting.

K. Induction of Tuberization

1. Tuberization in Potato Plants

The first sign of potato tuberization is a swelling at the subapical region of the stolon. The swelling at this early stage is mainly brought about by radial cell expansion (Koda and Okazawa, 1983b). Subsequently, vigorous thickening growth due to cell division and expansion occurs and starch accumulates in the cells. As described in Section II, we isolated TAG from potato leaves as a tuber-inducing substance (Koda et al., 1988). TA and TAG appear to trigger the tuberization by inducing cell expansion. The whole process of the tuberization seem to be regulated by the combined action of several plant hormones. A decrease in gibberellin level is a prerequisite for the tuberization. Cytokinin, which accumulates in the swollen part (Koda, 1982; Koda and Okazawa, 1983b), seems to induce vigorous cell division, and ABA, the level of which continues to increase during the course of tuberization, seems to induce dormancy of the meristems.

Since the chemical structure of TA is closely related to that of JA, we compared potato tuber-inducing activities of JA, CA, and their analogs (Koda et al., 1991a). The procedure for bioassay of the tuber-inducing activity is described in detail in Section V,B,1,c. All compounds used were racemic as well as diastereo mixtures (cis–trans with respect to the plane of the cyclopentane moiety). JA and JA-Me showed strong and almost equivalent tuber-inducing activities when present at concentrations greater than 10^{-7} M. The activities of CA and CA-Me were somewhat lower than those of JA and JA-Me. The appearance of typical tubers induced by JA-Me is shown in Fig. 12. These compounds also affected the negative geotropism of the lateral shoots at concentrations above 10^{-6} M. HJA, HJA-Me, dihydro-CA, and dihydro-CA-Me had almost no tuber-inducing activity. Reduction of the carboxyl group at the C-1 position to a hydroxyl group caused an almost complete loss of activity. Modification of the oxygen atom at the C-3 position caused a slight decrease in the activity. The chemical structures of prostaglandins (PGs), which are important animal hormones, resemble that of JA. However, PGB_2 and PGE_1 had no tuber-inducing activity. These results suggest that partial structures that are indispensable for the activity are the carboxyl group or its esters at the C-1 position, a double bond (pentenyl group) in the substituent at the C-2 position, and an oxygen atom at the C-3 position. These structural requirements for the tuber-inducing activity are different from those reported for the inhibitory activity of such compounds on the growth of rice seedlings (Yamane et al., 1980) and for the senescence-promoting activity assayed with oat leaves (Ueda et al., 1981). The most remarkable difference is that a double bond in the substituent at the C-2 position is not essential for the latter two activities. It appears that the mechanism of action of JA and related compounds in the induction of tubers is different from the mechanism for the growth inhibition and the promotion of senescence.

In our previous paper (Koda et al., 1991a), tuber-inducing activity of synthetic JA was compared to those of isolated TA and TAG, and their activities were found to be almost equivalent. However, comparison of tuber-inducing activities of the cis isomers of JA, TA, and TAG indicated that, among the three compounds, JA had the strongest activity (Fig. 13) (Y. Koda and Y. Kikuta, unpublished data).

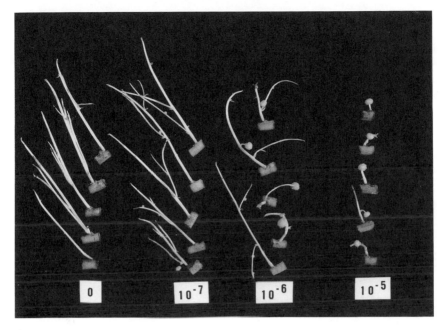

FIG. 12   Typical appearance of potato tubers induced by the cis isomers of JA-Me after 3 weeks in culture. Numbers indicate the molar concentrations of JA-Me in the medium. The concentration of sucrose in the medium was 2%.

2. Tuberization in Yam Plants

Yam plants (*Dioscorea* spp.) are monocots and usually form a single, large underground tuber at the lowest node of the main viny stem and many aerial tubers at axillary buds. Morphologically, the two kinds of tuber are identical. The tubers are an important source of food in many tropical countries. Yam tuberization is under the control of the photoperiod, as is potato tuberization (Garner and Allard, 1923). Short days stimulate the process, while long days inhibit it. Therefore, it is probable that TA, JA, and their related compounds are involved in yam tuberization. If these substances are involved in the process, they must be present in the leaves, and the level of these compounds should show some changes that are associated with tuberization. Furthermore, they must be capable of inducing tuberization.

JA was isolated from yam leaves (*Dioscorea batatas*) with the guidance of the assay for tuber-inducing activity, which was carried out on cultures of single-node segments of potato stems *in vitro,* and identified by HPLC and mass spectrometry (Koda and Kikuta, 1991). TA was not found and

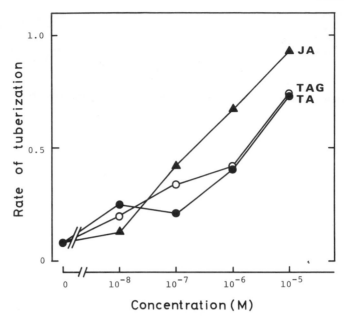

FIG. 13   Comparison of potato tuber-inducing activities of the cis isomers of JA, TA, and TAG. The concentration of sucrose in the medium was 2%.

only a small amount of JA-Me was detected. The level of endogenous JA in the leaves of the plants was 5×10^{-7} mol/kg fresh weight ($\sim 100\ \mu$g/kg) at an early stage of plant growth and increased continuously with the growth of the plant. The effect of exogenous JA on the tuberization of yam plants was examined in cultures of single-node segments of yam stem *in vitro*. The cultures were maintained either in the dark or under a photoperiod of 12 hr of light and 12 hr of darkness. Under the both sets of conditions, JA at concentrations above $10^{-7}\ M$ induced tuberization. The typical appearance of tubers induced by JA is shown in Fig. 14. These results strongly suggest that tuberization of yam plants is induced by JA or its derivatives.

3. Tuberization in Jerusalem Artichoke Plants

Since tuberization of Jerusalem artichoke plants (*Helianthus tuberosus* L.) is induced by short days (Hamner and Long, 1939), it is probable that JA-related compounds are also involved in tuberization of the plants. We found that JA is present in the leaves and this substance is capable of inducing tuberization *in vitro* (Y. Koda and Y. Kikuta, unpublished data), suggesting the involvement of JA or its derivatives in tuberization of the plants.

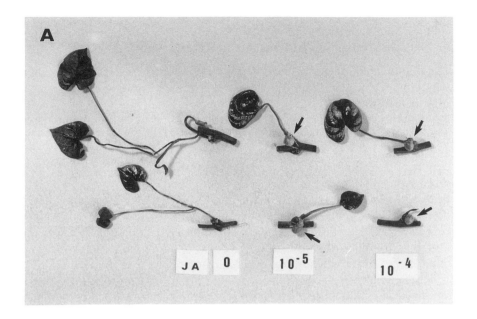

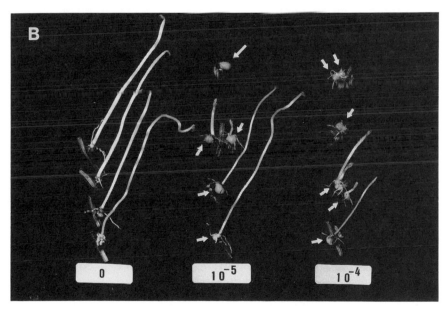

FIG. 14 Typical appearance of yam tubers induced by JA under (A) a photoperiod of 12 hr of light and 12 hr of darkness and (B) in the dark after 1 month. Numbers indicate the molar concentrations of JA in the medium. The sucrose concentration in the medium was 2%. Arrows indicate tubers. Attachments to the base of the lateral shoots of control segments are stipules and not tubers (Koda and Kikuta, 1991).

4. Possible Mechanisms of Induction of Tuberization

The involvement of JA-related compounds in tuberization in potato plants, which are dicots, and in yam plants, which are monocots, suggests that tuberization in many other tuber-forming plants may also be controlled by JA-related compounds. The fact that tuberization in both yams and potatoes is induced by short days (Ewing, 1985; Garner and Allard, 1923) suggests that the level of JA-related compounds are controlled by the length of the photoperiod. As mentioned in Section III,A, JA is synthesized from linolenic acid by a series of enzymes, the first of which is lipoxygenase (Vick and Zimmerman, 1984). Since phytochrome-mediated control of lipoxygenase activities in seedlings of some plants has been reported (Surrey, 1967; Oelze-Karow et al., 1970), it is possible that the length of the photoperiod affects the level of JA-related compounds via changes in the activities of lipoxygenase and related enzymes.

Since radial cell expansion occurs at an early stage of potato tuberization (Koda and Okazawa, 1983b), it is likely that JA-related compounds are capable of inducing radial cell expansion and the expansion triggers tuberization. The direction of cell expansion depends to a considerable extent on the orientation of cellulose microfibrils in the cell wall (Green et al., 1970). The orientation is considered to be controlled by cortical microtubules (Gunning and Hardham, 1982). Mita and Shibaoka (1983) found that bulb formation of onion was accompanied by the disruption of cortical microtubules in the cells of leaf sheath. They also found that microtubule-disrupting compounds, such as colchicine and cremat, cause lateral expansion of cells in the leaf sheath and postulated that the "bulbing hormone" may be a compound which is capable of disrupting microtubules. It is possible, therefore, that JA-related compounds may cause disorientation or disruption of the microtubules and, consequently, induce the radial expansion of cells which leads to tuberization.

Abe et al. (1990) examined whether JA-Me has microtubule-disrupting activity using tobacco BY-2 cells. They applied JA-Me to the cultured cells at the logarithmic phase of growth. The cells were treated with JA-Me for 30 min, fixed, and then stained for microtubules by an immunofluorescence method. Disruption of microtubules by JA-Me at concentrations above 10^{-5} M was observed only in a small number of cells. They speculated that microtubules in cells at a certain stage in the cell cycle may be sensitive to JA-Me. They applied JA-Me to cells in which the cell cycles were synchronized by treatment with aphidicolin, an inhibitor of DNA polymerase-α. Disruption of microtubules was found exclusively in cells at the S phase of the cell cycle. Since suppression of the efflux of protons from guard cells in Commelina by JA-Me was demonstrated (Raghavendra and Reddy, 1987), the authors speculated that JA-Me lowers the cytosolic pH by inhibiting the efflux of protons from the cytosol, and then the decrease in cytosolic pH causes the depolymerization of microtubules. In order to verify the hypothesis that the disruption of cortical microtubules by JA-related compounds is involved in cell expansion, similar studies using cells which are able to respond to JA, such as potato, are necessary.

Recently, we found that an application of JA at concentrations above 10^{-5} M to potato cells excised from tubers induced a rapid swelling of the cells (K. Takahashi, Y. Koda, and Y. Kikuta, unpublished data). An experiment is necessary to see whether or not disruption of microtubules is involved in the swelling. Since JA-related compounds seem to be able to affect proton efflux from cells, it is also probable that JA-related compounds affect cytosolic pH as well as pH in the cell wall. The changes in pH may affect the turgor of the cells and extensibility of the cell walls by changing activities of enzymes involved in breakdown of starch and cellulose; thereby finally inducing swelling of the cell.

VII. Concluding Remarks

In general, growth and development of a plant are regulated in part by environmental factors such as photoperiod and temperature. The factors are received by the plants as informational inputs and then the inputs are translated into biochemical signals which are transmitted from the site of perception to the target cells and tissues. The signals then induce metabolic cascades resulting in various physiological responses. The representative biochemical signals are plant hormones. However, at present, the events between the input from the environment and the physiological responses have not been completely determined. With regard to JA-related compounds, our knowledge about this area is very limited yet. However, it is possible that environmental stimulis, such as photoperiods and wounding, can easily be translated into the levels of JA-related compounds. As stated in Section III,A, one of the rate-limiting factors for JA biosynthesis seems to be the level of free linolenic acid. Much linolenic acid is contained in phospholipids in plasma membrane. If environmental factors can activate phospholipase A_2, which is able to liberate fatty acid from phospholipids, and a specific lipoxygenase, which is able to start biosynthesis of JA, then the level of JA in the cell can be enhanced easily.

The biosynthetic pathway of JA from linolenic acid is similar to that of prostaglandins from arachidonic acid in mammalian cells. Since many prostaglandins are formed via an arachidonic acid cascade, the similarity has led to the expectation that there are many unknown compounds which are formed from linolenic acid and have regulatory effects on plant development. The additional information that will result from further studies in this field will determine whether these speculations are right or wrong.

Studies of the role of JA in the control of plant development started only a decade ago. Although the number of researchers in this field is very limited at present, a lot of information about involvement of JA-related compounds in various phases of plant development has been accumulated during the past decade. The information indicates that JA-related compounds satisfy various criteria for plant hormones, such as ubiquitous presence in plant kingdom, various biological activities at low concentrations, and interaction with other hormones. However, to get general agreement that JA-related compounds are a new group of plant hormones, an increase in the number of researchers in this field and further accumulation of our knowledge appear to be necessary.

References

Abe, M., Shibaoka, H., Yamane, H., and Takahashi, N. (1990). *Protoplasma* **156,** 1–8.
Abeles, F. B., Dunn, L. J., Morgens, P., Callahan, A., Dinterman, R. E., and Schmidt, J. (1988). *Plant Physiol.* **87,** 609–615.
Abeles, F. B., Hershberger, W. L., and Dunn, L. J. (1989). *Plant Physiol.* **89,** 664–668.
Aldridge, D. C., Galt, S., Giles, D., and Turner, W. B. (1971). *J. Chem. Soc. C,* pp. 1623–1627.
Anderson, J. M. (1985). *J. Chromatogr.* **330,** 347–355.
Anderson, J. M. (1988). *J. Plant Growth Regul.* **7,** 203–211.
Anderson, J. M. (1991). *J. Plant Growth Regul.* **10,** 5–10.
Anderson, J. M., Spilatro, S. R., Klauer, S. F., and Franceschi, V. R. (1989). *Plant Sci.* **62,** 45–52.
Ansell, G. B., and Spanner, S. (1982). *In* "Phospholipid" (J. N. Hawthorne and G. B. Ansell, eds.), pp. 1–50. Elsevier, Amsterdam.
Astle, M. C., and Rubery, P. H. (1983). *Planta* **157,** 53–63.
Astle, M. C., and Rubery, P. H. (1985). *Planta* **166,** 252–258.
Baker, T. C., Nishida, R., and Roelofs, W. L. (1981). *Science* **214,** 1359–1361.
Bernard, R. L. (1972). *Crop Sci.* **12,** 235–239.
Blumenfeld, A., and Gazit, S. (1970). *Plant Physiol.* **45,** 535–536.
Bohlmann, F., Wegner, P., Jakupovic, J., and King, R. M. (1984). *Tetrahedron* **40,** 2537–2540.
Brückner, C., Kramell, R. Schneider, G., Knöfel, H.-D., Sembdner, G., and Schreiber, K. (1986). *Phytochemistry* **25,** 2236–2237.
Brückner, C., Kramell, R., Schneider, G., Schmidt, J., Preiss, A., Sembdner, G., and Schreiber, K. (1988). *Phytochemistry* **27,** 275–276.
Chapman, H. W. (1958). *Physiol. Plant.* **11,** 215–224.
Crabalona, L. (1967). *C. R. Hebd. Seances Acad. Sci.,* Ser. *C* **264,** 2074–2076.
Cuello, J., Quiles, M. J., Garcia, C., and Sabater, B. (1990). *Bot. Bull. Acad. Sin.* **31,** 107–111.
Dathe, W., Schneider, G., and Sembdner, G. (1978). *Phytochemistry* **17,** 963–966.
Dathe, W., Ronsch, H., Preiss, A., Schade, W., Sembdner, G., and Schreiber, K. (1981). *Planta* **153,** 530–535.

Demole, E., Lederer, E., and Mercier, D. (1962). *Helv. Chim. Acta* **45**, 675–685.

Engvild, K. C. (1989). *Physiol. Plant.* **77**, 282–285.

Eskin, N. A. M., Grossman, S., and Pinsky, A. (1977). *CRC Crit. Rev. Food Sci. Nutr.* **9**, 1–14.

Ewing, E. E. (1985). *In* "Potato Physiology" (Li H. Paul, ed.), pp. 153–207. Academic Press, Orlando, Florida.

Farmer, E. E., and Ryan, C. (1990). *Proc. Natl. Acad. Sci. U.S.A.* **87**, 7713–7716.

Fletcher, R. A., Venkatarayappa, T., and Kallidumbil, V. (1983). *Plant Cell Physiol.* **24**, 1057–1064.

Fukui, H., Koshimizu, K., Usuda, S., and Yamazaki, Y. (1977a). *Agric. Biol. Chem.* **41**, 175–180.

Fukui, H., Koshimizu, K., Yamazaki, Y., and Usuda, S. (1977b). *Agric. Biol. Chem.* **41**, 189–194.

Galliard, T., and Phillips, D. R. (1971). *Biochem. J.* **124**, 431–438.

Garner, W. W., and Allard, H. A. (1923). *J. Agric. Res.* **23**, 871–920.

Green, P. B., Erickson, R. O., and Richmond, P. A. (1970). *Ann. N.Y. Acad. Sci.* **175**, 712–731.

Gregory, L. (1956). *Am. J. Bot.* **43**, 281–288.

Grossman, S., and Leshem, Y. (1978). *Physiol. Plant.* **43**, 359–362.

Gunning, B. E. S., and Hardham, A. R. (1982). *Annu. Rev. Plant Physiol.* **33**, 651–698.

Hamberg, M. (1988). *Biochem. Biophys. Res. Commun.* **156**, 543–550.

Hamberg, M., Miersch, O., and Sembdner, G. (1988). *Lipids* **23**, 521–524.

Hamner, K. C., and Long, E. M. (1939). *Bot. Gaz. (Chicago)* **101**, 81–90.

Herrmann, G., Lehmann, J., Peterson, A., Sembdner, G., Weidhase, R. A., and Parthier, B. (1989). *J. Plant Physiol.* **134**, 703–709.

Kim, I.-S., and Grosch, W. (1981). *J. Agric. Food Chem.* **29**, 1220–1225.

Knöfel, H. D., Bruckner, C., Kramell, R., Sembdner, G., and Schreiber, K. (1984). *Biochem. Physiol. Pflanz.* **179**, 317–325.

Kockritz, A., Schwe, T., Hiek, B., and Hass, W. (1985). *Phytochemistry* **24**, 381–384.

Koda, Y. (1982). *Plant Cell Physiol.* **23**, 843–849.

Koda, Y., and Kikuta, Y. (1991). *Plant Cell Physiol.* **32**, 629–633.

Koda, Y., and Okazawa, Y. (1983a). *Jpn. J. Crop Sci.* **52**, 582–591.

Koda, Y., and Okazawa, Y. (1983b). *Jpn. J. Crop Sci.* **52**, 592–597.

Koda, Y., and Okazawa, Y. (1988). *Plant Cell Physiol.* **29**, 969–974.

Koda, Y., Omer, E. A., Yoshihara, T., Shibata, H., Sakamura, S., and Okazawa, Y. (1988). *Plant Cell Physiol.* **29**, 1047–1051.

Koda, Y., Yoshida, K., Gotoh, K. and Okazawa, Y. (1989). *Jpn. J. Crop Sci.* **58**, 111–113.

Koda, Y., Kikuta, Y., Tazaki, H., Tsujino, Y., Sakamura, S., and Yoshihara, T. (1991a). *Phytochemistry* **30**, 1435–1438.

Koda, Y., Yoshida, K., and Kikuta, Y. (1991b). *Physiol. Plant.* **83**, 22–26.

Letham, D. S. (1971). *Physiol. Plant.* **25**, 391–396.

Lofstedt, C., Vickers, N. J., Roelofs, W., and Baker, T. C. (1989). *Oikos* **55**, 402–408.

Lopez, R., Dathe, W., Bruckner, C., Miersch, O., and Sembdner, G. (1987). *Biochem. Physiol. Pflanz.* **182**, 195–201.

Madec, P. (1963). *In* "The Growth of the Potato" (J. D. Ivins and F. L. Milthorpe, eds.), pp. 121–131. Butterworth, London.

Maslenkova, L. T., Zanev, Y., and Popova, L. P. (1990). *Plant Physiol.* **93**, 1316–1320.

Mason, H. S., and Mullet, J. E. (1990). *Plant Cell* **2**, 569–579.

Matile, P. (1980). *Z. Pflanzenphysiol.* **99**, 475–478.

Melis, R. J. M., and Van Staden, J. (1984). *Z. Pflanzenphysiol.* **113**, 271–283.

Meyer, A., Miersch, O., Butter, C., Dathe, W., and Sembdner, G. (1984). *J. Plant Growth Regul.* **3**, 1–8.

Meyer, A., Gross, D., Volkefeld, S., Kummer, M., Schmidt, J., Sembdner, S., and Schreiber, K. (1989). *Phytochemistry* **28**, 1007–1011.

Meyer, A., Schmidt, J., Gross, D., Jensen, E., Rudolph, A., Volkefeld, S., and Sembdner, G. (1991). *J. Plant Growth Regul.* **10**, 17–25.

Miersch, O., Meyer, A., Volkefeld, S., and Sembdner, G. (1986). *J. Plant Growth Regul.* **5**, 91–100.

Miersch, O., Sembdner, G., and Schreiber, K. (1989). *Phytochemistry* **28**, 339–340.

Miller, B. L., and Huffaker, R. C. (1982). *Plant Physiol.* **69**, 58–62.

Mita, T., and Shibaoka, H. (1983). *Plant Cell Physiol.* **24**, 109–117.

Mueller-Uri, F., Parthier, B., and Nover, L. (1988). *Planta* **176**, 241–247.

Nooden, L. D., and Leopord, A. C. (1978). *In* "Phytohormones and Related Compounds: A Comprehensive Treatise" (D. S. Letham, P. B. Goodwin, and T. J. V. Higgins, eds.), Vol. 2, pp. 329–362. Elsevier, Amsterdam.

Nooden, L. D., Guiamet, J. J. Singh, S., Letham, D. S., Tsuji, J., and Schneider, M. J. (1990). *In* "Plant Growth Substances 1988" (P. P. Pharis and S. B. Rood, eds.), pp. 537–546. Springer-Verlag, Berlin.

Oelze-Karow, H., Schopfer, P., and Morh, H. (1970). *Proc. Natl. Acad. Sci. U.S.A.* **65**, 51–57.

Ohta, H., Ida, S., Mikami, B., and Morita, Y. (1986). *Plant Cell Physiol.* **27**, 911–918.

Parthier, B. (1990). *J. Plant Growth Regul.* **9**, 57–63.

Pleiss, U., Teich, A., Stock, M., Gross, D., and Schutte, H. R. (1983). *J. Labelled Compd. Radiopharm.* **20**, 205–211.

Raghavendra, A. S., and Reddy, K. B. (1987). *Plant Physiol.* **83**, 732–734.

Ravnikar, M., and Gogala, N. (1989). *Biol. Vestn.* **37**, 79–88.

Ravnikar, M., and Gogala, N. (1990). *J. Plant Growth Regul.* **9**, 233–236.

Rehm, M. M., and Cline, M. G. (1973). *Plant Physiol.* **51**, 93–96.

Saniewski, M., and Czapski, J. (1985). *Experientia* **41**, 256–257.

Saniewski, M., Czapski, J., Nowacki, J., and Lange, E. (1987a). *Biol. Plant.* **29**, 199–203.

Saniewski, N., Nowacki, J., and Czapski, J. (1987b). *J. Plant Physiol.* **129**, 175–180.

Satler, S. O., and Thimann, K. V. (1981). *C. R. Hebd. Seances Acad. Sci.* **293**, 735–740.

Satter, R. L., and Galston, A. W. (1981). *Annu. Rev. Plant Physiol.* **32**, 83–110.

Sekiya, J., Kajiwara, T., and Hatanaka, A. (1979). *Agric. Biol. chem.* **43**, 969–980.

Shibaoka, H., and Thimann, K. V. (1970). *Plant Physiol.* **46**, 212–220.

Simonds, N. W. (1965). *Eur. Potato J.* **8**, 92–97.

Staswick, P. E. (1990). *Plant Cell* **2**, 1–6.

Stoddart, J. L., and Thomas, H. (1982). *Encycl. Plant Physiol., New Ser.* **14A**, 592–636.

Surrey, K. (1967). *Plant Physiol.* **44**, 421–424.

Tsurumi, S., and Asahi, Y. (1985). *Physiol. Plant.* **64**, 207–211.

Tsurumi, S., Asahi, Y., and Suda, S. (1985). *Bot. Mag.* **98**, 87–95.

Ueda, J., and Kato, J. (1980). *Plant Physiol.* **66**, 246–249.

Ueda, J., and Kato, J. (1982a). *Agric. Biol. Chem.* **46**, 1975–1976.

Ueda, J., and Kato, J. (1982b). *Physiol. Plant.* **54**, 249–252.

Ueda, J., Kato, J., Yamane, H., and Takahashi, N. (1981). *Physiol. Plant.* **52**, 305–309.

Van den Bosch, H. (1982). *In* "Phospholipids" (J. N. Hawthorne and G. B. Ansell, eds.), pp. 313–358. Elsevier, Amsterdam.

Vick, B. A., and Zimmerman, D. C. (1976). *Plant Physiol.* **57**, 780–788.

Vick, B. A., and Zimmerman, D. C. (1981). *Plant Physiol.* **67**, 92–97.

Vick, B. A., and Zimmerman, D. C. (1982). *Plant Physiol.* **69**, 1103–1108.

Vick, B. A., and Zimmerman, D. C. (1983). *Biochem. Biophys. Res. Commun.* **111**, 470–477.

Vick, B. A., and Zimmerman, D. C. (1984). *Plant Physiol.* **75**, 458–461.

Vick, B. A., and Zimmerman, D. C. (1986). *Plant Physiol.* **80,** 202–205.
Vick, B. A., and Zimmerman, D. C. (1987). *Plant Physiol.* **85,** 1073–1078.
Vick, B. A., Zimmerman, D. C., and Weisleder, D. (1979). *Lipids* **14,** 734–740.
Weidhase, R. A., Lehmann, J., Kramell, H., Sembdner, G., and Parthier, B. (1987a). *Physiol. Plant.* **69,** 161–166.
Weidhase, R. A., Kramell, H., Lehman, J., Liebisch, H., Lerbs, W., and Parthier, B. (1987b). *Plant Sci.* **51,** 177–186.
Weiler, E. W. (1982). *Plant Physiol.* **54,** 230–234.
Wittenbach, V. A. (1983). *Plant Physiol.* **73,** 125–129.
Wittenbach, V. A., Lin, W., and Hebert, R. R. (1982). *Plant Physiol.* **69,** 98–102.
Woodworth, C. M. (1933). *J. Am. Soc. Agron.* **25,** 36–51.
Yamaguchi, N., and Minamide, T. (1985). *J. Jpn. Soc. Hortic. Sci.* **54,** 256–271.
Yamane, H., Sugawara, J., Suzuki, Y., Shimamura, E., and Takahashi, N. (1980). *Agric. Biol. Chem.* **44,** 2857–2864.
Yamane, H., Takagi, H., Abe, H., Yokota, T., and Takahashi, N. (1981). *Plant Cell Physiol.* **22,** 689–697.
Yamane, H., Abe, H., and Takahashi, N. (1982). *Plant Cell Physiol.* **23,** 1125–1127.
Yoshihara, T., Omer, E. A., Koshino, H., Sakamura, S., Kikuta, Y., and Koda, Y. (1989). *Agric. Biol. Chem.* **53,** 2835–2837.
Zeiger, E. (1983). *Annu. Rev. Plant Physiol.* **34,** 441–475.
Zimmerman, D. C., and Feng, P. (1978). *Lipids* **13,** 313–316.
Zimmerman, D. C., and Vick, B. A. (1970). *Plant Physiol.* **46,** 445–453.

Restriction Fragment Length Polymorphism Analysis of Plant Genomes and Its Application to Plant Breeding

C. Gebhardt and F. Salamini

Max-Planck-Institut für Züchtungsforschung, D-5000 Köln 30, Germany

I. Introduction

The insertion via transformation of specific foreign genes into plants, which acquire through this step new and heritable agronomic properties, is one contribution of molecular genetics to plant breeding which at present attracts much public attention. A second contribution, which is the subject of this review, makes use of molecular techniques as diagnostic tools to assist the conventional breeding process of combining parental genomes and selecting among the progeny of foundation crosses. Standard breeding procedures utilize the genetic variability present within the available gene pools of crop species to synthesize new cultivars. The characters selected are, besides yield, disease and stress resistance or specific qualities required for food use or processing. Most of these traits have a complex inheritance. Moreover, the phenotypic variability observed in segregating populations can be splitted into genetic and environmental components. The separation of heritable and environmental sources of phenotypic variability is the most time-consuming step in plant breeding. The availability of genetic markers diagnostic for a superior expression of a trait, which was easier to score than the trait itself and phenotypically neutral and environmentally independent, would greatly facilitate the selection process. Such markers would also help in reducing the handling of plant pathogens necessary for testing the inheritance of plant resistance genes.

DNA sequence variation is the basis for the genetic diversity within a species. In most cases, DNA sequence variation is phenotypically neutral because the majority of the mutations altering the order of the nucleotides are only maintained during evolution if they are silent. By definition, this type of genetic variation is environmentally independent; when variations are frequent enough and easy to score, all requirements for genetic markers of diagnostic value are present on which marker-assisted selection schemes can be built.

As restriction enzymes cut DNA at specific sequences, a point mutation within this site results in the loss or gain of a recognition site, giving rise in that region to restriction fragments of different length. Mutations caused by the insertion, deletion, or inversion of DNA stretches will also lead to a length variation of DNA restriction fragments. Genomic restriction fragments of differing lengths between genotypes can be detected on Southern blots (Southern, 1975). The genomic DNA is digested with an enzyme of choice, and the fragments are electrophoretically separated in a gel matrix, transferred to a carrier membrane, and hybridized against a suitable labeled probe.

The DNA sequence variation detected by this method was termed restriction fragment length polymorphism (RFLP) (Botstein *et al.,* 1980). The phenomenon was first described for mutant strains of adenoviruses by Grodzicker *et al.* (1974) and its Mendelian inheritance was demonstrated for ribosomal DNA of yeast by Petes and Botstein (1977). The potential of RFLPs as diagnostic markers became evident from studies of the human globin genes, where a direct correlation between the sickle cell mutation carried by a specific β-globin allele and the presence of certain RFLP fragments was evident (Kan and Dozy, 1978a,b; Jeffreys, 1979). The molecular basis of sickle cell anemia is precisely known. This is not the case for most heritable traits, particularly for those showing quantitative inheritance. For such traits, genetically closely linked markers which do not influence the phenotypic expression of the trait can be identified and serve as diagnostic tools. When a genetic linkage map saturated with marker loci is available, the genome of a species can be systematically scanned for markers cosegregating with a specific trait of interest. Because of their abundance, RFLPs are the first class of genetic markers allowing the construction of highly saturated linkage maps. This was first suggested for the human genome by Botstein *et al.* in 1980. The potential of this approach to plant breeding was recognized few years later (Beckmann and Soller, 1983, 1986; Tanksley, 1983), and its feasibility was experimentally tested in maize and tomato, two species which are genetically well characterized (Rivin *et al.,* 1983; Helentjaris *et al.,* 1985; Evola *et al.,* 1986). Since then, an impressive amount of RFLP data has been accumulated in several plant species, including important crops. Extensive RFLP linkage maps have been constructed and several genes contributing to agronomically relevant traits were mapped by RFLP markers. Both topics, RFLP linkage maps and genome localization of genes of interest, are very relevant to plant breeding; they are predominantly reviewed in this article. Other aspects of RFLP analysis in plants which also may have an impact on plant breeding in the future are the structural analysis of plant genomes, the measurement of genetic distances and evolutionary relationships, and the fingerprinting of plant genotypes with highly discriminative probes.

II. RFLP-Marker-Based Selection in Plant Breeding

The prerequisite for marker-based selection is the identification of RFLP markers tightly linked to a trait of agronomic interest. The most general approach to achieve this is the initial construction of a complete RFLP linkage map. Markers evenly spaced on the chromosomes are then used to scan the genome in single individuals of progenies segregating for the trait(s) of interest and for the RFLP markers. Linkage is detected by cosegregation of the trait with RFLP loci in a specific chromosomal region. In a third step, the linked markers may be used for the selection of the trait. As the field of RFLP analysis in plants has developed only over the past few years, research has mainly been concerned with the first two steps: construction of RFLP maps and mapping of traits. The third step, plant breeding based on marker selection, is still in its infancy.

A. Construction of RFLP Linkage Maps

1. The Genetic Material

Unlike human linkage maps, which are derived from segregation analysis in large family pedigrees (White *et al.*, 1985), plant geneticists and breeders establish inheritance of genetic loci in experimental populations. For the construction of RFLP linkage maps, parents are chosen that show the maximum of polymorphic loci in order to ensure the mapping of as many marker genes as possible per unit of resources available. In cases where the variability within a species was found to be very small, as in tomato (Helentjaris *et al.*, 1985; Miller and Tanksley, 1990a) or soybean (Apuya *et al.*, 1988; Keim *et al.*, 1989), interspecific crosses were used for obtaining the linkage maps (Helentjaris *et al.*, 1986a; Bernatzky and Tanksley, 1986; J. G. K. Williams *et al.*, 1990).

In self-fertilizing species, the populations analyzed with RFLP markers are the same as those used for segregation analysis of morphological markers. However, whereas the segregation of an unlimited number of RFLP markers can be followed in one segregating population, this is not the case when mapping morphological markers. The crossing scheme is outlined in Fig. 1A. Polymorphic, homozygous inbred lines are crossed to produce an heterozygous F_1 generation which is either selfed to give an F_2

A

Self-compatible
species

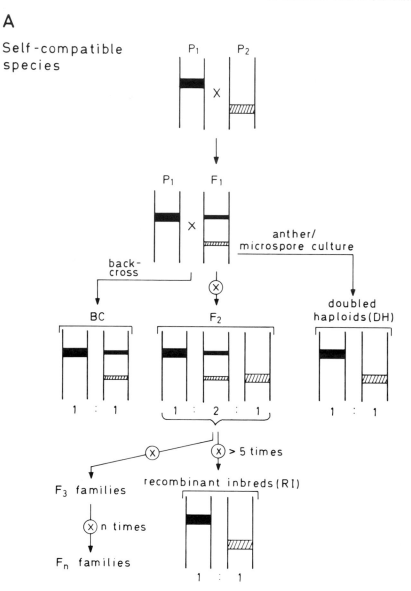

FIG. 1   Crossing schemes for segregation analysis in self-compatible inbreeding plant species (A) and self-incompatible, heterozygous species (B). See text for further explanations.

B

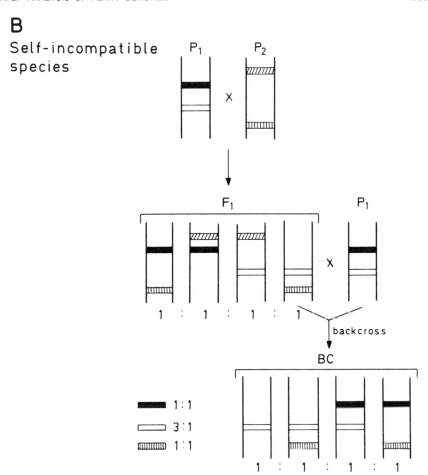

FIG. 1   (*continued*)

population segregating for the two parental alleles, or backcrossed to a recurrent parent to give a backcross progeny in which the alleles of the nonrecurrent parent segregate. Most RFLP maps in plants to date are based on segregation data from F_2 progenies (Helentjaris *et al.*, 1986a; Bernatzky and Tanksley, 1986; Landry *et al.*, 1987a; McCouch *et al.*, 1988; Chang *et al.*, 1988; Nam *et al.*, 1989; Slocum *et al.*, 1990; Chao *et al.*, 1989; J. G. K. Williams *et al.*, 1990; see also Table I). In parallel with an F_2 population, a backcross was also used for the tomato RFLP map of Bernatzky and Tanksley (1986). The genotypes of a valuable F_2 mapping population can be fixed and propagated as F_3 families derived from individual F_2 plants, as done for *Arabidopsis thaliana* (Nam *et al.*, 1989; Chang *et al.*, 1988).

As an alternative, recombinant inbred lines (RIs) can be developed by continuous selfing and utilized for mapping as demonstrated for maize (Burr *et al.*, 1988; Burr and Burr, 1991). Permanent RI populations can be used by different research groups to integrate segregation data into a single compound map, as with the human linkage studies based on the common use of cell lines derived from large kinships (Donis-Keller *et al.*, 1987). Due to the increased probability of chromosome pairing and recombination during an extended period of inbreeding, maps derived from RI populations show a higher resolution of closely linked loci as compared to F_2 populations (Burr and Burr, 1991). However, the development of RIs is time consuming and not readily accessible to self-incompatible species. A further possibility is the production of segregating populations of doubled haploid lines (DH) from anther or microspore culture of the F_1 (Fig. 1A). Here, the homozygosity is reached in a single step as compared to the five or more selfing generations required for RIs, although at a loss of the larger recombinational values of RIs (DHs fix the recombination events of meiosis taking place during the formation of the microsporocytes in the male parent).

Doubled haploids were used for linkage analysis in wheat (Chao *et al.*, 1989), in barley (Heun *et al.*, 1991; Graner *et al.*, 1991), and are available for mapping in rice (McCouch *et al.*, 1988). Although not restricted to self-compatible species, the practical use of doubled haploids for map construction and mapping of agronomically relevant genes is limited by the genotype-dependent response to *in vitro* culture conditions (for review, see Keller *et al.*, 1987; Sangwan and Sangwan-Norreel, 1990) and by the possible low vigor of the regenerated plants.

When homozygous inbred lines are not available, or when, as for the diploid self-incompatible potato, it is impractical to work with them, maps can still be constructed using F_1 or backcross progenies of partially heterozygous parents (Bonierbale *et al.*, 1988; Gebhardt *et al.*, 1989b, 1991). The crossing scheme outlined in Fig. 1B shows the inheritance of a tetraallelic RFLP locus as the most complex situation which occurs in allogamous diploid species. It must be considered that, in cases like the diploid potato, the parents differ in their allelic state at different loci. Besides loci with four or three different alleles as shown in Fig. 1B, combinations of two homozygous, two heterozygous, or one homozygous and another heterozygous RFLP allele occur (Fig. 1A) which segregate as in F_2s and backcross populations. At least in some species, like the potato, the allelic state of such progenies can be maintained indefinitely by vegetative propagation of the tubers.

All RFLP maps available for crop species have so far been derived from segregation data of diploid populations, although several important crops, like the cultivated potato or the winter wheat, are polyploid. Linkage analysis in tetraploids is theoretically possible (Bailey, 1961); it was performed for resistance traits (Cockerham, 1970) and might be useful as a basis for marker-based selection in breeding programs in the future. However, the effort required to establish a saturated RFLP linkage map would increase tremendously when using tetraploids, because it requires very large populations to be screened or a large marker number for establishing significant linkages.

In plant species extensively studied in the past decades by geneticists and cytogeneticists, well-characterized aneuploid stocks are available which are utilized for the assignment of RFLP loci to chromosomes and chromosomal arms. In maize, Helentjaris et al. (1986b) mapped RFLP markers to 8 of the 10 maize chromosomes by analyzing a set of monosomics where the $2n - 1$ condition was easily identified by the absence of a known morphological marker. An RFLP marker located, for example, on chromosome four was detected by the absence of one parental RFLP allele in the monosomic individuals. The same approach was applied to data from B–A translocations which carry either one or three copies of a specific chromosomal arm (Weber and Helentjaris, 1989). Monosomics and B–A translocations were also useful for the integration of the cytological and the genetic maps of maize with RFLP-based maps. In tomato and rice, primary trisomics helped in the efficient chromosome allocation of RFLP loci (Young et al., 1987; McCouch et al., 1988). Each primary trisomic line harbors three instead of two copies of one particular chromosome. Therefore, an RFLP marker revealing a locus on that chromosome will be identified from a 3:2 ratio of band intensity on autoradiograms when compared to normal disomics. A fragment polymorphism is not required in this case and more than one marker can be combined in one hybridization assay (Young et al., 1987).

In the hexaploid wheat ($2n = 6\times = 42$; AABBDD genomes) nullisomic–tetrasomic (NT) and ditelosomic (DT) stocks are available. In these strains, for an individual chromosome or chromosomal arm the two copies of one genome are replaced by two additional copies of a different genome, a situation which, for example, creates a BBBBDD genome for a particular chromosome. A marker of the corresponding A chromosome is therefore absent in the NT line. Using these genetic stocks, RFLP markers were mapped onto wheat chromosomes (Sharp et al., 1989; Chao et al., 1989; Kam-Morgan et al., 1989).

As the genomes of the Triticeae are homologous and as most markers show cross-hybridization within the family, alien addition lines in wheat of *Hordeum vulgare* (Barley), *Secale cereale* (rye), and *Aegilops umbellulata* chromosomes also allowed the assignment to chromosomes of RFLP loci whose position is, moreover, frequently conserved in all the species involved (Sharp et al., 1989; Chao et al., 1989; Kam-Morgan et al., 1989;

Graner *et al.*, 1990). The detection in the addition lines of polymorphic RFLP alleles contributed from the alien chromosome is highly probable because of the genetic distance existing among different species. However, the use of such markers for the scope of plant breeding within a species might be restricted due to the absence of intraspecific polymorphism.

The advantage of using aneuploid stocks or other cytogenetic variants for RFLP loci assignment is that only a small number of genotypes has to be analyzed, as compared to linkage analyses requiring populations of 50 to 100 genotypes. They, however, do not provide information on map distances and order of markers, with the exception of the translocation stocks available for maize (Weber and Helentjaris, 1989; Hoisington and Coe, 1990).

2. RFLP Markers

a. Sources and Selection of Markers All projects aimed at the construction of an RFLP linkage map in a given species begin with the collection of DNA probes suitable as RFLP markers. The criteria for selecting probes are (1) their capacity to reveal clear DNA polymorphisms within the germplasm to be used for mapping, and (2) the detection of one or only a few loci per probe. While it seems attractive to map many loci simultaneously with a single probe, criterion 1 limits to some extent the use of such probes (clear RFLP patterns). Ambiguities may also arise when the probe is applied to germplasms segregating for different alleles.

Probes have been obtained from cDNA as well as genomic sequences, most of them anonymous but also some of known identity. cDNA markers have been mapped in maize (Helentjaris, 1987; Burr *et al.*, 1988), tomato (Bernatzky and Tanksley, 1986), lettuce (Landry *et al.*, 1987a), potato (Gebhardt *et al.*, 1989b), and wheat (Chao *et al.*, 1989). In different species, cDNA clones revealed more polymorphisms than genomic clones (Landry *et al.*, 1987b; Miller and Tanksley, 1990a), but frequently showed weaker hybridization signals (Helentjaris *et al.*, 1986a; Miller and Tanksley, 1990a; also our own unpublished observations in potato). Early workers in the field had difficulties in finding suitable genomic probes due to dispersed repetitive DNA present in the cloned sequences (Helentjaris *et al.*, 1985, 1986a). The situation improved when methylation-sensitive restriction enzymes were chosen to construct partial genomic libraries to be used as the source of probes (Landry and Michelmore, 1985; Burr *et al.*, 1988). The rationale is that transcribed genomic regions, enriched in low-copy-number sequences, are hypomethylated. Random clones of such undermethylated regions might therefore be enriched for single or low-copy-number sequences (Burr *et al.*, 1988). In several plant species this expectation was fulfilled when genomic *Pst*I fragments were cloned and used as probes. Depending on the species, partial genomic *Pst* libraries containing inserts in the size range from 0.5 to 3 kb were found to be enriched between 50 and 90% for low-copy-number sequences (Burr *et al.*, 1988; McCouch *et al.*, 1988; Gebhardt *et al.*, 1989b; Figdore *et al.*, 1988; Miller and Tanksley, 1990b). The fraction of repetitive clones still present can be detected by proper hybridization experiments using the total genomic DNA of the species as a probe (Landry and Michelmore, 1985). Because random *Pst* clones are easily generated and produce clear RFLP patterns, they are the preferred source of RFLP markers in plants. The rice and *Brassica oleracea* RFLP maps (McCouch *et al.*, 1988; Slocum *et al.*, 1990) and a large part of the tomato and potato RFLP maps (Tanksley and Mutschler, 1990; Gebhardt *et al.*, 1991) are based on genomic *Pst* clones used as markers. At present, there is no indication of a bias in the genome coverage caused by this type of marker.

In *Arabidopsis thaliana*, a species with an exceptionally small amount of repetitive DNA, large genomic fragments (~12.5 kb) have an increased chance of revealing DNA polymorphisms (Chang *et al.*, 1988; Nam *et al.*, 1989).

b. Efficiency of RFLP Detection The selection of cloned sequences detecting DNA polymorphisms is a trial-and-error process which can be very tedious. The efficiency of this process is mainly hindered by the limited variation in the DNA sequence existing between genotypes. In several species, before map construction, studies have been conducted to assess the degree of polymorphism available within a germplasm pool (Helentjaris *et al.*, 1985; Evola *et al.*, 1986; Havey and Muehlbauer, 1989; Landry *et al.*, 1987b; Figdore *et al.*, 1988; Gebhardt *et al.*, 1989b; McCouch *et al.*, 1988; Graner *et al.*, 1990; Apuya *et al.*, 1988; Keim *et al.*, 1989; Tingey *et al.*, 1990; Van de Ven *et al.*, 1990; Jacobs *et al.*, 1990; Chase *et al.*, 1991). Although working groups considered different numbers of genotypes, probes, and restriction enzymes and used various experimental methods and data analyses, a picture emerges from these experiments which can be summarized as follows: maize, potato, and *Brassica oleracea* and *B. campestris* show very high levels of intraspecific polymorphism, whereas tomato, soybean, *Vicia faba*, wheat, and lentil species have very low levels. An intermediate level of polymorphism is found in rice, lettuce, and barley. The reasons for these differences have been discussed, comparing maize and tomato, by Helentjaris *et al.* (1985). With data available from more plant species, it appears evident that cross-pollinated species (maize, *Brassica,* potato) have a higher level of DNA polymorphism than self-pollinated ones (tomato, soybean, lentil, barley). In maize, the action of transposable elements may also contribute to the generation of variability.

A narrow genetic base due to the selection applied in breeding crop species and/or a recent evolutionary origin (wheat) could also be the reason for finding low variability. Compared to intraspecific polymorphism, interspecific comparisons usually reveal a sufficient level of polymorphism. For the efficient construction of saturated RFLP linkage maps it was therefore necessary, in some plants, to use progenies from interspecific crosses (see Section II,A,1). Besides the choice of highly divergent parents, experimental parameters like the number and the type of restriction enzymes employed and the conditions chosen to separate the genomic restriction fragments affect the frequency with which RFLPs are detected.

When point mutations generate DNA polymorphisms, the chance of detecting them increases with the number of different restriction enzymes tested per probe. Because of the higher frequency of sites, restriction enzymes recognizing a 4-base pair (bp) sequence score more nucleotides per unit length of DNA than 6-bp cutters (Kreitman and Aquade, 1986). Therefore, they are considered more efficient in detecting point mutations.

DNA insertions or deletions generate new RFLP alleles with all enzymes having a recognition site upstream and downstream from the DNA rearrangement. No correlation is expected in such cases between the number of restriction enzymes utilized and frequency of polymorphisms found. A positive correlation, on the contrary, should be found between frequency of polymorphic RFLP loci and the mean fragment size (averaged over a series of probes) created by a specific enzyme. This was indeed found in the germplasm surveys carried out with DNA probes in maize (Helentjaris *et al.*, 1985), rice (McCouch *et al.*, 1988; Wang and Tanksley, 1989), soybean (Apuya *et al.*, 1988), tomato (Miller and Tanksley, 1990a), and barley (Graner *et al.*, 1990). The evidence suggests that the RFLPs identified in these species are mostly the result of insertions or deletions detectable with 6-bp cutter enzymes which create fragments with a size range between ~2 and 20 kb.

Small DNA rearrangements, caused, for example, by "footprints" left behind by transposable elements after transposition (Schwarz-Sommer *et al.*, 1985; Saedler and Nevers, 1985), would be difficult to detect in the fragment separation range usually employed (Southern, 1975). Fragment size differences of 5 to 10 bp could, however, still be detected in the separation range between 250 and 2000 bp on denaturing polyacrylamide gels. Two 4-bp cutter restriction enzymes were sufficient for detecting from 50 to 90% (mean 76%) informative comparisons in a germplasm pool of 38 diploid potato genotypes (Gebhardt *et al.*, 1989b).

c. Other Methods for Detecting DNA Sequence Variation   The identification of DNA sequence variation with RFLP is, although successful, laborious and expensive. In its present form, it is not a laboratory routine which can be used economically by small breeding companies. It was suggested by Beckmann (1988) that 18-bp oligonucleotides could be designed from available DNA sequence information and used as allele-specific probes revealing a point mutation by the presence or absence of an hybridization signal. With short oligonucleotides, a single base pair mismatch between the template DNA and the probe leads to the loss of interaction at the appropriate hybridization stringency applied. Size separation of genomic restriction fragments would not be required in this case. Insertion/deletion polymorphisms, which seem to be prominent in at least some plant species (Section II,A,2,b), would be detected with this method only when the oligonucleotide probe maps directly in the insertion.

Primer-directed amplification of DNA fragments by the thermostable polymerase of *Thermophilus aquaticus,* the polymerase chain reaction (PCR; Saiki *et al.,* 1988), provides a novel method by which the detection of DNA sequence variation could become a relatively simple diagnostic tool in plant breeding. Two oligonucleotides in opposite orientation are annealed to their target sequence in the genomic DNA and used to prime the exponentially increasing synthesis by Taq polymerase of the DNA between them. The specificity of the amplification reaction is determined by the sequence of the two primers. No amplification should result when any one primer scans a point mutation or is deleted in the target DNA, a situation leading to a presence or absence type of response. An insertion/deletion event in the DNA between the two primers can be recognized as a fragment size polymorphism of the amplification products. Both types of events can easily be monitored using small quantities of DNA and without the preparation of genomic Southern blots and the use of radioactive detection methods.

When the sequence information for oligonucleotide preparation is obtained from previously mapped RFLP markers, DNA polymorphisms closely linked to agronomically relevant traits (see Section II,C) can be easily detected by PCR. Vice versa, sequences stored in data banks provide information for designing primers for PCR amplification, the product of which can define a polymorphic genetic marker. This was suggested by Olson *et al.* (1989) for the human genome and by Beckmann and Soller (1990) for all eukaryotes. The use of the PCR method was further extended by J. G. K. Williams *et al.* (1990), who "invented" primers from which random DNA segments were amplified in pro- and eukaryotic species. Some of the amplified fragments behaved as dominant genetic markers and were mapped in the framework of an existing soybean RFLP map based on an interspecific cross (J. G. K. Williams *et al.,* 1990). These markers were termed RAPDs (random amplified polymorphic DNA). RAPDs were shown to be applicable to interspecific comparisons in the genus *Lycopersicon* (Klein-Lankhorst *et al.,* 1991b). It remains to be shown whether

A   **B**

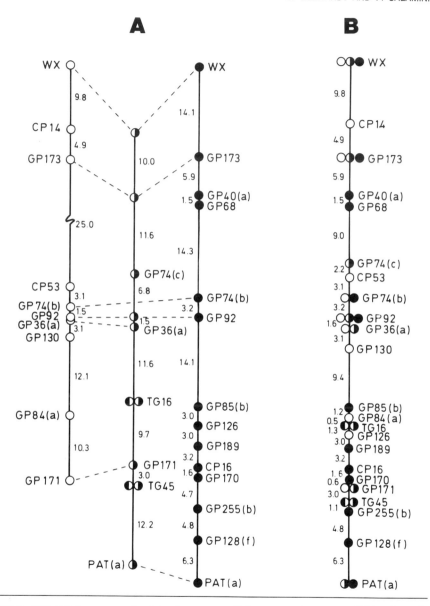

FIG. 2   RFLP linkage map of potato chromosome 8 (for nomenclature, see Gebhardt *et al.*, 1991). (A) Three linkage subgroups were obtained with RFLP alleles descending from parent P_1 (open circles), P_2 (solid circles) with an expected segregation ratio of 1 : 1 and from both parents (half open, half solid circles) with an expected segregation ratio of 3 : 1 (compare Fig. 1B). Alleles of the same RFLP locus are connected by dashed lines (Ritter *et al.*, 1990). (B) The same linkage groups as combined from the three subgroups. Recombination frequencies

RAPD markers will be useful in revealing intraspecific polymorphisms more easily than RFLPs and whether they will be informative in different genetic backgrounds.

In all cases discussed, PCR technology will have significant impact on the practical use of marker-based selection in plant breeding programs.

d. Data Processing and Linkage Analysis   RFLP maps in plants are a compendium of recombination frequencies among a large number of RFLP loci whose alleles segregate in progenies of parents having different allelic states (Section II,A,1). The size of the mapping populations in which RFLP (or RAPD) phenotypes are scored ranges from approximately 40 to 100 individuals. Linkage between any two loci is detected by a significant deviation of the experimental data from the values expected in the case of independent assortment of fragments. The degree of linkage, the recombination frequency, and the order of linked loci are estimated using the maximum likelihood equations (Fisher, 1921; Haldane and Smith, 1947; Morton, 1955). The mathematical theory of linkage was elaborated during the first half of this century (Fisher and Balmakund, 1928; Immer, 1930; Mather, 1938; Allard, 1956; Bailey, 1961), and, at present, the large amount of data generated by RFLP analysis makes the development of computer programs essential. The program LINKAGE-1 (Suiter *et al.*, 1983) was used to generate the linkage map of lettuce (Landry *et al.*, 1987a) and that of the group 7 chromosomes of wheat (Chao *et al.*, 1989). MAP-MAKER (Lander *et al.*, 1987) is now widely used for map construction in plants with F_2 segregation data (Nam *et al.*, 1989; Chang *et al.*, 1988; Hoisington and Coe, 1990). Available programs are not easily adaptable to mapping in allogamous species; because of this, alternative computation has been developed for linkage analysis in the diploid potato, including cases with distorted segregation ratios and several fragment configurations (Ritter *et al.*, 1990).

Figure 2 shows how the linkage map of chromosome 8 of potato (Gebhardt *et al.*, 1991) was compiled starting from linkage subgroups and using algorithms described in Ritter *et al.* (1990) to evaluate the recombination frequencies. As the mapping population was a backcross of a F_1 plant to one of the heterozygous parents, a maximum of three different alleles segregated per RFLP locus: one inherited from the F_1 parent (P_2), the

---

are given for each interval. Loci derived from anonymous genomic clones have GP numbers, and loci derived from anonymous cDNA clones have CP numbers. Multiple loci recognized by the same probe are indicated by the small letters in parentheses. TG16 and TG45 are reference RFLP markers of tomato obtained from S. D. Tanksley (Cornell University, Ithaca, New York; for details, see Gebhardt *et al.*, 1991). WX is the locus identified for granule-bound starch synthase and PAT is the patatin locus (see Gebhardt *et al.*, 1989b).

second from the recurrent parent P_1, and the third from both, F_1 and P_1 (see Fig. 1B). Alternatively, only one allele, either derived from F_1 or P_1 (backcross type segregation), or two alleles (F_2 type segregation) were scored per locus. As a consequence, three linkage subgroups were obtained: one for loci with alleles descending from F_1, a second from P_1, and the third for loci with alleles common to both parents. The subgroups were connected and oriented relative to each other via common alleles of three and two allelic loci and merged into a compound chromosome map (Fig. 2).

B. RFLP Maps Constructed in Plants

RFLP maps have been constructed in recent years for several plant species (summarized in Table I), including some of the world's most important crop species, e.g., maize, rice, soybean, and potato. However, the Triticeae, particularly wheat, has not yet been included in the list: this is due to its large and complex genome and the low degree of polymorphism found (Gale *et al.*, 1990). Nevertheless, a map of wheat homologous group 7 chromosomes has been published (Chao *et al.*, 1989). The most extensive maps have been developed for maize, with nearly 900 RFLP loci (Hoisington and Coe, 1990), and for tomato, with the number of loci approaching 1000 (Ganal *et al.*, 1991). For these two species, which are of worldwide importance as major crops, the availability of classical genetic maps with numerous morphological and isozyme markers, the large set of genetic material available, and the numerous polymorphisms found made them the first choice for RFLP mapping experiments in higher plants. The connection between RFLP linkage groups and the chromosomes of the existing genetic map was possible in maize by using monosomic lines and B–A translocations (Helentjaris *et al.*, 1986b; Weber and Helentjaris, 1989). Identified primary trisomics and isozyme loci were used in rice and tomato, respectively, to coordinate cytogenetic and RFLP maps (Young *et al.*, 1987; Bernatzky and Tanksley, 1986; McCouch *et al.*, 1988). The RFLP maps of *Arabidopsis thaliana* were related to the established genetic map by using morphological markers (Chang *et al.*, 1988; Nam *et al.*, 1989). In species like potato, soybean, and *Brassica oleracea,* in which comparatively few genetic data were available, the new RFLP maps provided the first extensive genetic analysis of the genome (Bonierbale *et al.*, 1988; Gebhardt *et al.*, 1989b; 1991; Tingey *et al.*, 1990; Slocum *et al.*, 1990).

As the genomes of *Solanum* and *Lycopersicon* species are largely homologous (Bonierbale *et al.*, 1988), the newly constructed potato RFLP maps were related to the established genetic map of the tomato by mapping

TABLE I

Linkage Maps in Plants Obtained with Molecular Markers (RFLP and Isozyme Markers)

Species	Crosses for mapping	Mapping population[a]	Number of loci mapped	References
Tomato (*Lycopersicon esculentum*)	*L. esculentum* × *L. pennellii*	46 F_2 46 BC	>350	Bernatzky and Tanksley (1986); Young *et al.* (1987); Zamir and Tanksley (1988); Tanksley and Mutschler (1990)
	L. esculentum × *L. hirsutum*	50 F_2	104	Helentjaris *et al.* (1986a)
Maize (*Zea mays*)	H427 × 761 (inbred lines)	46 F_2	338	Helentjaris *et al.* (1986a); Helentjaris (1987)
	T232 × CM37 CO159 × Tx303 (inbred lines)	46 RI 38 RI	334	Burr *et al.* (1988), Burr and Burr (1991)
	CO159 × Tx303 (inbred lines)	46 F_2	262	Hoisington and Coe (1990)
Lettuce (*Lactuca sativa*)	cv. Calmar × cv. Kordaat	66 F_2	46	Landry *et al.* (1987a)
Rice (*Oryza sativa*)	IR 34583 (indica type) × Bulu Dalam (javanica type)	50 F_2	123	McCouch *et al.* (1988)
Potato (*S. tuberosum*)	*S. phureja* × *S. tuberosum* × *S. chacoense*)	65 F_1	134	Bonierbale *et al.* (1988)
	S. tuberosum × *S. tuberosum*	67 BC	~300	Gebhardt *et al.* (1989b, 1991)
Pepper (*Capsicum* species)	*C. annuum* × *C. chinense*	46 BC	85	Tanksley *et al.* (1988)
Arabidopsis thaliana	Niederzenz (Nd-0) × Columbia (C), Niederzenz (Nd-0) × Landsberg (La-0) (ecotypes)	? F_2	90	Chang *et al.* (1988)
	Landsberg erecta (La) × Columbia (Col-0)	118 F_2	94	Nam *et al.* (1989)
Brassica oleracea	Broccoli cv. Packman × cabbage cv. Wisconsin Golden Acres	96 F_2	258	Slocum *et al.* (1990)

(*continued*)

TABLE I *(continued)*

Species	Crosses for mapping	Mapping population[a]	Number of loci mapped	References
Soybean (*Glycine max.*)	*G. maximum* × *G. soja*	68 F$_2$	~550	Tingey *et al.* (1990)
Barley (*Hordeum vulgare*)	cv. Proctor × cv. Nudinka	91 DH	155	Heun *et al.* (1991)
	cv. Igri × cv. Franka	71 DH	215	Graner *et al.* (1991)
	cv. Vada × *H. spontaneum*	135 F$_2$		
Cuphea lanceolata	LN43/1xLN68/1 (inbred lines)	140 F$_2$	37	Webb *et al.* (1991)

a BC, Backcross; RI, recombinant inbred lines; DH, doubled haploid lines.

chromosome-specific tomato markers in potato and vice versa (Gebhardt *et al.*, 1991).

C. The Use of RFLP Linkage Maps for Localizing Agronomic Traits

In principle, an RFLP linkage map is not essential for the identification of RFLP markers linked to a single locus trait. When genetic material is available which has been specifically bred for introgression of a single trait, as are nearly isogenic lines (NILS) (see Section II,C,1), then the screening of this material with large numbers of markers of unknown map position will result in the finding of linked markers by trial and error (Young *et al.*, 1988; Hinze *et al.*, 1991). An RFLP linkage map offers the possibility of systematically screening the genome with a minimum number of markers for cosegregation with one or more traits of interest of qualitative as well as quantitative inheritance. As the plant breeding process is aimed at the simultaneous improvement of several agronomic characters, integrated knowledge of the genomic positions of genes influencing those characters will be very valuable. In order to gain this knowledge, it is of practical importance that the RFLP linkage map, which is normally constructed with only two genetically distant parental lines, provides information on different genetic backgrounds carrying valuable agronomic traits. In crop species with low intraspecific variability, the use of the RFLP maps in breeding could therefore be limited to the introgression of traits from more distantly related wild species.

After the first phase of RFLP construction as reviewed in Sections II,A and B, the second phase, i.e., using the information for the mapping of genes influencing agronomic traits, is currently being pursued. Several single loci conferring resistance to plant pathogens, as well as QTLs, have so far been mapped but their number will certainly increase during the next few years.

1. Tagging of Monogenic Traits with RFLP Markers

In inbreeding species, the availability of nearly isogenic lines (NILs) facili-tated the detection of RFLP markers linked to monogenic disease resis-tance genes. NILs are obtained by crossing a donor parent carrying the resistance gene with a susceptible recipient parent, followed by repeated backcrosses to the recipient parent under continuous selection for the resistance trait. The donor genome is therefore progressively diluted in the introgressed lines until, in the ideal situation, only a short chromosomal segment is retained around the resistance gene. When the parents and the NILs derived from them are compared with molecular probes, in the NILs only those that are linked to the resistance locus are expected to be polymorphic. The linkage and the genetic distance between RFLP markers and the resistance locus are then verified by segregation analysis in normal mapping populations.

Using the appropriate NILs and testing 122 probes at random, Young et al. (1988) identified two RFLP markers tightly linked to the Tm-2a locus on chromosome 9 of tomato which confers resistance to tobacco mosaic virus. A similar approach was used by Hinze et al. (1991) for tagging the ml-o resistance locus on the barley chromosome 4 which is active against the fungus Erysiphe graminis f.sp. hordein (powdery mildew). Four out of approximately 1100 markers tested were finally shown to be linked to the resistance gene. Instead of NILs, Jung et al. (1990) used addition lines of Beta vulgaris carrying a small chromosomal fragment of the wild species Beta procumbens known to contain a resistance gene to the nematode Heterodera schachtii. In selecting for DNA sequences specific for Beta procumbens, they were able to identify markers indicative for the presence of the resistance gene.

In a few other cases, the chromosome allocation of the resistance locus was known from the established genetic maps. Knowing the position of the Fusarium oxysporum resistance gene I2 on chromosome 11 of tomato reduced to 14 the number of chromosome 11 RFLP markers which were used on NILs. One of them was found to be closely associated with the resistance gene (Sarfatti et al., 1989). The association was not, however, confirmed by segregation analysis. Similarly, four mapped RFLP markers of maize chromosome 6 were sufficient to map more precisely in backcross populations a major gene for resistance to maize dwarf mosaic virus (McMullen and Louie, 1989). Tomato introgression lines nearly isogenic for the gene *Mi* conferring resistance to the root knot nematode *Meloidogyne incognita*, known to be located on tomato chromosome 6, were analyzed with 21 tomato RFLP markers of chromosome 6 and with 7 potato markers of the homologous potato chromosome 6. One potato RFLP marker was found to be highly diagnostic for the *Mi* resistance allele (Klein-Lankhorst *et al.*, 1991a). The latter three examples show how the integration of classical linkage maps with RFLP maps, as well as the integration of RFLP maps of homologous genomes like potato and tomato (Bonierbale *et al.*, 1988; Gebhardt *et al.*, 1991), can help in the efficient tagging of resistance genes with RFLP markers.

In many cases, however, the chromosomal allocation of a resistance gene is not known. The RFLP linkage map is then used to select a set of equidistant informative markers covering all chromosomes. NILs or progenies segregating for the resistance gene are tested with the marker set for polymorphisms (in the case of NILs) or for cosegregation with the resistance allele indicating linkage (an example is given in Fig. 3). The latter approach was used for mapping the resistance locus *Gro1* against the root cyst nematode *Globodera rostochiensis* onto chromosome 7 (Barone *et al.*, 1990) and the locus *Rx1* conferring extreme resistance to potato virus X onto chromosome 12 of potato (Ritter *et al.*, 1991). A second resistance locus *Rx2* having the same phenotype as *Rx1* was identified on chromosome 5 of potato via a short cut: knowing that many resistance genes in potato were introduced from wild *Solanum* species, it was assumed that rare RFLP alleles within a germplasm pool of *S. tuberosum* breeding lines might indicate the presence of "foreign" chromosomal segments introgressed together with the resistance gene and maintained during the breeding process. Indeed, the first marker of this type tested was linked to a PVX resistance allele (Fig. 3; Ritter *et al.*, 1991; Debener *et al.*, 1991). With the statistical approach, assuming a genome length of 1000 centiMorgan (cM) and a 20cM distance between marker loci, approximately 50 informative markers would have to be assayed instead of one.

Another resistance locus, *I1*, conferring resistance to race 1 of the fungus *Fusarium oxysporum* f.sp. *lycopersici*, was identified on chromosome 7 of tomato (Sarfatti *et al.*, 1991), in the same region homologous to potato chromosome 7 to which the nematode resistance locus *Gro1* has been mapped (Barone *et al.*, 1990). A similar observation was made for the virus resistance locus *Rx2* (Ritter *et al.*, 1991) and a locus conferring race-specific resistance to *Phytophthora infestans* (unpublished results from this laboratory) which map to the same chromosomal region in

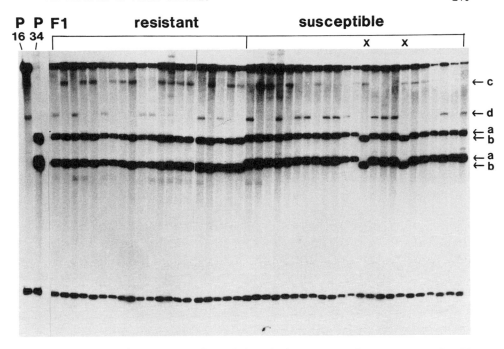

FIG. 3   Cosegregation between the locus *Rx2* conferring extreme resistance to potato virus X and the potato RFLP marker GP21. A Southern blot was prepared with genomic DNA of the susceptible (P16) and the resistant (P34) parents, and 20 resistant and 23 susceptible F_1 progeny. The blot was hybridized to the marker probe GP21. The RFLP alleles *a* and *b* descended from P34, the alleles *c* and *d* from P16. Allele *b* was linked to the resistance locus. Two recombinants are indicated with x. Reproduced with permission from Ritter *et al.* (1991).

potato. It may be speculated from coincidences like this that molecular disease resistance mechanisms might turn out to be based on common mechanisms for different plant pathogens.

The RFLP map constructed in rice has, at last, been utilized together with a set of NILs for mapping two blast resistance loci (*Pyricularia oryzae* Cav.) to rice chromosomes 6 and 12 (Yu *et al.*, 1991).

The first genetic locus not involving a pathogen resistance trait but controlling the supernodulation phenotype in soybean was tagged with RFLP markers by Landau-Ellis *et al.* (1991).

2. Quantitative Trait Loci

Several characters of plant species, among which are traits of agronomic importance, are inherited quantitatively. This type of genetic variation is due to multiple factors acting collectively on the expression of a trait (Nilson-Ehle, 1909; Johannsen, 1909). In recent years, these loci have been designated with the acronym QTL (for *q*uantitative *t*rait *l*ocus; Geldermann, 1975). The essential feature which makes feasible the finding and characterization of a QTL is its linkage with a known marker locus segregating with Mendelian ratios. Moreover, it is necessary that, in the population considered, a high level of linkage disequilibrium exists between markers and QTLs; that is, not only alternative states of the marker locus must segregate, but also the linked QTL.

Sax (1923) was able to reveal the presence of a QTL in *Phaseolus vulgaris* following its linkage with a gene controlling the coat color of the seed; an analogous approach was adopted by Everson and Schaller (1955) which associated yield differences to awn barbing in barley. The use of morphological markers to locate QTLs has, however, serious drawbacks: frequently, this class of chromosomal markers induces at the phenotypic level strong pleiotropic changes which mimic the presence of QTLs linked to the marker locus. The situation improved when phenotypically more neutral markers became available. These were first recognized in isozymic variants whose loci can easily be mapped to chromosomes.

a. QTL Mapping: The Traditional Approach   In maize, the selection for increased grain yield was found to modify the allelic frequencies at 8 isozyme loci (Stuber *et al.*, 1980). Vice versa, the experimental modification by selection of allelic frequencies at 7 isozyme loci increased significantly the yield of grain (Stuber *et al.*, 1982). Both experiments were interpreted as supporting linkage between isozymic loci and QTLs for yield. Tanksley *et al.* (1982) used 12 isozyme loci to locate QTLs influencing 4 quantitative characters in a *Lycopersicon esculentum*[2] × *L. pennellii* backcross population. The 48 comparisons between marker loci and quantitative trait expression frequently proved significant, with a minimum of 5 QTLs detected per trait and 21 QTLs mapped to chromosomes. In this experiment, two pairs of linked enzyme markers made possible the application of a three-point mapping procedure: data for three QTLs fitted predictions for a QTL closer to one marker than to the other.

At the end of the 1970s, a number of isozyme marker loci, available together with proper biometrical procedures establishing and interpreting the associations of QTLs and marker loci (Jayakar, 1970; Mather and Jinks, 1971; McMillan and Robertson, 1974; Soller and Brody, 1976; Tanksley *et al.*, 1982; Soller and Beckmann, 1983), made possible more detailed studies on number, genomic distribution, and type of gene action of QTLs. Two contributions (Stuber *et al.*, 1987; Edwards *et al.*, 1987) provided further biometrical refinements of what later has been referred to as the "traditional approach" to QTL analysis (Lander and Botstein, 1989).

They also represented the highest level of contribution of isozyme research to theoretical breeding of an important crop such as maize.

In the experiment of Edwards et $al.$ (1987), two F_2 maize populations derived from crosses among inbred lines were analyzed with 17 and 20, respectively, segregating marker loci. Forty characters were measured, resulting in the construction of 82 plant traits. The statistical method adopted was a simple factor analysis of variance of each pairwise combination of a quantitative trait and a marker locus. When significant variation was associated with differences in genotypic classes of a marker locus, a QTL was supposed to exist on the chromosomal region marked. Details of this study were as follows. In the F_1 a marker is heterozygous (M/m); also, a QTL linked to it by r (recombination frequency) may be heterozygous (Q/q). F_2 progenies will consist of nine genotypic classes with frequencies expressed as functions of r. Genotypes at the QTL, however, cannot be discriminated and their effect is only calculated via association with the genotype at the marker locus M. When $r = 0$, the mean expression for MM, Mm, and mm equals the assigned values of QTL genotypes $+a$, d, and $-a$, respectively (a is the additive and d the dominance effect at the QTL). If genotypes $QQMM$, $QQMm$, $QQmm$ have a value of $+a$, $qqMM$, $qqMm$, and $qqmm$ have a value of $-a$, and $QqMM$, $QqMm$, and $Qqmm$ have a value of d, the expressions for resolving additive and dominance effects (knowing the mean expression for the QTL of marker classes MM, Mm, and mm) are additivity = $(MM - mm)/2 = a(1 - 2r)$ and dominance $= Mm - (MM + mm)/2 = d(1 - 2r)^2$. The dominance/additive ratio has the value $(1 - 2r)d/a$. The ratio among effects of dominance is, however, biased at high values of r.

When using this model, failure to detect a QTL by a marker does not mean that the corresponding chromosomal region is not hosting such a locus: the two inbreds, in fact, may have identical alleles at a QTL, or different alleles with equivalent expression of the trait. Also, effects of more than one but linked QTLs cannot be excluded; the existence of such cases has, moreover, an influence on the estimation of the type of gene action at the QTL. The results of Edwards et $al.$ (1987) indicate that, for each of the traits, QTLs were detected and allocated to genomic sites. Most of the associations between a QTL and a marker were significant. The cumulative single effects of marker-linked regions explained between 8 and 40% of phenotypic variation. Single loci accounted for 0.3–16% of the phenotypic variation of traits. The overdominance type of gene action appeared frequently; this may reflect the effects of multiple QTLs within the chromosomal regions next to the markers. Digenic epistasis was found to be of no importance in the expression of the traits considered.

In an accompanying paper, the same authors (Stuber et $al.$, 1987) considered maize yield and yield-related traits and found that about two-thirds of the associations among isozyme loci and 25 QTLs were significant. The proportion of variation accounted for by QTLs associated to marker loci varied from 1 to 11%. Occasionally, differences between the means of the *MM* and *mm* classes were more than 16% of the population mean. For grain yield the gene action was predominantly dominant or overdominant. It was additive for several ear characteristics.

b. QTL Mapping: The Advent of RFLP   With the advent of RFLP, markers as suitable as isozymes in QTL studies, but unlimited in number, became available. It was soon realized that more precise methods of QTL mapping were necessary to exploit the new tool at the basic and applied level. In this respect, Soller and Beckmann (1983) introduced a clear distinction between what can be done in selfing or in outcrossing species.

In selfers, the "traditional approach" used in isoenzyme studies can be applied fruitfully. Estimates of main effect and relative dominance of the QTLs are, however, attenuated by the value of *r* between marker loci and QTLs. When QTL traits are bracketed between two markers, the situation improves (Thoday, 1961). Better estimates of existence, location, and characteristics of QTLs in self-pollinated species have been developed (Lander and Botstein, 1989). Lander and Botstein have elaborated a method for mapping QTLs using RFLP maps that was aimed at identifying promising crosses for QTL mapping, exploiting the full power of complete linkage maps by the approach of interval mapping of QTLs, and decreasing the number of progenies to be genotyped. For a description of the method, which can be applied to a backcross generation of pure lines, the reader has to consult the original publication; in this review, we simply underline its basic assumptions.

When the estimated phenotypic effect *b* of a single allele substitution at a putative QTL is known, maximum likelihood equations (MLEs) can be calculated which maximize the probability that the observed data will have occurred. These MLEs are compared to MLEs obtained under the assumptions that no QTL is linked ($b = 0$), and the evidence that a QTL is existing is indicated by the LOD score. The LOD score is the ratio of the probability that the data have arisen assuming the presence of a QTL to the probability that no QTL is present. If the LOD score exceeds a predetermined threshold T, a QTL is accepted as present. Restricting the study to the effects at a single marker locus performed in a backcross population of inbred lines, and allowing for a 5% error rate (Soller and Brody, 1976), the threshold is $T = 0.83$. Moreover, for a QTL contributing σ^2 exp to the backcross variance, the expected LOD score per progeny (ELOD) is equal to $0.22(\sigma^2 \exp/\sigma^2 \mathrm{res})$, where σ^2 exp is the variance explained by the QTL and σ^2 res is the residual variance. This formula (similar to the one known from the traditional method) is useful to calculate the number of progenies required if the LOD score is expected to exceed T.

The method can be extended to more complex situations: if markers are available distributed along all the genome, the MLE equations can generate estimates of phenotypic effects and LOD scores for a QTL at any chromosomal location. The solution of appropriate likelihood functions using a MAPMAKER-QTL computer program (Lincoln and Lander, unpublished; Lander and Botstein, 1989) leads to LOD score values usable in a QTL likelihood map and showing how the LOD score varies through the genome. The method also describes how to increase the power of QTL mapping by selective genotyping of the extreme progenies of a population for the expression of a trait, and by decreasing the environmental variance via progeny testing. The method proposed by Lander and Botstein (1989) was applied to resolve quantitative traits into discrete Mendelian factors in tomato (Paterson *et al.*, 1988). Tomato chromosomes six and twelve are shown as examples for this analysis in Fig. 4. At least six QTLs were mapped controlling fruit mass, four QTLs were mapped for soluble solids content, and five QTLs for fruit pH in a backcross of *Lycopersicon esculentum*[2] and *L. chmielewskii* as parents.

The mapping of QTLs already localized to chromosomal regions can be improved by introducing an approach based on selected overlapping recombinant chromosomes (Paterson *et al.*, 1990). This "substitution mapping" makes use of an RFLP which is located near the QTL. Several chromosome segments are then identified which have this RFLP, and their overlapping regions are revealed by the use of additional markers. Phenotypic effects of each chromosome segment are then established in segregating populations, and effects specific for one segment are attributed to the QTLs mapping on that segment. Six genetic stocks were developed carrying chromosome segments of *L. chmielewskii* in the background of *L. esculentum*. The stocks were heterozygous for two to four chromosome fragments hosting QTLs increasing soluble solid concentration. The effects of the different genetic situations were analyzed in the field; it was assessed that QTLs affecting the high sugar phenotype(s) lie in intervals of as little as 3 cM. This case illustrates well how RFLPs can assist gene introgression from wild to cultivated species and, in general, how the gap between QTLs and linkage maps can be reduced.

In self-pollinating species, besides the cases presented here in detail, association of RFLP and QTLs has been shown for water use efficiency in tomato (Martin *et al.*, 1989) and for soluble solids in the already cited interspecific cross of *L. esculentum*[2] $\times$ *L. chmielewskii* (Osborn *et al.*, 1987; Tanksley and Hewitt, 1988). Similar methods have been used to map a QTL for plant height and thermotolerance in maize (Helentjaris, 1987;

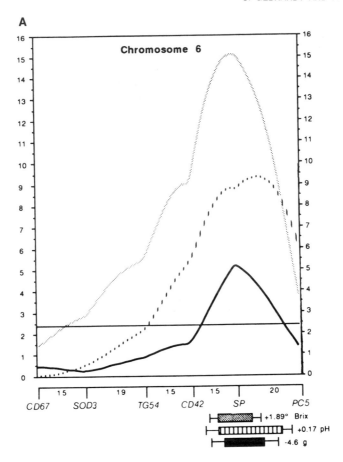

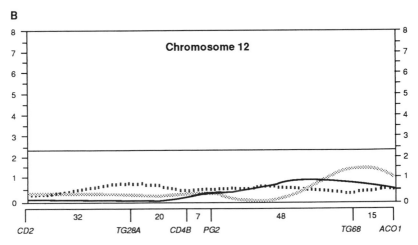

Ottaviano *et al.*, 1991), an allogamous species where the availability of inbred lines allows the same type of approach to QTL analysis as in self-pollinating species.

In outcrossers, the use of molecular markers for genetic analysis of QTLs is more problematic than in self-fertilizing plants (Soller and Beckmann, 1983). In such populations, under linkage equilibrium the sign of the QTL allele associated with the marker allele will be different in different homologous chromosomes. Soller and Beckmann (1983) summarize the possible approach to the solution of this problem, which depends critically, however, on the total number of polymorphic markers available and on large F_1 and F_2 populations. A valid alternative can be, in such cases, to score with DNA markers only individuals showing extreme phenotypes for the trait of interest.

c. Applications of QTL Mapping   QTL mapping in crop species has obvious practical applications. Two groups of species are separately discussed by Soller and Beckmann (1983). The selfers, which have a small genomic size (~10 Morgans), do not require many polymorphisms per chromosome; the markers can be used in the frame of already well-described breeding schemes. Among the applications suggested for such crops, the selection of superior pure lines out of a cross is a particularly attractive case. The analysis of the progeny of a cross showing hybrid vigor may lead to the identification of QTLs for heterosis and to the characterization of type and magnitude of genetic effects involved. New inbred lines can then be screened out of the cross for the presence of QTLs which are additional, in terms of number and effects, to those of each single parent.

In outcrossers (genome size longer than 30 Morgans) marker-assisted methods are not of general applicability (Soller and Beckmann, 1983). They can, however, also help in these species by identifying those chromosomal segments which have a relevant joint contribution to phenotypic effects. Biometric equations defining expected progress for RFLP-assisted selection in outcrossers are presented by Soller and Beckmann (1983).

FIG. 4   QTL likelihood maps for tomato chromosomes 6 (A) and 12 (B) indicating lod scores for fruit mass (solid lines and bars), soluble-solids concentration (dotted lines and bars), and pH (hatched lines and bars). The RFLP linkage map used in the analysis is presented along the abscissa. The height of the curves indicates the strength of the evidence ($\log_{10}$ of the odds ratio) for the presence of a QTL at each location. The lod scores were determined according to the method of QTL mapping developed by Lander and Botstein (1989). The horizontal line at a height of 2.4 indicates the stringent threshold that the lod score must cross to allow the presence of a QTL to be inferred. Therefore, on chromosome 6, a QTL was localized for all three phenotypes measured, whereas on chromosome 12, no QTL was detected. Reprinted by permission from Paterson *et al.* (1988). Copyright 1988 Macmillan Magazines Ltd.

The efficiency of *marker-assisted* selection (MAS) in the improvement of traits by QTL determination has been treated and critically discussed by Lande and Thompson (1990). The authors derive selection indices maximizing the rate of genetic gain under different MAS schemes. They also analyze limitations to the efficiency of MAS, including problems in the detections of QTL linkage with marker loci or due to sampling errors in estimating selection indices. Integration of MAS with phenotypic information depends on the heritability of the characters selected, on the additive genetic variance associable to the marker locus, and on the selection schemes. Among the practical constraints which limit the potential utility of MAS are the number of markers necessary to detect linkage disequilibria, the sample size needed to detect QTLs for low heritability traits, and the sampling errors in the estimation of weights of components in selection indices. The number of markers will possibly be reduced in the future, at least in selected cases: three repetitive probes of mouse, for instance, collectively identify 28 loci dispersed in 16 of the 19 autosomes (Siracusa *et al.*, 1991). Similar probes have been found in potato which have the capacity to reveal more than 9 loci mapping in the 12 chromosomes of this plant (Gebhardt *et al.*, 1989a; Görg *et al.*, 1991).

III. RFLP Analysis of Plant Genomes

When an RFLP map is available, the genetic architecture of a plant genome is amenable to detailed analysis that is impossible to carry out with only morphological or cytological markers. This has been demonstrated in several cases. Relevant to backcross breeding, for example, is the possibility of minimizing the number of backcrosses necessary for reducing the donor parent geome to a small segment around the introgressed locus, avoiding the introgression of undesirable genes from the donor. Young and Tanksley (1989) have shown with RFLP markers that tomato lines introgressed for the *Tm-2* gene of *Lycopersicon peruvianum* retained between 4 and 51 cM of the donor genome after up to 20 backcross cycles. They proposed to monitor, during backcrossing, the recombination around the locus of interest with closely linked RFLP markers, therefore increasing the efficiency of selection by an estimated factor of 50.

The hybrid nature of plants regenerated after protoplast fusion can also be demonstrated with RFLPs by the combined occurrence in the hybrid of restriction fragments specific to the two fused genomes (C. E. Williams *et al.*, 1990). The fate in their sexual progeny of individual chromosomes of somatic hybrids or of addition lines was followed using markers of known map position (C. E. Williams *et al.*, 1990; Hu and Quiros, 1991). The approach permitted the identification and mapping of deletions. Somaclonal variation induced during tissue culture was detected as changed RFLP patterns in regenerated plants when compared to the original genotype (Brown *et al.*, 1991; Müller *et al.*, 1990).

Plants derived from *in vitro* anther culture may have a different genetic architecture depending on their cellular origin. This is an important component of breeding schemes based on doubled haploid lines (DHs; see Section II,A,1). Haploids and DHs originating from reduced microspores are truly homozygous, whereas the diploid products of first and second division restitution (FDR and SDR) microspore nuclei give rise to 2n plants heterozygous to different degrees (Hermsen, 1984). Also, *in vitro* regenerated plants derived from somatic tissue and not from microspores are heterozygous, having the paternal genotype. Monitoring in the progeny RFLP loci which were heterozygous in the parent, Rivard *et al.* (1989) could classify the plants produced by *in vitro* anther culture of *Solanum chacoense* according to their origin from reduced microspores, FDR microspores, or somatic tissue.

In constructing an RFLP linkage map and depending on the species under study, a certain percentage of duplicated RFLP loci is usually observed. In maize, 29% of all probes analyzed detected duplicated loci (Helentjaris *et al.*, 1988). Determining their map positions, it was found that groups of linked duplicated loci have retained their linkage relationship after the duplication event. These observations indicated that, during the evolution of maize, large chromosomal segments have been duplicated, a finding which supports an old polyploid origin of this species, as already suggested by earlier cytological observations (cited in Helentjaris *et al.*, 1986a). The duplicated chromosome segments are not, however, arranged in five pairs of homologous chromosomes. Helentjaris *et al.* (1988) concluded that either extensive rearrangements have taken place after polyploidization or, as an alternative explanation, that the present genome composition is the result of interchromosomal duplications. Similar conclusions were drawn in considering the RFLP linkage map of *Brassica oleracea*, in which at least 35% of the loci are duplicated (Slocum *et al.*, 1990). Contrasting examples are tomato and potato: RFLP loci duplications are also found, but with little evidence of an extensive duplication of groups of linked loci (Bernatzky and Tanksley, 1986; Gebhardt *et al.*, 1989b, 1991). Therefore, in view of the available RFLP data, an allopolyploid structure of the 12 chromosomes of potato and tomato originating from ancestors with 6 chromosomes seems highly unlikely.

Using the same set of chromosome-specific RFLP markers, the construction and comparison of RFLP maps in related but sexually incompatible species offer the unique opportunity of comparing plant genomes in terms of linkage arrangements between homologous loci. The first examples of such a comparison in plants are the RFLP maps of tomato, pepper, and potato, all Solanaceae species, which were constructed with tomato markers (Bernatzky and Tanksley, 1986; Tanksley *et al.*, 1988; Bonierbale *et al.*, 1988). The basic chromosome number of the three species is 12. The DNA content is similar for tomato and potato and fourfold higher in pepper than in tomato and potato. The genomes of tomato and potato are largely homologous (Bonierbale *et al.*, 1988). The order of the loci tested is almost identical in the two species, with the exception of four paracentric inversions in three chromosomes. In contrast, extensive chromosome rearrangements were observed in pepper as compared to tomato (Tanksley *et al.*, 1988). The longest linkage block conserved in pepper was a 63 cM piece of tomato chromosome 2. The gene repertoire, however, was highly conserved between all three species, as nearly all tomato probes cross-hybridized to potato as well as pepper sequences. A small proportion of markers also showed differences in copy number (see also Gebhardt *et al.*, 1991). Further comparisons between maps derived from different crosses of potato revealed no differences in order of genes but differences in map length, depending of whether an inter- or intraspecific cross was used (Bonierbale *et al.*, 1988; Gebhardt *et al.*, 1991).

IV. Measurement of Genetic Distances with RFLP

A. Assessment of Genetic Distances within Species and Their Relationship to Heterosis

One of the applications of RFLP analysis consists of the assessment of the genetic variation within and between natural populations or varieties of crop species. Several studies already available indicate the existence of RFLP variants whose frequencies can be used to calculate genetic distances. A discussion has been provided in this review (see Section II,A,2,b) that outlines that cross-pollinated species show, a higher degree of intraspecific polymorphism than inbreeders. An evaluation of genetic distances within the self-pollinating species rice, however, shows that RFLP polymorphisms are also present within varieties, mainly in the form of homozygous variant alleles (Wang and Tanksley, 1989). The cited study indicates that, based on genetic distance calculations in rice, the ratio of genetic variation between versus within varieties is near 12 : 1.

These studies are also interesting when applied to cross-breeders. Here, one of the main objectives of RFLP analysis is the prediction of superior heterotic combinations within a set of genotypes. For this purpose, the calculation of genetic distances among genotypes has a central significance. Methods have been developed to transform RFLP data into indices of pairwise similarity (or dissimilarity). Detailed analyses have been carried out in maize, a species for which both heterotic patterns of combination and pedigrees of inbred lines are available (Smith *et al.*, 1990). The final aim is (1) the estimation of genetic similarities between pairs of inbred lines, (2) the comparison of genetic similarities based on RFLP with pedigree associations, and (3) the examination of the association between genetic diversities and grain yield heterosis.

In the experiment described by Smith *et al.* (1990), 37 inbred lines with clear pedigree relationships and with known performances in crosses were considered. The DNAs of the lines were tested with 257 probe–enzyme combinations and an index of similarity between pairs of lines was established. Lines were also clustered, based on pedigree relationships. Regression analyses were carried out between genetic distance indexes and F_1 yield and grain heterosis. A cluster diagram based on RFLP data grouped the lines into families consistent with their breeding history. Moreover, the distances based on DNA data had a higher correlation with yield performances and grain yield heterosis than any other measure of genetic similarity. The authors express their belief that RFLP measures of similarity, in combination with the knowledge of pedigrees and with the use of DNA markers capable of locating QTLs on chromosomes, will, in future, allow maize breeders to predict high-yielding hybrid pedigrees.

Similar experiments were carried out by Lee *et al.* (1989) and Melchinger *et al.* (1990b). Here, three sets of maize inbred lines were studied. The genetic distances based on RFLP data among all possible pairs of inbreds were estimated from the modified Rogers' distance (MRD; Rogers, 1972). In the first set of lines, grain yield and specific combining ability effects for this trait were significantly correlated with the MRD. Again, the genetic similarities of inbreds based on their RFLP profiles were consistent with expectations based on known pedigrees. The conclusions put forward by Lee *et al.* (1989), however, were to some extent different from those reached by Smith *et al.* (1990). Lee *et al.* (1989) support the use of RFLP analysis only as a potential alternative to field testing when the target is the assignment of inbred lines to heterotic groups. The analysis of the other two sets of inbred lines again supported the use of RFLPs in investigations of relationships among maize inbreds. Molecular probes, on the contrary, were credited with limited usefulness for predicting the heterotic performances of single crosses between unrelated lines (Melchinger *et al.*, 1990b). This last conclusion also emerges from further studies published on this subject which mainly considered lines belonging to different heterotic patterns (Melchinger *et al.*, 1990a; Godshalk *et al.*, 1990).

As evident from our discussion, the relevance of RFLP analysis for predicting high-yielding pedigrees still has to wait a more definitive assessment. We emphasize, nevertheless, that the positive aptitude toward this possibility expressed by Smith *et al.* (1990) was based on the use of a special type of RFLP probes for the prediction, that is, of markers linked to known QTLs.

A specific use of genetic distances based on RFLPs is the finding, within the available gene pool of a species, of germplasm introgressed from alien sources. This situation was investigated by Debener *et al.* (1991) for diploid and tetraploid clones of the cultivated potato (*Solanum tuberosum* ssp. *tuberosum*). Starting at the beginning of this century, the germplasm of this species was modified by breeders who introgressed genes from wild species and from cultivated forms of *Solanum tuberosum* ssp. *andigena* and *stenotomum*. Several potato clones bred in different eras and accessions of wild and cultivated relatives of potato were compared. RFLP data collected from many restriction enzymes and more than one probe for the same genomic region were used for the computation of locus-specific phenograms based on distance matrix methods. Several potato lines deviated significantly from the cluster represented by *S. tuberosum* ssp. *tuberosum* and its close relatives, indicating the presence of exotic germplasm at particular loci. Detecting such genomic regions has implications for the mapping of agronomic traits introgressed from wild germplasm which may still be linked to foreign chromosomal fragments.

B. Genetic Distance Among Species

Plant phylogenies can be studied by comparative analysis of DNA sequences of the same gene isolated from different species. This, however, provides information on genes which are to some extent arbitrarily considered as descriptors of genomes (Felsenstein, 1988; Nei, 1987). Up to now, the sequence data provided for the same gene cloned from several species are, in fact, limited. Moreover, evolution of single genes may not correspond to that of genomes (Nei, 1987). As an alternative, RFLP data can be accummulated from several genotypes for computing genetic distances among closely or loosely related species.

An example of such an approach to plant systematics is the genus *Brassica*. For the diploid species, RFLP data from 38 populations tested with 30 random RFLP probes indicate that two basic evolutionary pathways exist (Song *et al.*, 1988a, 1990). One includes *Brassica rapa* (syn. *campestris*) and *B. oleracea;* the second includes *B. nigra* together with *Raphanus* sp. and *Synapis* sp. The data support the possibility of tracing the cultivated forms back to specific wild relatives. In the context of such studies (Song *et al.*, 1988b), it was also established that, for *B. rapa,* two centers of diversity exist, one in central Europe and the second in South China. The hybrid origin of some diploid or amphidiploid *Brassica* species was also assessed using RFLP-based genetic distances. *Raphanus sativus* was found to derive from crosses between *B. nigra* and a species related to *B. oleracea* (Song *et al.*, 1990). Possible ancestors for the amphidiploid species *B. juncea*, *B. carinata*, and *B. napus* were also identified (Song *et al.*, 1988a). The work done with the genus *Brassica* indicates that RFLP data match quite well with other cytological or morphological criteria used in species taxonomy. They not only provide hypotheses on phylogeny, but also can assess the relative time of emergence of species: *B. oleracea*, for example, is, based on RFLP data, evolutionarily older than *B. rapa* (Song *et al.*, 1988a).

Within the Solanaceae, tomato and potato taxonomies have been studied by computing genetic distances with RFLP. In tomato (Miller and Tanksley, 1990b), 156 accessions belonging to 8 species have been analyzed using 40 DNA probes. The study confirmed the current taxonomical classification. One accession of *Lycopersicon peruvianum* was, however, sufficiently distinct to be considered as a separate species; this highlights the potential of molecular probes to reveal errors in species assignment or to support the putative existence of new species. *L. esculentum* and *L. pimpinellifolium* were the two species less clearly differentiated and this may, in part, reflect recent introgressive hybridizations. Two major dichotomies were found in grouping the 8 *Lycopersicon* species: one corresponds to the division between self-incompatible and self-compatible species, the second between red and green fruit color. The authors interpret their RFLP data suggesting that color and style–pollen compatibility reactions are monophyletic in origin.

In potato, with two sets of 18 and 12 DNA probes, the phylogenetic relationships among 14 wild and 3 cultivated species were studied (Debener *et al.*, 1990). Phenetic trees were obtained where the reliability of tree topology increased with the number of polymorphic fragments scored. At least four main tree branches were found which group the 18 species considered. With minor exceptions. RFLP phenetic trees supported the taxonomic description among *Solanum* species based on earlier biosystematic studies. This conclusion, which is also valid for the cited cases of *Brassica* and tomato, should not be taken as an indication that RFLP does not contribute substantially to plant phylogeny. Having established that molecular probes are reliable measures of interspecific distances, their use can now be proposed also for species and genera with a difficult or insufficient taxonomic status. In this context, RFLP studies in *Lens culinaris* and its wild relative *L. nigricans* (Havey and Muehlbauer, 1989) and the molecular characterization of the indica versus japonica types of rice varieties (Wang and Tanksley, 1989) clearly indicate the potential of the technique to recognize separation of groups or races within species, a situation which may be taken as a proof of incipient speciation.

V. Fingerprinting with RFLPs

According to Bailey (1983), the basic criteria to be fulfilled by a character usable for variety identification are (1) distinguishable intervarietal variation, (2) minimal intravarietal differences, (3) environmental stability, and (4) experimental reproducibility. Available data largely support the conclusion that molecular probes are highly suitable to varietal identification. From this point of view, it is also easy to understand that the use of a sufficient number of RFLP probes can solve every problem of varietal fingerprinting. This is the case in the studies reported for several agricultural plant species in Section II,A. More interesting for RFLP fingerprinting is, however, the finding of RFLP probes which reduce drastically the number of experiments needed for varietal identification. Such molecular markers have been identified and cloned from the human genome. They are based on minisatellite DNAs which are highly efficient in detecting genetic differences among individuals (Jeffreys et al., 1985a,b). Probes equivalent in resolution to human minisatellites have not been reported in plants. The direct use on plant DNA of human minisatellite probes has, nevertheless, been proposed (Dallas, 1988). However, as shown in the cited experiment which considered eight rice varieties, they generated RFLP patterns with a poor resolution. These are nevertheless specific for the cultivars tested and remain unchanged after plant regeneration from *in vitro* culture.

The four conditions listed by Bailey, on the other hand, were fulfilled by the probes used for potato by Gebhardt et al. (1989a) and Görg et al. (1992). These probes, in combination with a sensitive detection system for small restriction fragment length differences, made it possible to distinguish with a minimum of two probe–enzyme combinations all 20 tetraploid and all 38 diploid lines of potato considered (Gebhardt et al., 1989a). The analysis with those potato probes was later extended to a set of 136 German tetraploid varieties of potato (Görg et al., 1992). The RFLP marker GP35 was the most effective: it revealed 122 unique patterns out of 134 evaluated. The remaining 12 varieties were grouped into 6 pairs, each one having the same RFLP pattern. Using additional markers, the two genotypes of three equal pairs could be distinguished from each other. For the three remaining pairs, the evidence, based on probability estimates, suggested similarity by origin. The RFLP marker GP35 was sequenced; it is not related to human minisatellite DNA (Görg et al., 1992). Using M13 phage DNA as probe has also been proposed for DNA fingerprinting (Ryskov et al., 1988). For plants, this probe was tested only on a single individual each of cotton, soybean, and orange. The two barley varieties tested gave different RFLP patterns.

Finding probes highly suitable in varietal fingerprinting may help to solve other needs typical of plant breeding projects. For example, Hillel *et al.* (1990) proposed the use of RFLP fingerprinting in plants to reduce the number of generations in backcross programs (see also Section III).

VI. Concluding Remarks

We have attempted to show how RFLP analysis has contributed to plant science, with the emphasis put on the implications for plant breeding. In terms of scientific progress, the old disciplines of quantitative genetics and plant taxonomy have been revived by the RFLP approach, and the young discipline of plant tissue culture has gained a valuable analytical tool. The results obtained so far are mainly descriptive. However, detailed description may lead in the future to explanations of, for example, what the structural components of a QTL are.

Whether marker-assisted selection will improve the efficiency in terms of time and costs of breeding new and adapted cultivars, and whether the approach will also be of help to developing countries remain to be demonstrated. This will depend on the collaborative efforts of plant breeders and molecular geneticists to identify and to map the agronomically important genes and to simplify the experimental techniques involved. The results reported from several crop species within the past 3 or 4 years support the expectation that this may be possible.

References

Allard, R. W. (1956). *Hilgardia* **24**, 235–278.

Apuya, N. R., Frazier, B. L., Keim, P., Jill Roth, E., and Lark, K. G. (1988). *Theor. Appl. Genet.* **75**, 889–901.

Bailey, D. C. (1983). *In* "Isozymes in Plant Genetics and Breeding" (S. D. Tanksley and T. J. Orton, eds.), Part A. pp. 425–440. Elsevier, Amsterdam.

Bailey, N. T. J. (1961). "Introduction to the Mathematical Theory of Linkage." Oxford Univ. Press (Clarendon), London and New York.

Barone, A., Ritter, E., Schachtschabel, U., Debener, T., Salamini, F., and Gebhardt, C. (1990). *Mol. Gen. Genet.* **224**, 177–182.

Beckmann, J. S. (1988). *Bio/Technology* **6**, 1061–1064.

Beckmann, J. S., and Soller, M. (1983). *Theor. Appl. Genet.* **67**, 35–43.

Beckmann, J. S., and Soller, M. (1986). *Oxford Surv. Plant Mol. Cell Biol.* **3**, 196–250.

Beckmann, J. S., and Soller, M. (1990). *Bio/Technology* **8**, 930–932.

Bernatzky, R., and Tanksley, S. D. (1986). *Genetics* **112**, 887–898.

Bonierbale, M. W., Plaisted, R. L., and Tanksley, S. D. (1988). *Genetics* **120**, 1095–1103.

Botstein, D., White, R. L., Skolnick, M., and Davis, R. W. (1980). *Am. J. Hum. Genet.* **32,** 314–331.

Brown, P. T. H., Göbel, E., and Lörz, H. (1991). *Theor. Appl. Genet.* **81,** 227–232.

Burr, B., and Burr, F. A. (1991). *Trends Genet.* **7,** 55–60.

Burr, B., Burr, F. A., Thompson, K. H., Albertson, M. C., and Stuber, C. W. (1988). *Genetics* **118,** 519–526.

Chang, C., Bowman, J. L., De John, A. W., Lander, E. S., and Meyerowitz, E. M. (1988). *Proc. Natl. Acad. Sci. U.S.A.* **85,** 6856–6860.

Chao, S., Sharp, P. J., Worland, A. J., Warham, E. J., Koebner, R. M. D., and Gale, M. D. (1989). *Theor. Appl. Genet.* **78,** 495–504.

Chase, C. D., Ortega, V. M., and Vallejos, C. E. (1991). *Theor. Appl. Genet.* **81,** 806–811.

Cockerham, G. (1970). *Heredity* **25,** 309–348.

Dallas, J. F. (1988). *Proc. Natl. Acad. Sci. U.S.A.* **85,** 6831–6835.

Debener, T., Salamini, F., and Gebhardt, C. (1990). *Theor. Appl. Genet.* **79,** 360–368.

Debener, T., Salamini, F., and Gebhardt, C. (1991). *Plant Breed.* **106,** 173–181.

Donis-Keller, H., Green, P., Helms, C., Cartinhour, S., Weiffenbach, B., Stephens, K., Keith, T. P., Bowden, D. W., Smith, D. R., Lander, E. S., Botstein, D., Akots, G., Rediker, K. S., Gravius, T., Brown, V. A., Rising, M. B., Parker, C., Powers, J. A., Watt, D. W., Kauffman, E. R., Bricker, A., Phipps, P., Müller-Kahle, H., Fulton, T. R., Ng, S., Schumm, J. W., Braman, J. C., Knowlton, R. G., Barker, D. F., Grooks, S. M., Lincoln, S. E., Daly, M. J., and Abrahamson, J. (1987). *Cell (Cambridge, Mass.)* **51,** 319–337.

Edwards, M. D., Stuber, C. W., and Wendel, J. F. (1987). *Genetics* **116,** 113–125.

Everson, E. H., and Schaller, C. W. (1955). *Agron. J.* **47,** 276–286.

Evola, S. V., Burr, F. A., and Burr, B. (1986). *Theor. Appl. Genet.* **71,** 765–771.

Felsenstein, J. (1988). *Annu. Rev. Genet.* **22,** 521–565.

Figdore, S. S., Kennard, W. C., Song, K. M., Slocum, M. K., and Osborn, T. C. (1988). *Theor. Appl. Genet.* **75,** 833–840.

Fisher, R. A. (1921). *Philos. Trans. R. Soc. London, Ser. A* **122,** 309–368.

Fisher, R. A., and Balmakund, B. (1928). *J. Genet.* **20,** 79–92.

Gale, M. D., Chao, S., and Sharp, P. J. (1990). *In* "Gene Manipulation in Plant Improvement II" (J. P. Gustafson, ed.), pp. 353–363. Plenum, New York.

Ganal, M. W. *et al.* (1992). In preparation.

Gebhardt, C., Blomendahl, C., Schachtschabel, U., Debener, T., Salamini, F., and Ritter, E. (1989a). *Theor. Appl. Genet.* **78,** 16–22.

Gebhardt, C., Ritter, E., Debener, T., Schachtschabel, U., Walkemier, B., Uhrig, H., and Salamini, F. (1989b). *Theor. Appl. Genet.* **78,** 65–75.

Gebhardt, C., Ritter, E., Barone, A., Debener, T., Walkemeier, B., Schachtschabel, U., Kaufmann, H., Thompson, R. D., Bonierbale, M. W., Ganal, M. W., Tanksley, S. D., and Salamini, F. (1991). *Theor. Appl. Genet.* **83,** 49–57.

Geldermann, H. (1975). *Theor. Appl. Genet.* **46,** 319–330.

Godshalk, E. B., Lee, M., and Lamkey, K. R. (1990). *Theor. Appl. Genet.* **80,** 273–280.

Görg, R., Schachtschabel, U., Ritter, E., Salamini, F., and Gebhardt, C. (1992). *Crop Sci.* (in press).

Graner, A., Siedler, H., Jahoor, A., Herrmann, R. G., and Wenzel, G. (1990). *Theor. Appl. Genet.* **80,** 826–832.

Graner, A., Jahoor, A., Schondelmaier, J., Siedler, H., Pillen, K., Fischbeck, A., Wenzel, A., and Herrmann, R. A. (1991). *Theor. Appl. Genet.* **83,** 250–256.

Grodzicker, T., Williams, J., Sharp, P., and Sambrook, J. (1974). *Cold Spring Harbor Symp. Quant. Biol.* **39,** 439–446.

Haldane, J. B. S., and Smith, C. A. B. (1947). *Ann. Eugen.* **14,** 10–31.

Havey, M. J., and Muehlbauer, F. J. (1989). *Theor. Appl. Genet.* **77,** 839–843.

Helentjaris, T. (1987). *Trends Genet.* **3**, 217–221.

Helentjaris, T., King, G., Slocum, M., Siedenstrang, C., and Wegman, S. (1985). *Plant Mol. Biol.* **5**, 109–118.

Helentjaris, T., Slocum, M., Wright, S., Schaefer, A., and Nienhuis, J. (1986a). *Theor. Appl. Genet.* **72**, 761–769.

Helentjaris, T., Weber, D. F., and Wright, S. (1986b). *Proc. Natl. Acad. Sci. U.S.A.* **83**, 6035–6039.

Helentjaris, T., Weber, D., and Wright, S. (1988). *Genetics* **118**, 353–363.

Hermsen, J. G. T. (1984). *Iowa State J. Res.* **58**, 421–434.

Heun, M., Kennedy, A. E., Anderson, J. A., Lapitan, N. L. V., Sorrells, M. E., and Tanksley, S. D. (1991). *Genome* **34**, 437–447.

Hillel, J., Schaap, T., Haberfeld, A., Jeffreys, A. J., Plotzky, Y., Cahaner, A., and Lavi, V. (1990). *Genetics* **124**, 783–789.

Hinze, K., Thompson, R. D., Ritter, E., Salamini, F., and Schulze-Lefert, P. (1991). *Proc. Natl. Acad. Sci. U.S.A.* **88**, 3857–3861.

Hoisington, D. A., and Coe, E. H. (1990). In "Gene Manipulation in Plant Improvement II" (J. P. Gustafson, ed.), pp. 331–352. Plenum, New York.

Hu, J., and Quiros, C. F. (1991). *Theor. Appl. Genet.* **81**, 221–226.

Immer, F. R. (1930). *Genetics* **15**, 81–98.

Jacobs, J. J. M. R., Krens, F. A., Stiekema, W. J., Van Spanje, M., and Wagenvoort, M. (1990). *Potato Res.* **33**, 171–180.

Jayakar, S. D. (1970). *Biometrics* **26**, 451–464.

Jeffreys, A. J. (1979). *Cell (Cambridge, Mass.)* **18**, 1–10.

Jeffreys, A. J., Wilson, V., and Thein, S. L. (1985a). *Nature (London)* **314**, 67–73.

Jeffreys, A. J., Wilson, V., and Thein, S. L. (1985b). *Nature (London)* **316**, 76–79.

Johannsen, W. (1909). "Elemente der exakten Erblichkeitslehre." Fischer, Jena.

Jung, C., Kleine, M., Fischer, F., and Herrmann, R. G. (1990). *Theor. Appl. Genet.* **79**, 663–672.

Kam-Morgan, L. N. W., Gill, B. S., and Muthukrishnan, S. (1989). *Genome* **32**, 724–732.

Kan, Y. W., and Dozy, M. (1978a). *Proc. Natl. Acad. Sci. U.S.A.* **75**, 5631–5635.

Kan, Y. W., and Dozy, M. (1978b). *Lancet* **2**, 910–912.

Keim, P., Shoemaker, R. C., and Palmer, R. G. (1989). *Theor. Appl. Genet.* **77**, 786–792.

Keller, W. A., Arnison, P. G., and Cardy, B. J. (1987). In "Plant Tissue and Cell Culture" (C. E. Green, D. A. Somers, W. P. Hackett, and D. D. Biesboer, eds.). Alan R. Liss, New York.

Klein-Lankhorst, R., Rietveld, P., Machiels, B., Verkerk, R., Weide, R., Gebhardt, C., Koornneef, M., and Zabel, P. (1991a). *Theor. Appl. Genet.* **81**, 661–667.

Klein-Lankhorst, R. M., Vermunt, A., Weide, R., Liharska, T., and Zabel, P. (1991b). *Theor. Appl. Genet.* **83**, 108–114.

Kreitman, M., and Aquade, M. (1986). *Proc. Natl. Acad. Sci. U.S.A.* **83**, 3562–3566.

Landau-Ellis, D., Shoemaker, R., Angermüller, S., and Gresshoff, P. M. (1991). *Mol. Gen. Genet.* **228**, 221–226.

Lande, R., and Thompson, R. (1990). *Genetics* **124**, 743–756.

Lander, E. S., and Botstein, D. (1989). *Genetics* **121**, 185–199.

Lander, E. S., Green, P., Abrahamson, J., Barlow, A., Daly, M. J., Lincoln, S. E., and Newburg, L. (1987). *Genomics* **1**, 174–181.

Landry, B. S., and Michelmore, R. W. (1985). *Plant Mol. Biol. Rep.* **3**, 174–179.

Landry, B. S., Kesseli, R. V., Farrara, B., and Michelmore, R. W. (1987a). *Genetics* **116**, 331–337.

Landry, B. S., Kesseli, R., Leung, H., and Michelmore, R. W. (1987b). *Theor. Appl. Genet.* **74**, 646–653.

Lee, M., Godshalk, E. B., Lamkey, K. R., and Woodman, W. L. (1989). *Crop Sci.* **29**, 1067–1071.

Martin, B., Nienhuis, J., King, G., and Schaefer, A. (1989). *Science* **243**, 1725–1728.

Mather, K. (1938). "The Measurement of Linkage in Heredity." Methuen, London.

Mather, K., and Jinks, J. L. (1971). "Biometrical Genetics." Cornell Univ. Press, Ithaca, New York.

McCouch, S. R., Kochert, G., Yu, Z. H., Wang, Z. Y., Khush, G. S., Coffman, W. R., and Tanksley, S. D. (1988). *Theor. Appl. Genet.* **76**, 815–829.

McMillan, J., and Robertson, A. (1974). *Heredity* **32**, 349–356.

McMullen, M. D., and Louie, R. (1989). *Mol. Plant-Microbe Interact.* **2**, 309–314.

Melchinger, A. E., Lee, M., Lamkey, K. R., and Woodman, W. L. (1990a). *Crop Sci.* **30**, 1033–1040.

Melchinger, A. E., Lee, M., Lamkey, K. R., Hallauer, A. R., and Woodman, W. L. (1990b). *Theor. Appl. Genet.* **80**, 488–496.

Miller, J. C., and Tanksley, S. D. (1990a). *Theor. Appl. Genet.* **80**, 385–389.

Miller, J. C., and Tanksley, S. D. (1990b). *Theor. Appl. Genet.* **80**, 437–448.

Morton, N. (1955). *Am. J. Hum. Genet.* **7**, 277–318.

Müller, E., Brown, P. T. H., Hartke, S., and Lörz, H. (1990). *Theor. Appl. Genet.* **80**, 673–679.

Nam, H.-G., Giraudat, J., den Boer, B., Moonau, F., Loos, W. D. B., Hauge, B. M., and Goodman, H. M. (1989). *Plant Cell* **1**, 699–705.

Nei, M. (1987). "Molecular Evolutionary Genetics." Columbia Univ. Press, New York.

Nilsson-Ehle, H. (1909). "Kreuzungsuntersuchungen an Hafer und Weizen." Lund.

Olson, M., Hood, L., Cantor, C., and Botstein, D. (1989). *Science* **245**, 1434–1435.

Osborn, T. C., Alexander, D. C., and Fobes, J. F. (1987). *Theor. Appl. Genet.* **73**, 350–356.

Ottaviano, E., Goria, M. S., Pe, E., and Frova, C. (1991). *Theor. Appl. Genet.* **81**, 713–719.

Paterson, A. H., Lander, E. S., Hewitt, J. D., Peterson, S., Lincoln, S. E., and Tanksley, S. D. (1988). *Nature (London)* **335**, 721–726.

Paterson, A. H., De Verna, J. W., Lanini, B., and Tanksley, S. D. (1990). *Genetics* **124**, 735–742.

Petes, T. K., and Botstein, D. (1977). *Proc. Natl. Acad. Sci. U.S.A.* **74**, 5091–5095.

Ritter, E., Gebhardt, C., and Salamini, F. (1990). *genetics* **125**, 645–654.

Ritter, E., Debener, T., Barone, A., Salamini, F., and Gebhardt, C. (1991). *Mol. Gen. Genet.* **227**, 81–85.

Rivard, S. R., Cappadocia, M., Vincent, G., Brisson, N., and Landry, B. S. (1989). *Theor. Appl. Genet.* **78**, 49–56.

Rivin, C. J., Zimmer, E. A., Cullis, C. A., and Walbot, V. (1983). *Plant Mol. Biol. Rep.* **1**, 9–16.

Rogers, J. S. (1972). *Univ. Tex. Publ.* **7213**, 145–153.

Ryskov, A. P., Jincharadze, A. G., Prosnyak, M. I., Ivanov, P. L., and Limborska, S. A. (1988). *FEBS Lett.* **233**, 388–392.

Saedler, H., and Nevers, P. (1985). *EMBO J.* **4**, 585–590.

Saiki, R. K., Gelfand, D. H., Stoffel, S., Scharf, S., Higuchi, R., Horn, G. T., Mullis, K. B., and Ehrlich, H. A. (1988). *Science* **239**, 487–491.

Sangwan, R. S., and Sangwan-Norreel, B. S. (1990). *In* "Plant Tissue Culture: Applications and Limitations" (S. S. Bhojwani, ed.), pp. 220–241. Elsevier, Amsterdam.

Sarfatti, M., Katan, J., Fluhr, R., and Zamir, D. (1989). *Theor. Appl. Genet.* **78**, 755–759.

Sarfatti, M., Abu-Abied, M., Katan, J., and Zamir, D. (1991). *Theor. Appl. Genet.* **82**, 22–26.

Sax, K. (1923). *Genetics* **8**, 552–560.

Schwarz-Sommer, Z., Gierl, A., Cuypers, H., Peterson, P. A., and Saedler, H. (1985). *EMBO J.* **4**, 591–597.

Sharp, P. J., Chao, S., Desai, S., and Gale, M. D. (1989). *Theor. Appl. Genet.* **78,** 342–348.

Siracusa, L. D., Jenkins, N. A., and Copeland, N. G. (1991). *Genetics* **127,** 169–179.

Slocum, M. K., Figdore, S. S., Kennard, W. C., Suzuki, J. Y., and Osborn, T. C. (1990). *Theor. Appl. Genet.* **80,** 57–64.

Smith, O. S., Smith, J. S. C., Bowen, S. L., Tenborg, R. A., and Wall, S. J. (1990). *Theor. Appl. Genet.* **80,** 833–840.

Soller, M., and Beckmann, J. S. (1983). *Theor. Appl. Genet.* **67,** 25–33.

Soller, M., and Brody, T. (1976). *Theor. Appl. Genet.* **47,** 35–39.

Song, K. M., Osborn, T. C., and Williams, P. H. (1988a). *Theor. Appl. Genet.* **75,** 784–794.

Song, K. M., Osborn, T. C., and Williams, P. H. (1988b). *Theor. Appl. Genet.* **76,** 593–600.

Song, K., Osborn, T. C., and Williams, P. H. (1990). *Theor. Appl. Genet.* **79,** 497–506.

Southern, E. M. (1975). *J. Mol. Biol.* **98,** 503–517.

Stuber, C. W., Moll, R. H., Goodman, M. M., Schaffer, H. E., and Weir, B. S. (1980). *Genetics* **95,** 225–236.

Stuber, C. W., Goodman, M. M., and Moll, R. H. (1982). *Crop Sci.* **22,** 737–740.

Stuber, C. W., Edwards, M. D., and Wendel, J. F. (1987). *Crop Sci.* **27,** 639–648.

Suiter, K. A., Wendel, J. F., and Case, J. S. (1983). *J. Hered.* **74,** 203–204.

Tanksley, S. D. (1983). *Plant Mol. Biol. Rep.* **1,** 3–8.

Tanksley, S. D., and Hewitt, J. (1988). *Theor. Appl. Genet.* **75,** 811–823.

Tanksley, S. D., and Mutschler, M. A. (1990). *In* "Genetic Maps" (S. J. O'Brien, ed.), pp. 6.3–6.15. Cold Spring Harbor Lab., Cold Spring Harbor, New York.

Tanksley, S. D., Medina-Filho, H., and Rick, C. M. (1982). *Heredity* **49,** 11–25.

Tanksley, S. D., Bernatzky, R., Lapitan, N. L., and Prince, J. P. (1988). *Proc. Natl. Acad. Sci. U.S.A.* **85,** 6419–6423.

Thoday, J. M. (1961). *Nature (London)* **191,** 368–370.

Tingey, S. V., Rafalski, J. A., Williams, J. G. K., and Sebastian, S. (1990). *In* "Proceedings of the Sixth NATO Advanced Study Institute, Plant Molecular Biology." Schloss Elmau, Bavaria, Germany (in press).

Van de Ven, M., Powell, W., Ramsay, G., and Waugh, R. (1990). *Heredity* **65,** 329–342.

Wang, Z. Y., and Tanksley, S. D. (1989). *Genome* **32,** 1113–1118.

Webb, D. M., Knapp, S. J., and Tagliani, L. A. (1992). *Theor. Appl. Genet.* (in press).

Weber, D., and Helentjaris, T. (1989). *Genetics* **121,** 583–590.

White, R., Leppert, M., Bishop, D. T., Barker, D., Berkowitz, J., Brown, C., Callahan, P., Hohn, T., and Jerominski, L. (1985). *Nature (London)* **313,** 101–105.

Williams, C. E., Hunt, G. J., and Helgeson, J. P. (1990). *Theor. Appl. Genet.* **80,** 545–551.

Williams, J. G. K., Kubelik, A. R., Livak, K. J., Rafalski, J. A., and Tingey, S. V. (1990). *Nucleic Acids Res.* **18,** 6531–6535.

Young, N. D., and Tanksley, S. D. (1989). *Theor. Appl. Genet.* **77,** 353–359.

Young, N. D., Miller, J. C., and Tanksley, S. D. (1987). *Nucleic Acids Res.* **15,** 9339–9348.

Young, N. D., Zamir, D., Ganal, M. W., and Tanksley, S. D. (1988). *Genetics* **120,** 579–585.

Yu, Z. H., Mackill, D. J., Bonman, J. M., and Tanksley, S. D. (1991). *Theor. Appl. Genet.* **81,** 471–476.

Zamir, D., and Tanksley, S. D. (1988). *Mol. Gen. Genet.* **213,** 254–261.

Regulatory Functions of Soluble Auxin-Binding Proteins

Shingo Sakai

Institute of Biological Sciences, University of Tsukuba,
Tsukuba, Ibaraki 305, Japan

I. Introduction

The physiological effects of auxin are numerous and varied (Evans, 1974; Davies, 1987), but the biochemical events behind the physiological manifestations are poorly understood at the molecular level. It is well documented that a long-term (12–18 hr) exposure of intact plants or plant tissues to exogenous auxin has dramatic effects on nucleic acid and protein synthesis (Key, 1969; J. V. Jacobsen 1977). However, auxin-promoted incorporation of labeled precursors into RNA and protein was usually detected only after a period of at least 1 hr. Any effect of auxin on the synthesis of mRNA or specific proteins in the short term (within 30 min) could not be detected by techniques available in the 1970s.

The development of *in vitro* translation systems for mRNA and two-dimensional gel electrophoresis of translated proteins has made possible the identification of selective mRNA changes in pea and soybean seedlings within 10–20 min after exposure to auxin (Theologis, 1986). Walker and Key (1982), Zurfluh and Guilfoyle (1982a,b,c), and Theologis and Ray (1982) demonstrated that treatment of plant tissues with auxin triggers the transcription of genes which are otherwise unexpressed or expressed at low levels. However, the ways in which signals generated by auxin in cells are transmitted to nuclei, where RNA synthesis occurs, have not been clarified.

It has been thought that putative receptors for auxin localized in the cellular membranes generated signals that are transmitted to nuclei (Hardin *et al.*, 1972), but the mechanism remains unknown. Ettlinger and Lehle (1988) reported an increase in levels of inositol triphosphate as a result of treatment with auxin. These results may suggest the involvement of hydrolysis of inositol phosphate-containing phospholipids, with subsequent mobilization of Ca ions, as proposed in the case of animal peptide hormones

(Berridge, 1987). Another possibility is that soluble receptors activated by auxin then activate a specific gene, the product of which triggers a cascade of gene expression which leads to the formation of key proteins necessary for a physiological process to occur (Rubery, 1981; Theologis, 1986; Mennes et al., 1986; Libbenga and Mennes, 1987).

During the past 20 years, auxin-binding studies have been performed with membrane preparations and soluble proteins from various tissues and species in attempts to clarify the nature of auxin receptors. Many studies have indicated the presence of auxin-binding sites, notably, those on the membranes of maize coleoptiles, but also in membrane preparations from various other tissues and species (Hertel et al., 1972; Libbenga et al., 1986). In membranes isolated from maize, auxin-binding sites are located on the endoplasmic reticulum (ER), the tonoplast, and the plasmalemma (Dohrmann et al., 1978). However, it remains unclear whether these auxin-binding sites represent functional receptors. Soluble proteins with in vitro auxin-binding activity have been extracted from many plant sources (Libbenga et al., 1986). These studies were reviewed by Kende and Gardner (1976), Venis (1977), Stoddart and Venis (1980), Rubery (1981), and Libbenga et al. (1986). Short evaluations have been given by Libbenga (1978), Lamb (1978), Bogers and Libbenga (1981), Venis (1981), Mennes et al. (1986), and Libbenga and Mennes (1987).

One area in which there has been remarkable progress in determining the function of auxin-binding proteins is the stimulation of RNA synthesis in isolated nuclei by soluble auxin-binding proteins. Our group has made several contributions toward this. Therefore, in this article, I discuss recent topics in connection with this subject based both on our experience and that of other workers. The first half of this article reviews briefly the soluble auxin-binding proteins. The second half of the article focuses on the regulatory function of auxin-binding proteins on RNA synthesis in isolated nuclei. The findings of our group are presented, as well as certain other unpublished results.

II. Problems in the Isolation of Auxin-Binding Protein

A. Use of the Affinity Column

Several attempts have been made to isolate soluble auxin-binding proteins by affinity chromatography, using the lysine derivatives of indole-3-acetic acid (IAA) or 2,4-dichlorophenoxyacetic acid (2,4-D) coupled to cyanogen bromide-activated Sepharose (Venis, 1971; Rizzo et al., 1977). The isolated proteins, however, had no auxin-binding activity (Venis, 1971), and affinity chromatography on 2,4-D-linked Sepharose 4B has not been considered suitable for isolation of active auxin-binding proteins (Kende and Gardner, 1976; Rubery, 1981). Venis (1971) eluted auxin-binding proteins from an affinity column principally with 2 mM KOH (pH 11.2) and adjusted the pH of the eluates to pH 8.1 with 1 M Tris-HCl buffer (pH 7.5). As we thought that the auxin-binding proteins retained on the affinity column might be denatured during KOH elution at pH 11.2, we used 1 M NaCl in 50 mM Tris-HCl buffer (pH 9.2) as an eluant. The proteins eluted had auxin-binding activity (Sakai and Hanagata, 1983). Affinity chromtography on IAA-linked Sepharose 4B (Roy and Biswas, 1977), [5-OH]IAA-epoxy-activated Sepharose 6B (Van der Linde et al., 1985), 2-OH-3,5-diiodobenzoic acid-linked Sepharose 4B (Löbler and Klämbt, 1985), and naphthalene-1-acetic acid (NAA)-linked AH Sepharose 4B (Shimomura et al., 1986) has been widely used for the isolation of auxin-binding proteins from some plant tissues.

B. Binding Assay

A variety of methods have been used for measuring bound radioactivity to soluble auxin-binding proteins. Soluble ligand-binding protein complexes are analyzed by equilibrium dialysis (Mondal et al., 1972; Wardrop and Polya, 1980a; Sakai, 1985), or by adsorption of free ligand to dextran-coated charcoal (Likholat et al., 1974; Oostrom et al., 1975; Ihl, 1976; Roy and Biswas, 1977, Bogers et al., 1980; Van der Linde et al., 1984) or by precipitation at high ammonium sulfate concentration (Wardrop and Polya, 1977, 1980a; Murphy, 1980; Jacobsen, 1981, 1982; Sakai and Hanagata, 1983; Sakai, 1985).

Venis (1984) could not confirm specific auxin binding in dwarf bean seedlings (Wardrop and Polya, 1977, 1980a) and pea epicotyls (Jacobsen, 1981, 1982) by the equilibrium dialysis binding assay, and questioned the reliability of binding data generated by the ammonium sulfate precipitation assay. However, using different and independent procedures for assaying soluble auxin-binding in etiolated pea epicotyls, Jacobsen and Hajek (1985) could prove the reliability of the ammonium sulfate precipitation assay both for crude cytosols and for specific protein fractions obtained after chromatofocusing. The affinity of auxin-binding protein II (ABP-II) purified from etiolated mung bean seedlings for 2,4-D was determined using the two procedures of ammonium sulfate precipitation and equilibrium dialysis (Sakai, 1985). The dissociation constant (K_d) for 2,4-D, determined by either procedure, is about 10^{-5} M and the binding stoichiometry at saturated 2,4-D concentrations is of the order of 1 mol/mol protein. Recently, Kaur and Kapoor (1990) also have reported similar binding data with respect to the dissociation constants and site concentration obtained by the three types of binding assays, namely, ammonium sulfate precipitation, charcoal adsorption, and equilibrium dialysis.

III. Soluble Auxin-Binding Proteins

The group of Biswas and co-workers obtained two IAA-binding proteins from a nucleoplasmic fraction (designated n-IRP) and the nonhistone chromatin protein (c-IRP) of isolated nuclei from immature coconut endosperm (Mondal *et al.*, 1972; Roy and Biswas, 1977). The n-IRP was purified by chromatography on CM-cellulose and affinity chromatography on IAA-linked Sepharose. The molecular weight of this protein, estimated by means of SDS-polyacrylamide gel electrophoresis (SDS-PAGE), was 94,000 and the K_d for IAA was 7.5 μM. The other IAA-binding protein (c-IRP) was extracted with 2 M NaCl from chromatin and purified to apparent homogeneity by chromatography on DEAE-cellulose and affinity chromatography on IAA-linked Sepharose. The molecular weight was estimated to be 70,000 by SDS-PAGE. The c-IRP exhibited high- (K_d: 58 nM) and low- (K_d: 8.2 μM) affinity binding for IAA. The latter was referred to as "nonspecific binding," though the dissociation constant was in fact very similar to that of n-IRP (7.5 μM). The [^{14}C]-IAA–nIRP complex was reported to bind to coconut chromatin and to cause a twofold increase in chromatin transcription by *Escherichia coli* RNA polymerase (Section VII).

Likholat *et al.* (1974) demonstrated IAA binding to a cytosol fraction extracted from wheat coleoptiles. By addition of a 50-fold excess of unlabeled indole-3-propionic acid (IPA) or IAA to the incubation mixture, they found that IPA competed weakly while IAA inhibited the binding of labeled auxin by 50%. The binding of [^{14}C]-IAA to the cytosol fraction was about five times higher in extracts from 36-hr-old than from 72-hr-old plants. No evidence supporting a proteinaceous nature for the binding site was presented.

In vivo labeling was attempted by Ihl (1976), who incubated soybean cotyledons with [^{14}C]-IAA at 25°C for 1 hr, and showed IAA binding to a 100,000 g supernatant fraction from the cotyledons. The radioactivity associated with macromolecular material eluted in the void volume of Sephadex G-25 or G-50 column was reduced if a 275-fold excess of unlabeled auxin had been supplied during the incubation. The IAA-binding was sensitive to pronase and heat, but the chemical nature of the bound radioactivity was not determined.

Murphy (1979) demonstrated NAA binding to nonplant protein, bovine serum albumin, using the ammonium sulfate precipitation method. The binding was auxin-specific, reversible, and saturable. He pointed out that identification of auxin receptor activity based soley on these experiments should be treated with caution. Nevertheless, in a subsequent paper (Murphy, 1980), he used the ammonium sulfate precipitation method and demonstrated specific NAA binding at pH 4.75 to fractionated cytosol, prepared from 4- to 7-day-old dark-grown maize coleoptiles. The binding site was partially purified by conventional methods. The apparent molecular weight was estimated to be 38,700. The K_d for NAA was about 3.5 μM. Evidently, this binding site behaved like a macromolecule, but a proteinaceous nature for the macromolecule was not examined.

Wardrop and Polya (1977) reported the presence of a soluble, high-affinity-binding protein in 6- to 8-day-old seedlings of dwarf bean. Optimal binding occurred at pH 8.5. The molecular weight estimated by gel filtration was 315,000 and the K_d for IAA was 0.09 μM.

In two subsequent publications (Wardrop and Polya, 1980a,b), they identified this protein as ribulose-1,5-bisphosphate carboxylase (RuBP-Case) with a molecular weight of 550,000. The K_d for IAA was 0.8 μM and the stoichiometry of binding was about one IAA/RuBPCase. However, purified spinach and silver beet RuBPCase had no IAA-binding activity. Many criticisms have been made of these results (Libbenga et al., 1986).

Kaur and Kapoor (1990) demonstrated [^{14}C]-IAA binding to cytosolic proteins isolated from etiolated chickpea epicotyls using three types of binding assays. Similar binding data with respect to the dissociation constants (K_d 0.9–5 × 10^{-7} M) and site concentration (0.8–1.0 pmol/mg total protein) have been obtained by ammonium sulfate precipitation, dextran-coated chacoal adsorption, and equilibrium dialysis.

Jacobsen (1981) described the presence of an auxin-binding protein in epicotyls of 7-day-old etiolated pea seedlings. The K_d values for NAA and 2,4-D were 0.23 and 0.21 μM, respectively. In a subsequent paper (Jacobsen, 1982), the proteinaceous nature was demonstrated by sensitivity to proteinase K digestion. The binding protein was sensitive to freezing and thawing and prolonged storage at 0°C. The binding was specific for the auxins NAA, IAA, and 2,4-D with a K_d in the range of 0.1–0.4 μM.

Partial purification by a preparative chromatofocusing column revealed a second soluble auxin-binding site (sABP$_2$) in etiolated pea epicotyls (Jacobsen, 1984; Jacobsen and Hajek, 1985). While in epicotyls of 8- to 9-day-old seedlings high-affinity binding could be detected in the protein fraction with an isoelectric point (pI) range of 5.0–6.0 (sABP$_1$), in later stages (10–12 days), an additional site appeared distinctly in the peak fraction characterized by a pI range of 6.0–6.8 (sABP$_2$). K_d values for the two different binding sites were 0.55 × 10^{-7} M for sABP$_1$ and 2.24 × 10^{-7} M for sABP$_2$, respectively. In pea seedlings it was observed that the occurrence of sABP appears to be growth-stage dependent and tissue specific (Jacobsen and Hajek, 1987). The binding site(s) for $sABP_1$ in pea was predominantly localized in internodes of etiolated epicotyls from young seedlings, while a second site ($sABP_2$), which was not detectable in binding assays prior to day 9 of germination, seemed to be restricted to nodal tissues of the seedlings.

Subsequently, Jacobsen's group started the study to analyze the presence of cytoplasmic high-affinity binding sites for auxins using rapidly growing cell suspensions of soybean (Jacobsen et al., 1987; Herber et al., 1988). Cytosol preparations were studied in lag, log, and early stationary phases of the growth cycle. Two classes of binding sites were detected, which showed some similarities with binding sites reported from pea epicotyls (Jacobsen, 1984). While the number of both sites declined in the cytoplasm during the growth cycle, the number of one of the sites increased at the onset of rapid cell divisions. In parallel, both sites exhibited an increase in binding affinity during the growth cycle. The dissociation constants for NAA of the two sites were 1.99×10^{-8} and $4.02 \times 10^{-8} M$, respectively, at 0 days of culture.

Oostrom et al. (1975) detected a soluble IAA-binding protein in the cytosol of cultured tobacco pith explants. The binding site with a K_d of $10 \mu M$ was found in extracts from cultured pith explants, whereas extracts derived from freshly excised pith showed only low-affinity binding. A soluble IAA-binding protein was isolated from cultured tobacco pith explants (Oostrom et al., 1980). A molecular weight of about 300,000 was calculated from the Sepharose-CL-6B elution profile. The K_d for IAA was $0.01 \mu M$. The protein was not present in freshly excised stem pith, but it was detectable after one day of culture on medium with or without IAA and kinetin (Bogers et al., 1980).

Van der Linde et al. (1984) improved the purification of the IAA-binding protein. By using boric acid buffer (pH 6.8) instead of Tris buffer, contamination of cytosol by polyphenols was substantially reduced, but then the IAA-binding site was eluted in fractions with a molecular weight of 150,000–200,000. The K_d for IAA was about $1.6 \times 10^{-8} M$ and the number of binding sites varied from 0 to 2×10^{-13} mol mg^{-1} protein. These fractions stimulated RNA synthesis in isolated nuclei in the presence of IAA (Section VII). Similar IAA-binding proteins were detected in soluble fractions from cell-suspension cultures from tobacco and sycamore and $0.5 M$ KCl extracts from nuclei isolated from sycamore cell-suspension cultures.

Subsequently, Van der Linde et al. (1985) reported that the amount of specific IAA-binding protein in crude protein extracts could be significantly increased by adding MgATP and/or excess artificial substrate for phosphatases (p-nitropheynyl phosphate) to the binding assay medium. They suggested that the auxin-binding protein might be liable to affinity modulation by ATP-dependent phosphorylation and by dephosphorylation, transforming the binding proteins into a high- or low-affinity form, respectively (Libbenga and Mennes, 1987).

Bailey et al. (1985) confirmed the presence of essentially the same auxin-binding protein in both cytosol preparations and salt extracts of isolated nuclei from tobacco cell-suspension cultures. They purified the binding protein 50- to 60-fold by affinity chromatography. The molecular weight of the binding protein was estimated to be approximately 175,000 by gel filtration. They observed that the number of binding sites varied according to the growth stage of culture. In the cytosol preparations high-affinity binding could be detected only during the lag and stationary phases of culture, whereas salt extracts from nuclei contained detectable binding sites only in the exponential phase. They suggested the possibility that, during the growth cycle, the binding protein was recycling between the cytosol and the nucleus. Libbenga and Mennes (1987) reported that cytosol and high-salt nuclear extracts from tobacco cells contained a high-affinity auxin-binding protein. The K_d for IAA was 10^{-8}–10^{-9} M.

We have purified two buffer soluble auxin-binding proteins (ABP-I and ABP-II) from etiolated mung been seedlings by affinity chromatograhy on 2,4-D-linked Sepaharose 4B and by gel filtration on Sepharose 4B and Sephacryl S-200 (Sakai and Hanagata, 1983; Sakai, 1984, 1985) These proteins had a relatively high affinity for auxins and the binding was auxin-specific, reversible, and saturable (Table I). The purified auxin-binding proteins stimulated RNA synthesis in isolated nuclei (Section VII).

IV. Effects of Indole-3-Acetic Acid on RNA Synthesis in Isolated Nuclei and Chromatin

Although some reports indicated that auxin might enhance RNA synthesis by isolated nuclei (Roychoudhry and Sen, 1964, 1965; Maheshwari et al., 1966; Cherry, 1967), the reported effects of auxin on precursor incorporation into RNA of nuclei varied from no enhancement in some experiments to one- or twofold increases in other experiments. However, nuclei isolated from auxin-treated soybean hypocotyls showed higher levels of transcription than those from untreated hypocotyls (Chen et al., 1975).

Isolated chromatin did not show enhanced nucleic acid synthesis in responce to auxin in vitro (Schwimmer, 1968). However, chromatin iso-

TABLE I

Characteristics of ABP-I and ABP-II

ABP	Molecular weight	Subunit structure	Dissociation constant (μM)	2,4-D binding to ABP				
				Optimum pH	Inhibitor	Stoichiometry (mol/mol)	Reversibility	Ref.
I	390,000	(47,000)[6] (15,000)[6]	3.2 (IAA) 9.3 (2.4-D)	7.0–7.6	IAA NAA PCIB[a]	2	Yes	Sakai and Hanagata (1983) Sakai (1984)
II	190,000	(48,000)[4]	9.5 (2,4-D)	5–6	IAA NAA PCIB 2, 4, 5-T[b]	1	Yes	Sakai (1985)

[a] p-Chlorophenoxyisobutylic acid.
[b] 2, 4, 5-Trichlorophenoxyacetic acid.

lated from auxin-treated soybean hypocotyl tissues showed higher RNA synthetic activity *in vitro* than chromatin isolated from control (O'Brien *et al.*, 1968).

These results suggest that some factors are required for the stimulation of RNA synthesis in isolated nuclei or chromatin by auxins.

V. Factors Which Mediate Auxin Action

Matthysse and Phillips (1969) found that nuclei of cultured tobacco cells showed an increase in RNA synthesis if they were isolated in the presence of 2,4-D. If the nuclei were sedimented from a medium lacking 2,4-D, they no longer showed this stimulation. However, addition of the supernatant fraction from the centrifugation caused the nuclei to become auxin-responsive. This factor was also isolated from pea buds. The factor from pea buds and tobacco nuclei stimulated RNA synthesis by exogenous polymerase with pea bud chromatin in the presence of 2,4-D. It was inferred that, in the absence of 2,4-D, some factor required for the auxin response was lost from the nuclei. Factors prepared from different tissues of peas showed somewhat greater activity on chromatin derived from the same tissue (Matthysse, 1970), but no further elaboration of this system has been described.

The protein fraction prepared from soybean cotyledons stimulated RNA polymerase, but not if prepared from 2,4-D-treated tissues (Hardin *et al.*, 1970). The protein also stimulated the activity of *E. coli* RNA polymerase added to the soybean chromatin system. Hardin *et al.* suggested that this protein might mediate the action of the hormone *in vivo*, so that polymerase from auxin-treated tissue, having been already activated by the hormone–receptor complex, would be unresponsive to *in vitro* addition of the factor.

Teissere *et al.* (1975) reported that four fractions of nonhistone chromosomal protein stimulated activity of lentil root nucleolar RNA polymerase. Two of these fractions were studied in more detail and appeared to be initiation factors.

Hardin *et al.* (1972) reported that plasma membrane fractions stimulated soybean RNA polymerase and that the active factor could be released from the membranes by incubation with IAA or 2,4-D, but not by the inactive analog 3,5-dichlorophenoxyacetic acid. The soybean factor was certainly not protein, was ethanol-soluble, and, judging by its chromatographic properties, was of low molecular weight (Clar *et al.*, 1976).

Likholat *et al.* (1974) observed that membrane preparations isolated from 72-hr-old wheat seedling increased the template activity of chromatin prepared from the same seedlings in the presence of IAA. They demonstrated the liberation of auxin-binding sites from membrane preparations, isolated from 72-hr-old seedlings, by incubation with IAA.

VI. Stimulation of RNA Synthesis by Proteins Retained on a 2,4-Dichlorophenoxyacetic Acid-Linked Sepharose Column

Venis (1971) attempted to isolate auxin-binding proteins by affinity chromatography using IAA- or 2,4-D-linked Sepharose. When crude supernatants prepared from pea or maize shoots were passed over such columns, small amounts of protein were retained and could be eluted successively with 1 M NaCl, water, and then 2 m M KOH. The fractions obtained were tested for their effects on DNA-dependent RNA synthesis, supported by $E.$ $coli$ polymerase. Fractions eluted by NaCl were invariably inhibitory, while KOH-eluted fractions promoted RNA synthesis by 40–200% in different preparations. The isolated proteins, however, had no auxin-binding activity. Activity of the eluted KOH fractions from peas was not dialyzable and was completely destroyed by freezing or brief heating. Increasing amounts of the factor stimulated RNA synthesis up to a maximum of 200% of control, and the effect of the factor could be blocked with rifampicin. It was suggested that the active material was a high-molecular-weight protein. Addition of auxin was not necessary for activity, and attempts to detect auxin binding to these fractions by equilibrium dialysis was not successful (Section II,A).

To account for the lack of an auxin requirement for stimulation of RNA synthesis by the protein, Venis (1971) discussed the possibility that passage through the affinity column may transform the factor to an active configuration in which further contact with auxin was not required or a regulatory subunit might be stripped off the protein during the passage through the column. He also speculated that auxin may simply permit activity to be expressed by transporting an inherently active regulatory protein from the cytoplasm to the nucleus.

Rizzo et $al.$ (1977) obtained a similar, though more active, factor from soybean hypocotyls using a 2,4-D-linked Sepharose 4B column and preparation methods identical to those outlined by Venis (1971). The KOH eluate contained a transcription factor that stimulated RNA synthesis 2- to 7-fold using $E.$ $coli$ RNA polymerase and native calf thymus DNA. Experiments using soybean multiple RNA polymerases separated by anion-exchange chromatography showed that the factor stimulated RNA polymerase I in the presence of 2,4-D, but did not stimulate polymerases II and III.

VII. Stimulation of RNA Synthesis by Auxin-Binding Proteins

Biswas and co-workers (Mondal *et al.*, 1972; Roy and Biswas, 1977) isolated two auxin-binding proteins (n-IRP and c-IRP) from nuclei of immature coconut endosperm (Section III). The [^{14}C]-IAA–nIRP complex was reported to bind to coconut chromatin and to cause a twofold increase in RNA synthesis in the chromatin by *E. coli* RNA polymerase (Roy and Biswas, 1977). The synthesis of 9–12S RNA was stimulated by the complex. RNA synthesis supported by chick erythrocyte chromatin was unaffected. Biswas and co-workers proposed the model that the high-affinity c-IRP in chromatin acts as an acceptor site for the n-IRP–IAA complex to trigger specific gene transcription.

Although the reported properties of the coconut proteins make them the best candidates in existence for genuine auxin receptors, it is unfortunate that their experimental details are unsatisfactory. Some criticisms of the data have been made by Kende and Gardner (1976), Stoddart and Venis (1980), and Libbenga *et al.* (1986).

Stimulation of RNA synthesis *in vitro* by a partially purified auxin-binding protein was described by Van der Linde *et al.* (1984). They used a transcription system consisting of nuclei isolated from tobacco callus tissue. Addition of partially purified protein preparations from tobacco callus to these nuclei resulted in an IAA-dependent stimulation of RNA polymerase activity, which was not observed with similar preparations that did not contain detectable amounts of specific IAA-binding sites. The average stimulation in the presence of 1 μM IAA was 42%. No stimulation was observed when α-amanitin was present in the transcription system at a concentration of 2 μg ml^{-1}, which is known to inhibit RNA polymerase II activity. By increasing the IAA concentration from 0 to 10^{-6} M, a good correlation was found between the IAA-binding occupancy and the relative stimulation.

Bailey *et al.* (1985) and Libbenga and Mennes (1987) reported similar results with partially purified auxin-binding proteins and nuclei isolated from tobacco cell-suspension cultures. When partially purified by affinity chromatography and allowed to preincubate with IAA, the binding proteins had a significant stimulation effect on total RNA synthesis in isolated nuclei.

Jacobsen *et al.* (1987) demonstrated that two auxin-binding proteins (sABP$_1$ and sABP$_2$) isolated from pea epicotyls stimulated transcription in isolated nuclei in the presence of 10^{-8} *M* IAA.

Kaur and Kapoor (1990) described that a fraction eluted from IAA-linked Sepharose 4B affinity column stimulated transcription *in vitro* up to sevenfold.

As shown in Table II, the auxin-binding proteins (ABP-I and ABP-II) purifed from etiolated mung bean seedlings stimulated RNA synthesis in isolated mung bean nuclei both in the presence and absence of $(NH_4)_2SO_4$ (Sakai *et al.*, 1986). However, in contrast to the results of Van der Linde *et al.* (1984), the stimulation of RNA synthesis by either ABP-I or ABP-II was not increased by the addition of 10^{-5} *M* IAA to the reaction mixture. As the stimulatory effects of both ABP-I and ABP-II on RNA synthesis were not further increased by IAA, it seemed possible that IAA present in the isolated nuclei interacted with ABPs. We therefore isolated nuclei from hypocotyl segments which had been incubated in a moist chamber for 24 hr at 25°C in the dark, a treatment which would deplete the level of endogenous auxin. However, as with nuclei from fresh tissue, ABP-I and ABP-II stimulated RNA synthesis of the nuclei prepared from auxin-depleted tissues, and further addition of IAA had no effect on the stimulation of RNA synthesis (Table III).

The stimulatory effects of ABP-I and ABP-II were completely abolished by the addition of α-amanitin (Fig. 1). These results suggested that the

TABLE II

Effects of ABP-I and ABP-II on RNA Synthesis by Isolated Nuclei[a]

| | [³H]UMP incorporated/5 × 10⁷ nuclei | | | |
| | No $(NH_4)_2SO_4$ | | 50 m*M* $(NH_4)_2SO_4$ | |
Addition	pmol	% of control	pmol	% of control
None	36.00 ± 0.85	100	65.00 ± 0.19	100
Bovine serum albumin (100 μg)	35.48 ± 1.00	99	65.83 ± 0.15	101
IAA (10^{-5} *M*)	36.02 ± 0.54	100	65.10 ± 0.34	100
ABP-I (100 μg)	44.78 ± 2.48	124	74.32 ± 1.70	114
ABP-I + IAA	44.82 ± 1.17	125	74.11 ± 0.62	114
Boiled ABP-I	36.28 ± 0.29	101	64.82 ± 0.78	100
ABP-II (100 μg)	49.94 ± 1.03	139	89.72 ± 0.94	138
ABP-II + IAA	49.42 ± 1.34	137	88.70 ± 1.58	136
Boiled ABP-II	36.52 ± 1.01	101	65.11 ± 1.00	100

[a] Values are means of three experiments ± standard error. From Sakai *et al.* (1986; Table 2, p. 639).

TABLE III

Effects of ABP-I and ABP-II on RNA Synthesis by Nuclei
Isolated from Auxin-Depleted Mung Bean Hypocotyls[a]

Addition	[³H]UMP incorporated/5 × 10⁷ nuclei	
	pmol	% of control
None	55.18 ± 1.60	100
IAA ($10^{-5} M$)	55.24 ± 0.89	100
ABP-I (100 μg)	60.70 ± 1.02	110
ABP-I + IAA	61.00 ± 1.12	110
ABP-II (100 μg)	72.28 ± 1.25	131
ABP-II + IAA	71.65 ± 0.94	130

[a] The assay was carried out in the presence of
50 mM $(NH_4)_2SO_4$ as described in Sakai *et al.*
(1986).

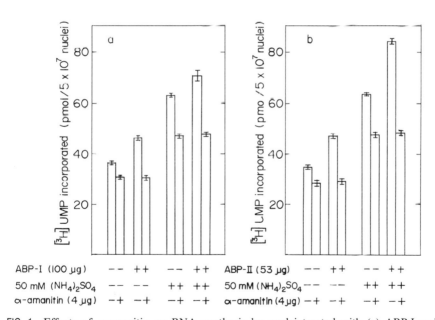

FIG. 1 Effects of α-amanitin on RNA synthesis by nuclei treated with (a) ABP-I and
(b) ABP-II. Data are averages of three experiments. The standard error for each set of values
is indicated by a vertical line. From Sakai *et al.* (1986; Fig. 7, p. 641).

stimulation of RNA synthesis by both ABP-I and ABP-II was due to the activation of RNA polymerase II.

VIII. Modulation of Gene Expression in Isolated Nuclei by Auxin-Binding Proteins

The auxin-specific alterations in gene expression within 15 min after application of auxin to plant tissues (Theologis, 1986) suggest that a net increase in RNA synthesis is not necessarily required for the short-term auxin action. To examine the possibility that, by interacting with ABPs, auxin may influence the synthesis of RNA qualitatively rather than quantitatively, we compared the products of the translation *in vitro* of poly(A)$^+$RNA extracted from isolated nuclei which were treated with ABP-I or ABP-II in the presence or absence of IAA.

To start this study, we first examined the integrity of transcription products in the preparation of isolated nuclei, because a good deal of evidence has accumulated to indicate that runoff transcription by isolated nuclei reflects the state of transcription of genes in the cell from which the nuclei were isolated, but among the transcription products there are few full-size transcripts and probably little or no re-initiation occurs. One-dimensional gel electrophoresis of RNA synthesized in the preparation of nuclei from mung bean hypocotyls revealed formation of RNAs that correspond to full-size rRNAs, i.e., 18S and 25S, and larger species of RNA (Kikuchi *et al.*, 1989). These results indicated that the preparation of isolated nuclei was capable of synthesizing transcripts of exact and appropriate sizes. Addition of either ABP-I or ABP-II alone stimulated synthesis of heterogeneous species of RNAs of between 4S and 15S in size, but IAA together with ABPs did not further alter the distribution of sizes of the newly synthesized RNA. Synthesis of these heterogeneous RNAs was inhibited by α-amanitin, indicating that these RNAs were synthesized by RNA polymerase II. Since pulse–chase experiments were not possible because of the lability of isolated nuclei, the heterogeneous RNA formed in the presence of ABPs and IAA should include RNA molecules in the course of synthesis.

To ascertain the effects of ABP and IAA on the initiation of transcription in the isolated nuclei, [γ-S]GTP was included in the reaction mixture. Although this method does not give accurate figures for the contribution of initiation to the total synthesis of RNA, the results (Table IV) indicated that the initiation of transcription by the isolated nuclei was increased by

TABLE IV

Initiation of Transcription in Isolated Nuclei[a]

Additions		Chromatography on Hg-Sepharose		
ABP (200 μg)[b]	IAA (10 μM)	[^{3}H]RNA applied ($\times$ 10^6 dpm)	[^{3}H]RNA eluted ($\times$ 10^4 dpm)	(%)
None	—	1.29	0.57	0.44
I	—	1.34	1.72	1.28
I	+	1.29	1.91	1.48
II	—	1.29	1.15	0.89
II	+	1.30	1.55	1.19

[a] From Kikuchi et al. (1989; Table 1, p. 767).
[b] Per 1 ml of reaction mixture.

ABP-I or ABP-II, and that the simultaneous addition of IAA and ABPs further increased the rate of initiation of transcription. These results suggested that the initiation of transcription occurred in the isolated nuclei and IAA-bound ABPs modulated the initiation of RNA synthesis.

Although addition of ABP-I or ABP-II alone to the isolated nuclei stimulated RNA synthesis (Sakai et al., 1986), the translatable transcripts in poly(A)$^+$ RNA from these nuclei did not show marked differences on isoelectric focusing (IEF)/SDS-PAGE and nonequilibrium pH gradient electrophoresis (NEPHGE)/SDS-PAGE from the control. From these results, it appeared that ABP-I or ABP-II alone stimulated the synthesis of the same species of mRNA that were synthesized in the control nuclei. However, the addition of IAA together with ABP-I or ABP-II to the nuclei resulted in the appearance of novel translatable transcripts that were not detected otherwise. As shown in Figs. 2 and 3, one peptide on an IEF/SDS gel (mw 79,000, pI 5.2) and another peptide on a NEPHGE/SDS gel (mw 18,000) were observed for which the relative amount of the corresponding mRNA increased significantly when the nuclei were treated with ABP-I plus IAA and ABP-II plus IAA, respectively.

The levels of two peptides which had the same molecular weights and isoelectric point as these specifically translated peptides were also increased in the translation products of poly(A)$^+$ RNA from IAA-treated or 2,4-D-treated sections of mung bean hypocotyls (Figs. 4 and 5). These results suggested that the mRNAs synthesized in the isolated nuclei in the presence of ABP plus IAA were also synthesized in tissues treated with auxin.

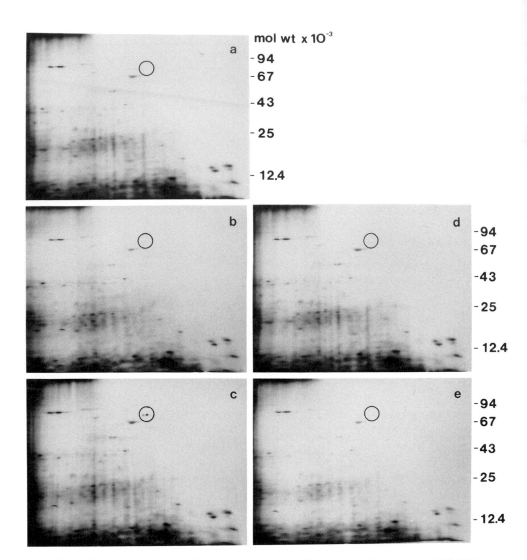

FIG. 2 Products of the translation (neutral and acidic polypeptides) of poly (A)$^+$ RNA prepared from isolated nuclei. (a) No addition, (b) +ABP-I; (c) +ABP-I + IAA, (d) +ABP-II, (e) +ABP-II + IAA. To aid in the comparison of data, all panels are marked with a circle at the position that corresponds to the polypeptide for which a difference was found in the samples. Such marking of a polypeptide was limited to the one that showed approximately the same difference in at least three separate experiments. The standard proteins of known molecular weight were phosphorylase b, 94,000; bovine serum albumin, 67,000; ovalbumin, 43,000; chymotrypsinogen A, 25,000; cytochrome *c*, 12,400. Their mobilities in only one direction are indicated. From Kikuchi *et al.* (1989; Fig. 2, p. 768).

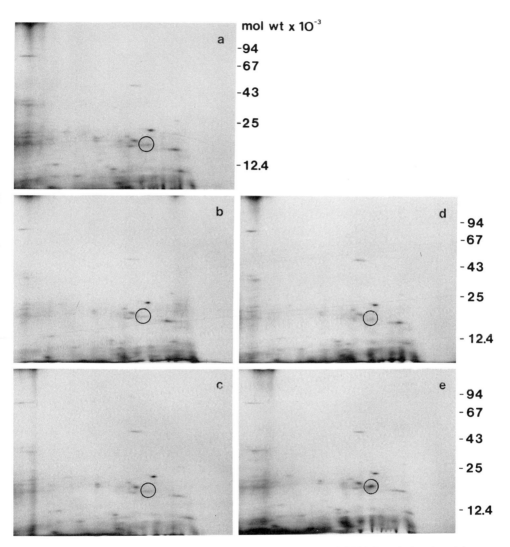

FIG. 3 Products of translation (basic polypeptides) of poly (A)$^{+}$ RNA from isolated nuclei. Conditions are as described in the legend of Fig. 2. From Kikuchi *et al.* (1989; Fig. 3, p. 769).

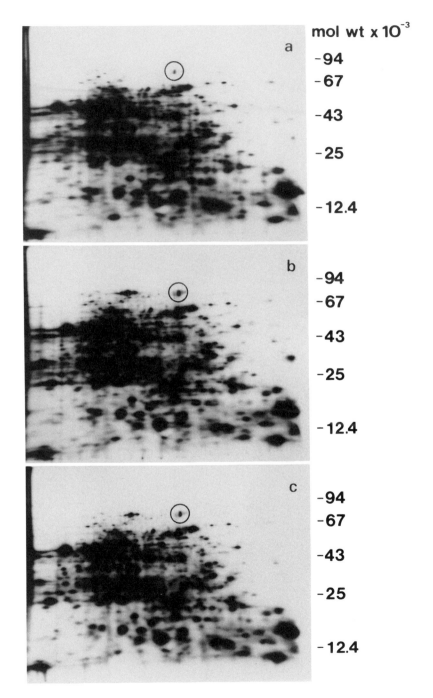

IX. Interaction of Auxin-Binding Proteins and RNA Polymerase II

As ABP-I and ABP-II stimulated mRNA synthesis and cause specific gene expression in the presence of IAA, interaction of ABPs and PNA polymerase II was examined using ABP-I. Nuclei were isolated from etiolated mung bean hypocotyls, and solubilized nuclear proteins were separated by a column of mono Q. α-Amanitin-insensitive RNA polymerase I and -sensitive RNA polymerase II were eluted at 0.1 and 0.25 M $(NH_4)_2SO_4$, respectively (Fig. 6). As the activity of RNA polymerase II was very labile and could not be purified, the fractions of RNA polymerase II were applied directly on a column of ABP-I-linked Sepharose 4B. As shown in Fig. 7, RNA polymerase II activity was completely retained on the affinity column and eluted by 0.3 M $(NH_4)_2SO_4$. Addition of 10 μM IAA to the buffer solution did not affect the retention of RNA polymerase II activity. From these results, it was considered that ABP-I had a binding affinity for RNA polymerase II. However, partially purified RNA polymerase II was used in this experiment, so we next examined this possibility using purified RNA polymerase II from wheat germ. RNA polymerase II was purified by the procedure reported by Jendrisak and Burgess (1975) and contaminants were removed by a column of DNA–agarose (Table V and Fig. 8).

As shown in Fig. 9, the purified RNA polymerase II was also retained on the column of ABP-I-linked Sepharose 4B and eluted by 0.3 M $(NH_4)_2SO_4$. However, the RNA polymerase II was not retained on a column of acid-denatured ABP-I-linked Sepharose 4B (Fig. 10).

The addition of ABP-I to the reaction mixture for *in vitro* RNA synthesis caused the stimulation of RNA polymerase II activity (Table VI). As ABP-I had no RNA polymerase activity (Sakai *et al.*, 1986), the stimulation was considered to be caused by the interaction of RNA polymerase II and ABP-I. The stimulation of RNA synthesis by ABP-I was not increased by the addition of 10 μM IAA. These results were very consistent with those of Venis (Section VI).

FIG. 4   Products of translation (neutral and acidic polypeptides) of poly (A)$^+$ RNA prepared from mung bean hypocotyl segments. (a) Control, (b) IAA-treated, (c) 2,4-D-treated. To aid in the comparison of data, the three panels are marked with a circle at the position that corresponds to the polypeptide for which a difference was found in Fig. 2. From Kikuchi *et al.* (1989; Fig. 4, p. 770).

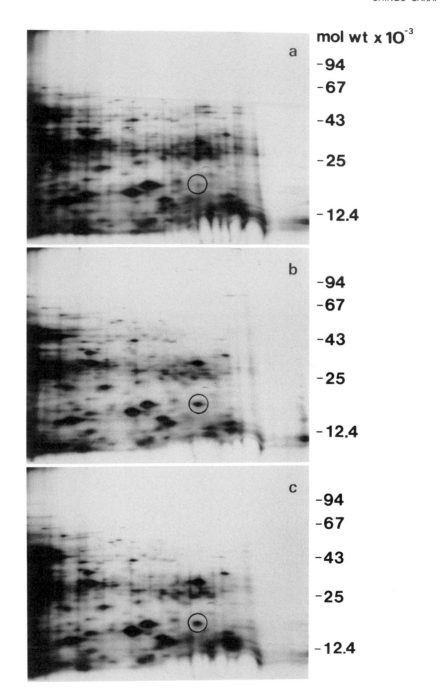

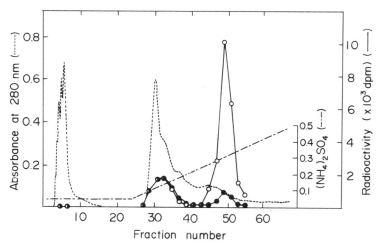

FIG. 6   Mono Q column profile of RNA polymerase solubilized from mung bean nuclei.
Nuclei were isolated from etiolated mung bean hypocotyls (150 g) as described (Sakai *et al.*,
1986). The nuclei were homogenized with 50 m*M* Tris-HCl buffer (pH 7.9) containing 0.5 *M*
(NH$_4$)$_2$SO$_4$, 0.1 m*M* EDTA, 5 m*M* MgCl$_2$, 10 m*M* 2-mercaptoethanol, and 25% (v/v)
glycerol. The homogenate was centrifuged at 27,000 g for 20 min. Solid (NH$_4$)$_2$SO$_4$ was added
to the supernatant to give 100% saturation, and the precipitate that formed was collected by
centrifugation at 27,000 g for 30 min. This precipitate was dissolved in 50 m*M* Tris-HCl buffer
(pH 7.9) containing 50 m*M* (NH$_4$)$_2$SO$_4$, 0.1 m*M* EDTA, 0.5 m*M* MgCl$_2$, 10 m*M* 2-
mercaptoethanol, and 25% (v/v) glycerol, and dialyzed against the same buffer for 18 hr with
several changes of the buffer. The precipitate formed was removed by centrifugation. A
portion of the supernatant (4 mg of protein) was loaded onto a column of mono Q and
fractionated with a linear gradient of (NH$_4$)$_2$SO$_4$ (50–500 m*M*). RNA polymerase activity was
assyed with 50 μl from each 0.5-ml fraction in 0.5 ml of a reaction mixture that contained
50 m*M* Tris-HCl buffer (pH 8.0), 10 m*M* dithiothreitol, 5 m*M* MgCl$_2$, 1 m*M* MnCl$_2$, 5% (v/v)
glycerol, 0.4 m*M* each of ATP, CTP, and GTP, 0.02 m*M* (5μCi) [5,6-³H]UTP, 50 m*M*
(NH$_4$)$_2$SO$_4$, and 40 μg of heat-denatured calf thymus DNA for 15 min at 30°C. The reaction
was terminated by the addition of 2 ml of 10% trichloroacetic acid containing 8 m*M* sodium
pyrophosphate. The precipitate was collected by centrifugation at 24,000 g for 30 min The
pellet was suspended in 1 ml of 10% trichloroacetic acid containing 8 m*M* sodium pyrophosphate and the suspension was centrifuged. This washing procedure was repeated two more
times. The precipitate was dissolved in 0.5 ml of 0.1 *M* KOH, the pH was adjusted with 0.5 ml
of 0.1 *M* Tris-HCl buffer (pH 7.6), and radioactivity of the solution was measured with a liquid
scintillation counter. ○, No α-amanitin; ●, 0.5 μg/ml α-amanitin.

FIG. 5   Products of translation (basic polypeptides) of poly (A)$^+$ RNA prepared from mung
bean hypocotyl segments. Conditions are as described in the legend to Fig. 4. To aid in the
comparison of data, the three panels are marked with a circle at the position that corresponds
to the polypeptide for which a difference was found in Fig. 3. From Kikuchi *et al.* (1989; Fig.
5, p. 770).

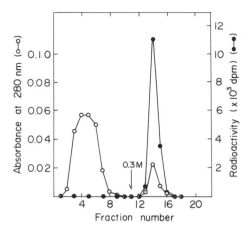

FIG. 7   Affinity chromatography on a column of ABP-I-linked Sepharose 4B of the partially purified mung bean RNA polymerase II. Purified ABP-I (4 mg) was coupled to 0.5 g of CNBr-activated Sepharose 4B (Pharmacia Fine Chemicals AB, Uppsala, Sweden) under the conditions suggested by the supplier. The fractions of RNA polymerase II (Fig. 6) were pooled and dialyzed against 50 mM Tris-HCl buffer (pH 7.9) containing 50 mM $(NH_4)_2SO_4$, 0.1 mM EDTA, 0.5 mM MgCl$_2$, 10 mM 2-mercaptoethanol, and 25% (v/v) glycerol. The fraction was then applied to the column of ABP-I-linked Sepharose 4B equilibrated with the same buffer. The column was eluted with 0.3 M $(NH_4)_2SO_4$ in the buffer. Fifty microliters from each 1-ml fraction was assayed as described in the legend to Fig. 6.

TABLE V

Purification of Wheat Germ RNA Polymerase II[a]

Purification step	Volume (ml)	Protein (mg)	Activity (units)[b]	Specific activity (units/mg)
Crude extract	3,400	127,840	1,512.5	0.012
Polyethyleneimine eluate	2,580	5,882	461.1	0.079
$(NH_4)_2SO_4$ precipitate	198	1,623	492.2	0.296
DEAE-Sephadex peak	290	78.3	257.6	3.29
Phosphocellulose peak	31	9.18	283.6	30.9
DNA-agarose peak	23	3.73	325.8	87.4

[a]RNA polymerase II was purified from 1 kg of wheat germ by the method of Jendrisak and Burgess (1975) and contaminants were removed by DNA-agarose chromatography.

[b]One unit is 1 nmol UMP incorporated/10min at 30°C.

(a) 5% gel   (b) 7.5 – 20% gel

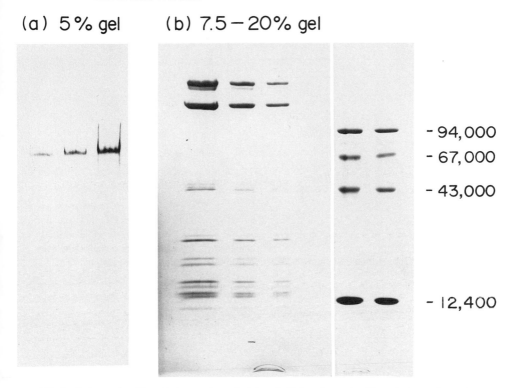

- 94,000
- 67,000
- 43,000
- 12,400

FIG. 8   Polyacrylamide gel electrophoresis (a) and SDS-polyacrylamide gel electrophoresis (b) of the purified wheat germ RNA polymerase II. The standard proteins of known molecular weight were phosphorylase b, 94,000; bovine serum albumin, 67,000; ovalbumin, 43,000, cytochrome *c*, 12,400.

X. Binding of Auxin-Binding Protein I to DNA

The results reported in Section VIII demonstrate that, in the transcription system composed of isolated nuclei, auxin interacts with soluble ABPs and stimulates the expression of specific genes which are also activated in tissues treated with auxin. There is a possibility that the auxin–ABP complex recognizes the promoter region of specific gene(s). Therefore, we examined interaction of ABP-I and DNA using affinity chromatography. Purified mung bean DNA was digested by *Sau*3AI, and the DNA fragments obtained were applied to a column of ABP-I-linked Sepharose 4B both in the presence and absence of 10 μM IAA. As shown in Fig. 11, most of the DNA fragments passed through the affinity column, but small amounts of DNA fragments were retained and could be eluted suc-

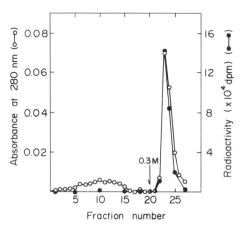

FIG. 9   Affinity chromatography on the column of ABP-I-linked Sepharose 4B of the purified wheat germ RNA polymerase II. The purified wheat germ RNA polymerase was applied to the column of ABP-I-linked Sepharose 4B. Conditions are as described in the legend to Fig. 7.

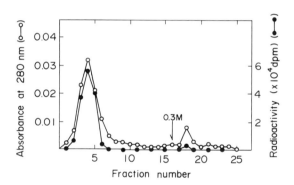

FIG. 10   Affinity chromatography on a column of acid-denatured ABP-I-linked Sepharose 4B of the purified wheat germ RNA polymerase II. The column of ABP-I-linked Sepharose 4B was treated with 0.1 M HCl for 10 min at 0°C and washed successively with distilled water and the same buffer used for the affinity chromatography. Conditions are as described in the legend to Fig. 7.

TABLE VI

Stimulation of Wheat Germ RNA Polymerase II Activity by ABP-I[a]

Addition	[³H]UMP incorporated (dpm)	%
None	9,750	100
ABP-I (19 µg)	19,890	204
ABP-I +IAA (10 µM)	19,830	203
Heat-denatured ABP-I	9,880	101

[a]The purified wheat germ RNA polymerase (1.7 µg) and mung bean DNA (34 µg) were used for the assay. Other conditions are as described in the legend to Fig. 6.

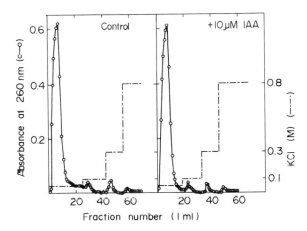

FIG. 11   Binding of DNA fragments on the column of ABP-I-linked Sepharose 4B. Mung bean DNA (1 mg) was purified from etiolated mung bean hypocotyls and digested by Sau3AI. The DNA fragments were precipitated by ethanol and dissolved in 10 ml of 10 mM Tris-HCl buffer (pH 7.6) containing 50 mM KCl and 10 mM 2-mercaptoethanol. The solution was passed through a column of Sepharose 4B (1 × 6 cm) previously washed with the same buffer. The DNA-fragment fractions were combined and a portion of the fractions (~0.5 mg) was applied to the column of ABP-I-linked Sepharose 4B in the presence or absence of 10⁻⁵ M IAA. The column was washed with the buffer and then eluted successively with 0.1, 0.3, and 0.8 M KCl.

cessively with 0.1, 0.3, and 0.8 M KCl. These results showed that ABP-I had a binding activity to DNA. We are now investigating whether ABP-I binds to different DNA fragments in the presence or absence of IAA.

XI. Auxin-Regulated Genes

Several cDNA clones to mRNA that are induced by auxin have been isolated and characterized (Baulcombe and Key, 1980; Walker and Key, 1982; Theologis et al., 1985; Hagen and Guilfoyle, 1985; McClure and Guilfoyle, 1987; Takahashi et al., 1989; Reddy and Poovaiah, 1990), and auxin-regulated genes also have been isolated (Theologis et al., 1985; Czarnecka et al., 1988; Hagen et al., 1988; Alliotte et al., 1989; Takahashi et al., 1990). Although several auxin-regulated genes have been isolated, regulatory sequences involved in hormonal activation of transcription of these genes have not been identified. Recently, Takahashi et al. (1990) have reported the existence of auxin-responsive elements in the par gene, which is expressed during the transition from G_0 phase to S phase in the early stage of tobacco mesophyll protoplasts cultured in vitro. When a chimeric gene, in which a reporter gene for bacterial β-glucuronidase (GUS) had been placed downstream of the 5′ flanking sequences of the par gene, was introduced into tobacco mesophyll protoplasts by electroporation, the chimeric gene elicited auxin-regulated expression of GUS activity. Because deletion of a 111-base pair (bp) direct repeat in the 5′ flanking sequences of the par gene abolished the auxin-induced GUS activity, they claimed that located in the 111-bp direct repeat of the par gene promoter is an auxin-responsive region which regulates auxin-mediated activation of transcription. In this context, the relationship between an auxin-binding protein and trans-acting factors that induce par gene expression should be interesting.

XII. Summary

Since the effects of auxin on plant tissues are complex, the mode of action of auxin at the molecular level may not depend on a single mechanism. There may be a mechanism by which the interaction of auxin with receptors localized in the cytoplasmic membranes activates certain enzymes which are necessary to generate the putative second messengers. On the other hand, soluble auxin-binding proteins have been isolated from a variety of plant tissues. Some of these proteins have a high affinity for auxins and the binding is auxin specific, reversible, and saturable, characteristics which suggest that these proteins may be auxin receptors. Although these criteria are often used to distinguish real receptors from nonfunctional binding proteins, it is necessary to clarify the biological function of the binding proteins to classify them as putative receptors.

The reported results on the function of soluble auxin-binding proteins demonstrate that, in the transcription system composed of isolated nuclei, auxin interacts with soluble auxin-binding proteins and stimulates the expression of specific genes. Thus, one of the mechanisms of action of auxin may involve a direct interaction with a soluble receptor protein, such that the resultant auxin–receptor complex, possibly together with other protein factors, can subsequently recognize the promoter region of specific gene(s) and interact with RNA polymerase II to cause a transcription of the gene(s).

Acknowledgments

I wish to express my thanks to Professor Darryl Macer (University of Tsukuba) for his critical reading of the manuscript. This work was supported in part by grants from the Ministry of Education, Science and Culture, Japan, and by a Grant-in-Aid (Bio Media Program) from the Ministry of Agriculture, Forestry and Fisheries, Japan.

References

Alliotte, T., Tiré, C., Engler, G., Peleman, J., Caplan, A., Van Montagu, M., and Inzé, D. (1989). *Plant Physiol.* **89,** 743–752.

Bailey, H. M., Barker, E. J. D., Libbenga, K. R., Van der Linde, P. C. G., Mennes, A. M., and Elliott, M. C. (1985). *Biol. Plant.* **27,** 105–109.

Baulcombe, D. C., and Key, J. L. (1980). *J. Biol. Chem.* **255,** 8907–8913.

Berridge, M. J. (1987). *Annu. Rev. Biochem.* **56,** 159–193.

Bogers, R. J., and Libbenga, K. R. (1981). *In* "Aspects and Prospects of Plant Growth Regulators" (B. Jeffcoat, ed.), Monogr. 6, pp. 177–185. Wessex Press, Wantage, Oxfordshire.

Bogers, R. J., Kulescha, Z., Quint, A., Van Vliet, T. B., and Libbenga, K. R. (1980). *Plant Sci. Lett.* **19,** 311–317.

Chen, Y., Lin, C., Chang, H., Guilfoyle, T. J., and Key, J. L. (1975). *Plant Physiol.* **56,** 78–82.

Cherry, J. H. (1967). *Ann. N. Y. Acad. Sci.* **144,** 154–168.

Clark, J. E., Morré, D. J., Cherry, J. H., and Yunhans, W. N. (1976). *Plant Sci. Lett.* **7,** 233–238.

Czarnecka, E., Nagao, R. T., Key, J. L., and Gurley, W. B. (1988). *Mol. Cell. Biol.* **8,** 1113–1122.

Davies, P. J. (1987). *In* "Plant Hormones and Their Role in Plant Growth and Development." Martinus Nijhoff Publishers, Kluwer Academic Publishers, Dordrecht, The Netherlands.

Dohrmann, D., Hertel, R., and Kowalik, H. (1978). *Planta* **140**, 97–106.
Ettlinger, C., and Lehle, L. (1988). *Nature (London)* **331**, 176–178.
Evans, M. L. (1974). *Annu. Rev. Plant Physiol.* **25**, 195–223.
Hagen, G., and Guilfoyle, T. J. (1985). *Mol. Cell. Biol.* **5**, 1197–1203.
Hagen, G., Uhrhammer, N., and Guilfoyle, T. J. (1988). *J. Biol. Chem.* **263**, 6442–6446.
Hardin, J. W., O'Brien, T. J., and Cherry, J. H. (1970). *Biochim. Biophys. Acta* **224**, 667–670.
Hardin, J. W., Cherry, J. H., Morré, D. J., and Lembi, C. (1972). *Proc. Natl. Acad. Sci. U.S.A.* **69**, 3146–3150.
Herber, B., Ulbrich, B., and Jacobsen, H.-J. (1988). *Plant Cell Rep.* **7**, 178–181.
Hertel, R., Thomson, K. St., and Russo, V. E. A. (1972). *Planta* **107**, 325–340.
Ihl, M. (1976). *Planta* **131**, 223–228.
Jacobsen, H.-J. (1981). *Cell Biol. Int. Rep.* **5**, 768.
Jacobsen, H.-J. (1982). *Physiol. Plant.* **56**, 161–167.
Jacobsen, H.-J. (1984). *Plant Cell Physiol.* **25**, 867–873.
Jacobsen, H.-J., and Hajek, K. (1985). *Biol. Plant.* **27**, 110–113.
Jacobsen, H.-J. and Hajek, K. (1987). *In* "Molecular Biology of Plant Growth Control" (J. E. Fox and M. Jacobs, eds.), pp. 257–266. Alan R. Liss, New York.
Jacobsen, H.-J., Hajek, K., Mayerbacher, R., and Herber, B. (1987). *In* "Plant Hormone Receptors" (D. Klämbt, ed.), pp. 63–69. Springer-Verlag, Berlin.
Jacobsen, J. V. (1977). *Annu. Rev. Plant Physiol.* **28**, 537–564.
Jendrisak, J. J., and Burgess, R. R. (1975). *Biochemistry* **14**, 4639–4645.
Kaur, S., and Kapoor, H. C. (1990). *Plant Sci.* **72**, 151–157,
Kende, H., and Gardner, G. (1976). *Annu. Rev. Plant Physiol.* **27**, 267–290.
Key, J. L. (1969). *Annu. Rev. Plant Physiol.* **20**, 449–474.
Kikuchi, M., Imaseki, H., and Sakai, S. (1989). *Plant Cell Physiol.* **30**, 765–773.
Lamb, C. J. (1978). *Nature (London)* **274**, 312–314.
Libbenga, K. R. (1978). *Proc. Int. Congr. Plant Tissue Cell Culture, 4th, Manitoba 1978* pp. 325–333.
Libbenga, K. R., and Mennes, A. M. (1987). *In* "Plant Hormones and Their Role in Plant Growth and Development" (P. J. Davies, ed.), pp. 194–217. Martinus Nijhoff Publishers, Kluwer Academic Publishers, Dordrecht, The Netherlands.
Libbenga, K. R., Van Telgen, H. J., Mennes, A. M., Van der Linde, P. C. G., and Van der Zaal, E. J. (1978). *In* "Molecular Biology of Plant Growth Control" (J. E. Fox and M. Jacobs, eds.), pp. 229–243. Alan R. Liss, New York.
Libbenga, K. R., Mann, A. C., Van der Linde, P. C. G., and Mennes, A. M. (1986). *In* "Hormones, Receptors and Cellular Interactions in Plants" (C. M. Chadwick and D. R. Garrod, eds.), pp. 1–68. Cambridge Univ. Press, Cambridge.
Likholat, T. V., Pospelov, V. A., Morozova, T. M., and Salganik, R. I. (1974). *Sov. Plant Physiol. (Engl. Transl.)* **21**, 779–784.
Löbler, M., and Klämbt, D. (1985). *J. Biol. Chem.* **260**, 9848–9853.
Maheshwari, S. C., Guha, S. and Gupta, S. (1966). *Biochim. Biophys. Acta* **117**, 470–472.
Matthysse, A. G. (1970). *Biochim. Biophys. Acta* **199**, 519–521.
Matthysse, A. G., and Phillips, C. (1969). *Proc. Natl. Acad. Sci. U.S.A.* **63**, 897–903.
McClure, B. A., and Guilfoyle, T. J. (1987). *Plant Mol. Biol.* **9**, 611–623.
Mennes, A. M., Nakamura, C., Van der Linde, P. C. G., Van der Zaal, E. J., Van Telgen, H. J., Quint, A., and Libbenga, K. R. (1986). *In* "Plant Hormone Receptors" (D. Klämbt, ed.), pp. 51–62. Springer-Verlag, Berlin.
Mondal, H., Mandal, R. K., and Biswas, B. B. (1972). *Nature (London)*, New Biol. **240**, 111–113.
Murphy, G. J. P. (1979). *Plant Sci. Lett.* **15**, 183–191.
Murphy, G. J. P. (1980). *Plant Sci. Lett.* **19**, 157–168.

O'Brien, T. J., Jarvis, B. C., Cherry, J. H., and Hanson, J. B. (1968). *Biochim. Biophys. Acta* **169**, 35–43.

Oostrom, H., Van Loopik-Detmers, M. A., and Libbenga, K. R. (1975). *FEBS Lett.* **59**, 194–197.

Oostrom, H., Kulescha, Z., Van Vliet, T. B., and Libbenga, K. R. (1980). *Planta* **149**, 44–47.

Reddy, A. S. N., and Poovaiah, B. W. (1990). *Plant Mol. Biol.* **14**, 127–136.

Rizzo, P. J., Pederson, K., and Cherry, J. H. (1977). *Plant Sci. Lett.* **8**, 205–211.

Roy, P., and Biswas, B. B. (1977). *Biochem. Biophys. Res. Commun.* **74**, 1597–1606.

Roychoudhry, R., and Sen, S. P. (1964). *Physiol. Plant.* **17**, 342–362.

Roychoudhry, R., and Sen, S. P. (1965). *Plant Cell Physiol.* **6**, 761–765.

Rubery, P. H. (1981). *Annu. Rev. Plant Physiol.* **32**, 569–596.

Sakai, S. (1984). *Agric. Biol. Chem.* **48**, 257–259.

Sakai, S. (1985). *Plant Cell Physiol.* **26**, 185–192.

Sakai, S., and Hanagata, T. (1983). *Plant Cell Physiol.* **24**, 685–693.

Sakai, S., Seki, J., and Imaseki, H. (1986). *Plant Cell Physiol.* **27**, 635–643.

Schwimmer, S. (1968). *Plant Physiol.* **43**, 1008–1010.

Shimomura, S., Sotobayashi, T., Futai, M., and Fukui, T. (1986). *J. Biochem. (Tokyo)* **99**, 1513–1524.

Stoddart, J. L., and Venis, M. A. (1980). *Encyclo. Plant Physiol., New Ser.* **9**, 445–510.

Takahashi, Y., Kuroda, H., Tanaka, T., Machida, Y., Takebe, I., and Nagata, T. (1989). *Proc. Natl. Acad. Sci. U.S.A.* **86**, 9279–9283.

Takahashi, T., Niwa, Y., Machida, Y., and Nagata, T. (1990). *Proc. Natl. Acad. Sci. U.S.A.* **87**, 8013–8016.

Teissere, M., Penon, P., Van Huystee, R. B., Azou, Y., and Ricard, J. (1975). *Biochim. Biophys. Acta* **402**, 391–402.

Theologis, A. (1986). *Annu. Rev. Plant Physiol.* **37**, 407–438.

Theologis, A., and Ray, P. M. (1982). *Proc. Natl. Acad. Sci. U.S.A.* **79**, 418–421.

Theologis, A., Huynh, T. V., and Davis, R. W. (1985). *J. Mol. Biol.* **183**, 53–68.

Van der Linde, P. C. G., Bouman, H., Mennes, A. M., and Libbenga, K. R. (1984). *Planta* **160**, 102–106.

Van der Linde, P. C. G., Maan, A. C., Mennes, A. M., and Libbenga, K. R. (1985). *Proc. FEBS Meet.* **16**, Part C, 397–403.

Venis, M. A. (1971). *Proc. Natl. Acad. Sci. U.S.A.* **68**, 1824–1827.

Venis, M. A. (1977). *Adv. Bot. Res.* **5**, 53–88.

Venis, M. A. (1981). *In* "Aspects and Prospects of Plant Growth Regulators" (B. Jeffcoat, ed.), Monogr. 6, pp. 187–195. Wessex Press, Wantage, Oxfordshire.

Venis, M. A. (1984). *Planta* **162**, 502–505.

Walker, J. C., and Key, J. L. (1982). *Proc. Natl. Acad. Sci. U.S.A.* **79**, 7185–7189.

Wardrop, A. J., and Polya, G. M. (1977). *Plant Sci. Lett.* **8**, 155–163.

Wardrop, A. J., and Polya, G. M. (1980a). *Plant Physiol.* **66**, 105–111.

Wardrop, A. J., and Polya, G. M. (1980b). *Plant Physiol.* **66**, 112–118.

Zurfluh, L. L., and Guilfoyle, T. J. (1982a). *Plant Physiol.* **69**, 332–337.

Zurfluh, L. L., and Guilfoyle, T. J. (1982b). *Plant Physiol.* **69**, 338–340.

Zurfluh, L. L., and Guilfoyle, T. J. (1982c). *Planta* **156**, 525–527.

Subcellular and Molecular Mechanisms of Bile Secretion

Susan Jo Burwen, Douglas L. Schmucker, and Albert L. Jones

Cell Biology and Aging Section, Veterans Administration Medical Center, San Francisco, California 94121 and the Departments of Medicine and Anatomy and the Liver Center, University of California, San Francisco, San Francisco, California 94143

I. Introduction

Bile is well known as a yellow pigmented fluid that is secreted by liver and delivered to the duodenum of the small intestine. Bile serves two major purposes: to assist in the digestion of dietary fats, by facilitating lipid absorption by the gut, and to provide a route, in addition to the kidney, for the removal and elimination of waste products, including drugs as well as natural metabolites. For example, bilirubin, which gives bile its yellow color, is taken up from plasma by hepatocytes, conjugated, and secreted into bile, from whence it is eliminated via the bowel.

Bile consists of water and inorganic ions (electrolytes) and contains many organic solutes, including bile acids, cholesterol, phospholipid, proteins and protein breakdown products, bile pigments, drug metabolites, and other organic anions and cations. Many of these constituents, such as bile acids, phospholipids, and some biliary proteins, are enriched in bile relative to their plasma concentrations. Bile is isosmotic with plasma. The osmotic activity of the organic constituents, particularly bile acids, phospholipids, and cholesterol, is reduced by micelle formation, and biliary proteins are at fairly low concentration (even though some are enriched above their plasma concentrations). The inorganic electrolytes account for nearly all of the osmotic activity of the bile.

Bile formation results from the active transport of solutes by hepatocytes into the canaliculus, followed by passive water flow (Boyer, 1980; Blitzer and Boyer, 1982; Scharschmidt and VanDyke, 1983). Electrolyte transport mechanisms are requisite for bile formation, since the uptake of many biliary constituents is sodium-dependent. The active transport of electrolytes also creates the osmotic gradients that result in passive water flow.

The composition of bile depends on the site of sampling. Bile formation starts at the level of the hepatocyte. The primary secretion from the hepatocyte into the bile canaliculus is modified by additions and subtractions (particularly water and electrolytes) during the course of its journey to the common bile duct. Its composition is further modified during storage in the gall bladder between meals.

The term "canalicular bile" refers to bile formed by the hepatocyte. It is technically impossible to sample the canalicular contents *in vivo*. Bile composition data have been obtained primarily by sampling bile collected from cannulas inserted into the common bile duct. Recently, an *in vitro* system that allows direct sampling of canalicular contents has been developed (Boyer *et al.*, 1988). Primary cultures of pairs of rat hepatocytes (couplets) form a sealed canalicular compartment, from which bile can be withdrawn by micropuncture. Isolated hepatocyte couplets serve as a model for the study of bile secretion, as well as of canalicular bile composition, although the composition of the bathing medium directly affects the composition of the canalicular contents. Since the original observations of microfilament-dependent bile canalicular contractility and its role in bile secretion, the potential of hepatocyte couplets has increased substantially and the model has been well characterized (Oshio and Phillips, 1981; Gautam *et al.*, 1987, 1989; Boyer *et al.*, 1988). Furthermore, primary hepatocyte cultures grown on gas-permeable membranes offer another *in vitro* model for the study of pericanalicular bile formation (Petzinger *et al.*, 1988).

Other techniques have contributed to our knowledge of the molecular mechanisms of bile formation. Isolation and separation of highly purified basolateral and canalicular plasma membrane vesicles from rat liver (Inoue *et al.*, 1983; Meier *et al.*, 1984; Blitzer and Donovan, 1984) initiated the investigation of polarized transport mechanisms at the membrane level. Photoaffinity labeling techniques have been successfully applied to identify domain-specific membrane proteins involved in sinusoidal uptake and canalicular secretion of bile acids and other amphipathic substances (Kramer *et al.*, 1982; von Dippe *et al.*, 1983; Wieland *et al.*, 1984; Fricker *et al.*, 1987b; Berk *et al.*, 1987; Frimmer and Ziegler, 1988). Other biochemical techniques, as well as immuno- and cytochemistry and electron microscopy, have also been applied to studies on the molecular mechanisms of bile formation.

This article discusses the source of the constituents of bile (some are made by liver and some are derived from plasma), their metabolism by liver, and the pathways by which they reach bile. The contribution of biliary constituents to bile formation and flow, and regulation of biliary secretion, are also addressed, with an emphasis on structure–function relationships.

II. Structure of the Bile Secretory Apparatus

A. Organization of the Liver Lobule

The hundreds of billions of cells in a normal liver are divided into small structural units known as lobules that are easily seen at the light microscope level, polyhedral in shape, and approximately 1 × 2 mm in size. The normal human liver consists of about one million of these lobules. Concepts of lobule configuration have been derived by considering specific functions of these small structural units. Historically, the first concept of lobular structure placed the bile ductules and portal veins at the center of the liver lobule, and the periphery of the lobule was defined by the central veins. Three central veins surrounding one portal area comprised a triangular-shaped lobule. Since this lobular configuration did not have any practical application to liver function or pathology, its use was abandoned.

In the more anatomical concept of lobule structure (Fig. 1), the classic triad structures (hepatic artery, portal vein, and bile ductules) (Fig. 2) are located around the periphery of the lobule. At the center of the lobule, the central vein is the beginning of the venous drainage of the liver. The central veins, like stems of a bunch of grapes, flow together and eventually become the hepatic veins which drain the blood directly into the inferior vena cava. In cross section, the lobule is a hexagonal structure, with the interlobular portal canals, containing the triad structures, located at the angles of the hexagons. The paralobular vessels and bile ductules branch out from the interlobular triads and surround the periphery of the lobule. Connective tissue and liver lymphatics are contained in these inter- and perilobular triads. In certain species, such as pig, raccoon, camel, and polar bear, the connective tissue is well developed and the outline of the classic hexagonal lobule is easy to delineate. Plates of liver, from 12 to 24 cells, extend radially from the central vein to the periphery of the lobule. These plates, or lacuni, do not communicate freely with the portal vein. A limiting plate of hepatic cells surrounds the circumference of the liver lobule, forming an almost continuous wall between the interior of the lobule and the space occupied by the portal canals. Openings in this limiting plate allow triad structures to come in contact with the hepatic parenchyma.

A more functional concept of lobule structure has come into widespread usage because it takes into account hepatic blood flow patterns and ob-

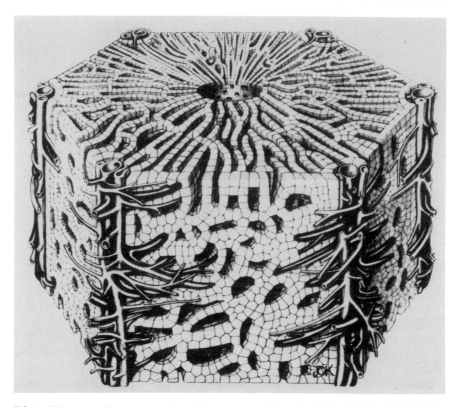

FIG. 1 Diagrammatic representation of the anatomy of the classic liver lobule. The lobule is hexagonal in shape, and the triad structures are located at the periphery of the lobule at the angles of the hexagon. The central vein, located at the center of the lobule, is the site where the venous drainage of the liver begins. Plates of hepatocytes, called lacuni, extend radially from the central vein to the periphery of the lobule. A limiting plate of hepatocytes forms a wall between the interior of the lobule and the space occupied by the triad structures.

served functional changes in certain liver diseases. In this concept, the area of entry of lobule blood flow, surrounding the portal vein, which is at the most peripheral part of the hexagonal lobule, is considered zone 1. The area at the termination of lobular blood flow, near the central vein (also known as the terminal hepatic venule) is designated as zone 3. Zone 1 cells are the first to receive blood and proliferate during liver regeneration (Grisham, 1962) and the last to undergo necrosis. Regions more distal to zone 1, i.e., zones 2 and 3, receive blood with fewer nutrients and less oxygen, and are the most susceptible to damage by certain hepatotoxins.

Goresky *et al.* (1973) were the first to propose a portal-to-central lobular concentration gradient for substances that are efficiently cleared from the

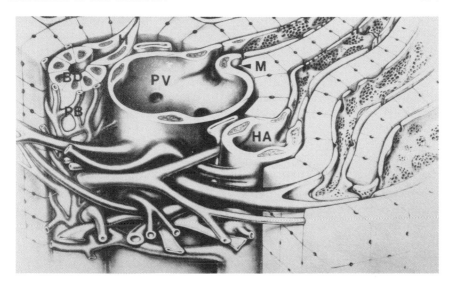

FIG. 2   Diagram of a classic liver triad structure, consisting of the portal vein (PV), hepatic artery (HA), and bile ductule (BD). Also shown are the peribiliary plexus (PB), the canal of Hering (H), and the space of Mall (M).

plasma by liver, by demonstrating that galactose was taken up exclusively by zone 1 cells. Subsequently, a number of substances have been shown to be taken up preferentially by the peripheral, or zone 1, cells: bile acids (Jones *et al.*, 1980; Suchy *et al.*, 1983), certain asialoglycoproteins (Daniels *et al.*, 1987), and epidermal growth factor (St. Hilaire *et al.*, 1983). Numerous studies undertaken to determine metabolic zonation of liver parenchyma have suggested that, in addition to the preferential uptake function mentioned above, periportal zone 1 cells are the most active in gluconeogenesis, fatty acid oxidation, and amino acid utilization. The more distal (zones 2 and 3) cells are the most active in glycolysis, drug transformation, and ammonia detoxification (Gumucio, 1983; Jungermann and Katz, 1982). The greater abundance of smooth endoplasmic reticulum in zone 3 cells as compared to zone 1 cells apparently accounts for the higher rate of drug metabolism by zone 3 cells and their selective damage by hepatotoxins whose metabolism results in the formation of toxic free radicals.

Bile salt uptake by liver is very efficient; >90% of an injected dose of taurocholate is taken up in a single pass. Jones *et al.* (1980) demonstrated a lobular concentration gradient for bile salt uptake using ^{125}I-labeled cholylglycylhistamine. Initial evidence supporting the formation of lobular concentration gradients by bile salts was based on the observation that canalicular lumen size, which is enlarged in the presence of physiological concentrations of bile salts, follows a lobular gradient, from larger diameters in zone 1 to smaller diameters in zone 3 cells. Infusion of supraphysiological concentrations of bile salts abolishes this gradient and all lumena are enlarged (Jones *et al.,* 1978), indicating that, under basal conditions, portal hepatocytes transport most of the bile salts, but, with increased bile salt loads, central hepatocytes are recruited (Blitzer and Boyer, 1982). The expansion of the bile acid pool is able to change the bile salt secretory characteristics of zone 3 hepatocytes toward those of zone 1 cells (Reichen and Le, 1989).

Selective damage of different areas of the liver lobule by pharmacological agents provides information as to the relative contributions of hepatocytes in different zones to bile formation. Allyl alcohol, which selectively damages portal areas, causes decreased bile salt concentration in bile; bromobenzene, which selectively damages central areas, causes increased bile salt concentration in bile (Gumucio *et al.,* 1978). These results suggest that, under basal conditions, pericentral (zone 3) hepatocytes contribute most to bile acid-independent flow (very low bile acid secretion) whereas periportal (zone 1) hepatocytes contribute most to bile acid-dependent flow (high bile acid secretion).

B. The Biliary Conduit

The smallest branches of the biliary tree are the bile canaliculi, which are minute extracellular areas located between two or more adjacent hepatocytes (Fig. 3). The bile canaliculi are isolated from the remainder of the intercellular space, and thus the sinusoidal lumen, by junctional complexes (Figs. 4 and 7b). The most important component of the junctional complex is the tight junction, which is always located immediately adjacent to and completely surrounding the canalicular lumen. Under physiological conditions, this junction has been considered relatively impervious to even small molecules. Tight junctions are negatively charged (Bradley and Herz, 1978) and are less permeable to anions than to uncharged solutes. Therefore, organic anions, including bile acids, are inhibited from crossing tight junctions in either direction (blood-to-bile or bile-to-blood) (Blitzer and Boyer, 1982), ruling out the paracellular pathway (across tight junctions) as physiologically significant for biliary organic constituents. The area of cytoplasm immediately surrounding the bile canaliculus is designated as the pericanalicular cytoplasm. During times of enhanced bile secretion, this area of cytoplasm contains many vesicles, indicating a role for vesicles in bile secretion.

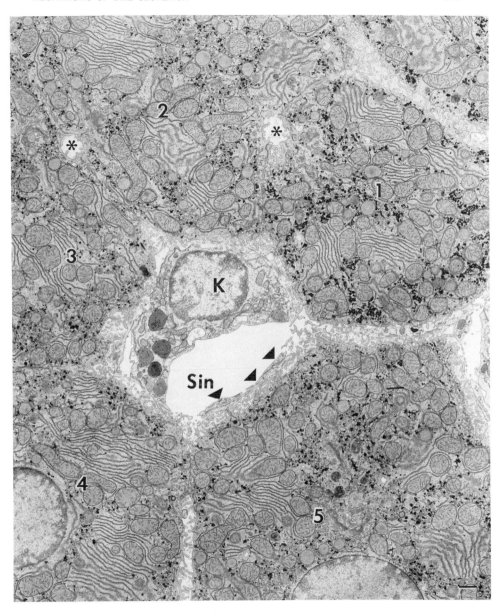

FIG. 3   Electron micrographic overview of five hepatocytes (labeled 1–5) and a sinusoidal area (Sin). Hepatocytes are rich in mitochondria and endoplasmic reticulum. Asterisks designate the bile canaliculi between adjacent hepatocytes. A Kupffer cell (K) lies within the sinusoidal space. An endothelial cell with numerous fenestrae is indicated by arrowheads. Bar = 1 μm.

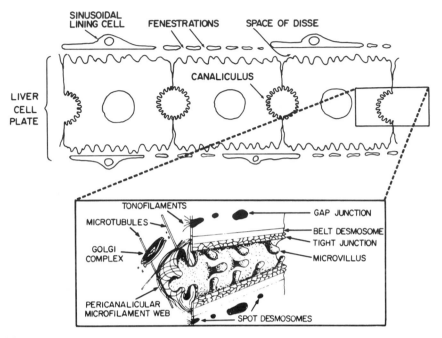

FIG. 4 Diagram of junctional complexes surrounding the bile canaliculus. The tight junction, which completely surrounds the canalicular lumen, is negatively charged and relatively impermeable to even small molecules.

Canalicular bile proceeds down the bile canaliculi, moving in the opposite direction of blood flow, from zone 3 toward zone 1 of the liver lobule. The canalicular bile enters the small terminal bile ductules, which initially consist of a few fusiform cells held together by junctional complexes in close association with the adjacent liver cells (canals of Hering) (Fig. 5) (Jones and Spring-Mills, 1983). The bile conveyed through the limiting plate of hepatocytes moves into larger perilobular ductules and interlobular ducts. Although in most mammals, with the exception of ruminants, there is no basement membrane visible by electron microscopy within the hepatic parenchyma, beginning at the canals of Hering and throughout the intra- and extrahepatic duct system, all lining epithelial cells contain a distinct basement membrane. Bile duct and ductule epithelial cells contain a prominent Golgi complex and numerous vesicles that undoubtedly are related to the transport of substances between bile and plasma (Fig. 6). Smooth muscle cells surrounding the larger ductules may play a role in the flow of the bile. In larger ducts, areas of mucous-secreting epithelium are surrounded by a rich vascular network. The large extrahepatic ducts are lined by columnar cells and their walls are similar to those of the intestine.

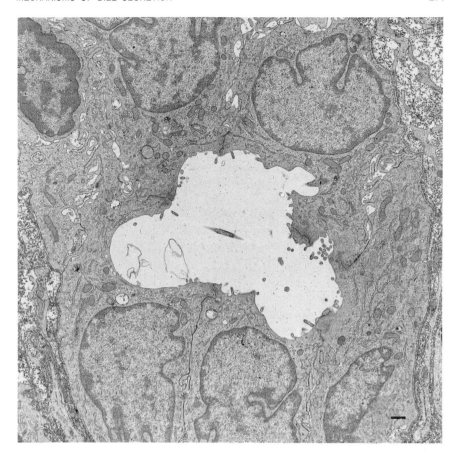

FIG. 5 Electron micrograph of a canal of Hering, consisting of epithelial cells held together by tight junctions. The electron-opaque material between adjacent cells is the cytochemical reaction product of horseradish peroxidase, which was injected into the bloodstream for the purpose of demonstrating the integrity of the tight junctions. Bar = 1 μm.

The gallbladder is a dispensable bag for the storage of bile. The human gallbladder holds 30–50 ml of bile. The epithelial cells lining the gallbladder are columnar cells whose apices contain numerous microvilli, lateral junctional complexes, and associated tonofilaments. The Golgi apparatus is supranuclear and there are well-developed endoplasmic reticulum and mitochondria. Glands are occasionally found in the lamina propria of the human gallbladder. Rokitansky-Aschoff crypts are seen in the surface epithelium. The cystic duct allows bile to flow freely from the common hepatic duct into the gallbladder and back again via a spiral valve of mucosa which holds the duct open at all times. At the distal end of the

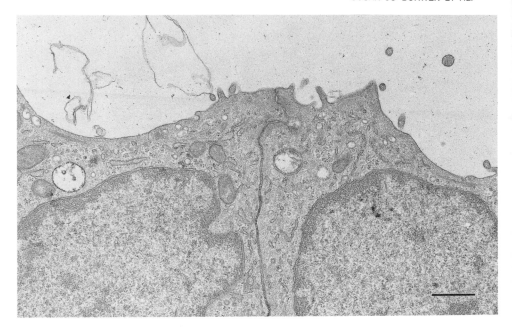

FIG. 6   Electron micrograph showing numerous cytoplasmic vesicles near the apical surface of bile ductule epithelial cells. These vesicles are presumed to play a role in the transport of substances between bile and blood. Bar = 1 μm.

common bile duct, at the site of its entrance into the intestine, a complex junctional region, referred to as the choledochoduodenal junction, regulates both the flow of bile into the intestine and the filling of the gallbladder, under hormonal and autonomic nervous system control.

C. Hepatocytes and Their Organelles

The formation of bile hinges on one major concept universal to all epithelia, which is that separation of two fluid compartments by epithelial cells permits the different composition of fluid (e.g., blood versus bile) in each compartment. The major epithelial cell in the liver is the hepatic parenchymal cell, or hepatocyte (Fig. 3), which accounts for 80% of the liver's volume. The cells range from 20 to 30 μm in diameter, are polyhedral in shape, and are arranged in plates dispersed between blood vascular channels or sinusoids. The bile canaliculi are centrally located between adjacent hepatocytes. Tight junctions between adjacent hepatocytes separate the bile canalicular (apical) from the sinusoidal (basolateral) hepatocyte surface and isolate the canalicular lumen (biliary or mucosal compartment) from the sinusoids (vascular or serosal compartment).

Hepatocytes are unrivaled by any other parenchymal cell type in functional diversity and complexity. They secrete bile salts into the intestine, thereby facilitating emulsification and absorption of dietary fat. They take up digested material from the afferent blood and store carbohydrates, protein, vitamins, and lipid, releasing these compounds often in association with a carrier (e.g., triglycerides complexed to protein). They synthesize albumin and other plasma proteins, glucose, fatty acid, cholesterol, and phospholipid. They metabolize, detoxify, and inactivate exogenous compounds, such as drugs and insecticides, and endogenous compounds, such as steroids and other hormones. Also, hepatocytes convert endogenous substances into more active forms (e.g., T4 to T3). Furthermore, in certain rodents, they play a major role in the intestinal immune system by sequestering circulating immunoglobulin A and secreting it into the intestine via the bile.

Due to their wide diversity of function, hepatocytes contain almost every organelle found in the animal kingdom. Mitochondria are especially numerous in liver cells (800/cell) (Fig. 3) and participate in oxidative phosphorylation and fatty acid oxidation. Changes in mitochrondrial structure, i.e., swelling or loss of intramitochondrial granules (calcium), are early nonspecific indications of hepatic injury. The general function of lysosomes appears to be the disposition and catabolism of exogenous and endogenous substances. Microbodies, which occupy 4% of the total cell volume, are specialized for β-oxidation of long-chain fatty acids. In most mammalian hepatocytes, the Golgi complex is composed of three to five closely packed, parallel, smooth-surfaced cisternae, variable numbers of large and small vesicles, and, occasionally, lysosomal elements. The word complex has been used because of the apparent "complexity" of this organelle with regard to its multifunctional capabilities, including very-low-density lipoprotein production, albumin secretion, and terminal glycosylation of many proteins, including plasma membrane receptors.

The endoplasmic reticulum is composed of a minute system of membrane-limited canals which form a continuous network of tubulocisternae that stretches from the plasma membrane to the bile canaliculus (Moller et al., 1983). Unlike cells in most tissues, in hepatocytes, both the rough- and smooth-surfaced varieties are well developed. In transmission electron micrographs, the rough endoplasmic reticulum usually appears as aggregates of ribosome-studded, flattened cisternae arranged in parallel profiles scattered randomly throughout the cytoplasm, whereas the smooth endoplasmic reticulum appears as minute membrane-limited tubules without ribosomes. The endoplasmic reticulum (microsomes) probably is involved, either directly or indirectly, in essentially every function of hepatocytes. Hepatic drug-metabolizing enzymes and enzymes involved in cholesterol biosynthesis and the conversion of cholesterol to bile acids are located in the smooth endoplasmic reticulum. The surface area of this membrane system is extraordinary. Morphometric data indicate that the liver of a 70 kg man contains enough smooth endoplasmic reticulum to cover a football field. The liver can rapidly increase the amount of smooth endoplasmic reticulum in response to the need for metabolism of xenobiotics.

Microfilaments and microtubules, which comprise the cytoskeleton, play an essential role in bile secretion. As a generalization, microfilaments are required for hepatocellular bile salt secretion and microtubules are involved in vesicle-dependent hepatocellular transport processes (such as for protein and lipid, see Section V).

Evidence for a role of microfilaments in bile salt secretion was first provided by a correlated structural and functional assessment of phalloidin-induced cholestasis in rats (Dubin *et al.*, 1978; 1980). Treatment with this actin-polymerizing agent resulted in (1) marked declines in basal and taurocholate-induced bile flow (>50%) and [^{14}C]erythritol and bile salt secretion, (2) dilatation of the bile canaliculi, and (3) a thickening of the pericanalicular microfilament network. Other studies using inhibitors of microfilament function substantiate these observations and support a role for this cytoskeletal element in bile secretion. For example, disruption of microfilaments with cytochalasin B or phalloidin resulted in cholestasis (Reichen *et al.*, 1981), and cytochalasin D was found to diminish significantly bile flow (50%) and [^{14}C]taurocholate secretion in rats, while not affecting the hepatobiliary transport of horseradish peroxidase (Kacich *et al.*, 1983). Oda *et al.* (1974) demonstrated that the pericanalicular microfilamentous network, consisting largely of actin filaments, remains attached to bile canalicular-enriched plasma membranes following their isolation. The loss of microfilament integrity is associated with the disruption of bile canalicular contractions in isolated rat hepatocyte couplets (Phillips *et al.*, 1975; Oshio and Phillips, 1981). These observations suggest that the pericanalicular microfilaments (1) provide and maintain bile canalicular tonus, (2) exert a contractile force that facilitates bile flow, and (3) maintain the high pressure found within the biliary tree.

In contrast with the established role for microfilaments, the role of microtubules in effecting bile salt secretion is controversial. Vinblastine and colchicine, agents which block tubulin polymerization, reputedly inhibit both biliary lipid secretion and bile flow in the perfused rat liver, suggesting that microtubule integrity is essential for these functions (Gregory *et al.*, 1978). Crawford and Gollan (1988) have speculated that the cotransport of bilirubin and bile salts from the endoplasmic reticulum to the bile canaliculus is microtubule-dependent. However, the importance of microtubules in facilitating bile secretion is questionable since colchicine at concentrations that significantly inhibit biliary secretion of plasma-derived proteins (Goldman *et al.*, 1983; Kacich *et al.*, 1983) has virtually no effect on basal bile flow or secretion of bile acids (Barnwell *et al.*, 1984; Lowe *et al.*, 1985; Coleman, 1987), although it does diminish taurocholate-induced secretion (Dubin *et al.*, 1980; Crawford *et al.*, 1988). Furthermore, colchicine has a minimal effect on taurocholate uptake by isolated rat hepatocytes, suggesting that the contribution of microtubules to bile salt transport is not quantitatively significant (Reichen *et al.*, 1981). These data suggest that bile salt secretion is independent of microtubule-associated vesicular transport.

Whereas microtubules do not appear to be important for bile salt secretion, several studies have shown that microtubules are absolutely essential for the vectorial movement of transport vesicles within hepatocytes. Biliary proteins and lipids both utilize vesicular transport mechanisms, and their secretion into bile requires microtubule integrity (Goldman *et al.*, 1983; Kacich *et al.*, 1983; Coleman, 1987; Hofmann, 1989; see Section V).

D. Role of Bile Duct Epithelial Cells

Canalicular bile composition may be substantially modified by bile ducts and the gallbladder en route to the small intestine. Despite the many years of liver research, little is known and perhaps appreciated of the role that intrahepatic bile duct epithelial cells play in the formation of the final secretory product that enters the duodenum through the common bile duct. In fact, much of what we describe in this review article on bile secretion relies on data which, for the most part, have not taken into account these important epithelial cells. *In vivo* bile secretion data are obtained from bile collected from the extrahepatic bile ducts or gallbladder. Analysis of bile from this source, though of obvious importance, cannot be assumed to be representative of bile that is secreted by the hepatic parenchymal cells into the bile canaliculi. To do so would be similar to a nephrologist attempting to equate urine from the urinary bladder with kidney glomerular filtrate. Insulin, for example, enters the glomerular filtrate intact but is subsequently reabsorbed by the kidney tubules. In the liver, 95% of the insulin and epidermal growth factor recovered in the common bile duct is degraded (Renston *et al.*, 1980b; Burwen *et al.*, 1984), although, by analogy with the kidney, they may have initially been secreted intact. As yet, there is no evidence that intact hormones are secreted by hepatocytes and subsequently taken up by bile duct epithelial cells as a method of biological conservation. However, this concept is certainly not unreasonable, given that the larger bile ducts, if not the ductules, are derived from an epithelial precursor common to the absorptive cells of the foregut.

The hepatocyte couplet model (Boyer *et al.*, 1988), which permits sampling of canalicular contents, provides one method for comparing canalicular with common bile duct bile. In addition, two separate laboratories have now developed methods for isolating intracellular bile ductule and duct epithelial cells for biochemical analyses or cell culture (Ishii *et al.*, 1989; Alpini *et al.*, 1989). Confirmation that the preparations are enriched in bile duct epithelial cells is based on γ-glutamyltranspeptidase activity and the presence of cytokerotines, as well as on recognition by specific monoclonal antibodies directed against *in situ* bile duct epithelial cells. These preparations should prove to be very useful in determining the role of bile duct epithelial cells in bile secretion as well as in mucosal immunity.

E. The Stem Cell

The identity of a stem cell from which bile duct epithelium as well as hepatocytes can be derived is as yet unresolved. Liver parenchymal cells live for approximately 1 year whereas bile duct epithelial cells live for almost 3 years. One possible mechanism of cell replacement is that each individual cell type divides and produces another exact copy of itself, obviating the need for a common stem cell (Slott *et al.*, 1990). Two different experimental models of liver injury provide contradictory evidence. In the first model, surgical removal of a large amount of liver (partial hepatectomy) results in regeneration of the liver to exactly its original size and function. During regeneration, the liver remnant produces normal albumin, suggesting that a common stem cell is not involved in liver growth. In the second model, exposure of livers to certain toxins results in substantial liver damage. The dead and dying cells are removed and the remaining viable liver also regenerates. However, these regenerating liver cells appear to be undifferentiated and produce α-fetoprotein instead of albumin, supporting the existence of a common stem cell. Our laboratory has found that, in normal adult rat liver, DNA synthesis as detected by [³H-] thymidine incorporation has no precise localization; sparse label is found scattered at random throughout the liver lobule and bile duct epithelium. On the other hand, Zajicek *et al.* (1988) have shown that, if labeled cells are followed for some period of time, there is normal streaming of liver parenchymal cells from zone 1 to zone 3, and bile duct epithelial cells appear to stream from minute ductules (canals of Hering) to larger ducts. We are of the opinion that both theories are correct, in that epithelial cells within the liver do have the ability to replace effete cells in any given locality, but that there are also reserved stem cells which can be recruited for the repair of large areas of damage. These stem cells may be located in the canals of Hering.

III. General Physiology of Bile Acid Secretion

A. Bile Acid/Salt Structure

Conjugated bile acids comprise the majority of organic compounds in bile. Biliary bile acid concentration ranges from 2 to 45 m M (Strange, 1984). Bile acids are amphipathic, acidic (anionic) sterols. By their very nature, they are not very water soluble. In mammals, bile acid structure is based on C_{24} cholic acid (Strange, 1984). Bile acids differ in the number of hydroxyl groups on the cholic acid ring, which can have profound effects on their physicochemical (hence physiological) properties. Bile acids form sodium salts and conjugates with amino acids, primarily taurine and glycine. In general, amino acid conjugation, sodium salt formation, and the presence of hydroxyl groups all tend to increase bile acid solubility.

Bile acids are derived from cholesterol. The carboxy group on the side chain is usually conjugated with glycine or taurine. Under physiological conditions, they are generally unprotonated and are often referred to as bile salts (Coleman, 1987). The hydrophobic properties of these molecules are derived from the steroid ring structure. The hydrophilic region results from the juxtaposition of ring hydroxy groups and the charge on the carboxylate or conjugate ion. The number, location, and orientation of the ring hydroxy groups are the factors that most directly affect the properties of bile acids (Strange, 1981; Hofmann and Roda, 1984).

B. Bile Acid Synthesis

Bile acids are synthesized from cholesterol by hepatocytes. Bile acid biosynthesis accounts for approximately half of cholesterol degradation. The hepatocyte utilizes both extracellular (dietary) sources of cholesterol and cholesterol stored within the cell for bile acid synthesis. The amount of cholesterol in the diet of rats has been shown to play a role in the regulation of bile acid synthesis (Sutton and Botham, 1989). Receptor-mediated uptake of low-density lipoprotein by cultured rat hepatocytes also stimulates bile acid synthesis, suggesting that uptake of low-density lipoprotein by the liver is intimately linked to a process activating bile acid synthesis (Junker and Davis, 1989).

Two enzymes play a key role in the bile acid biosynthetic pathway: 3-hydroxy-3-methylglutaryl-coenzyme A reductase is the rate-limiting enzyme for cholesterol synthesis, and cholesterol 17α-hydroxylase catalyzes the initial and rate-limiting step in bile acid synthesis (Heuman *et al.*, 1989). Factors which influence the activities of these enzymes thereby regulate the biosynthesis of bile acids by hepatocytes.

The interrelationship between the cholesterol and bile acid biosynthetic pathways is shown by the linkage between the rates of cholesterol synthesis and the activity of 17α-hydroxylase. Newly synthesized cholesterol regulates 17α-hydroxylase activity (Pandak *et al.*, 1990).

Hepatic bile acid synthesis is under negative feedback control by bile salts in the enterohepatic circulation, which suppress the activity of both cholesterol 17α-hydroxylase and 3-hydroxy-3-methylglutaryl-coenzyme A reductase (Heuman *et al.*, 1989). Heuman *et al.* (1989) studied the relationship between biliary bile salt hydrophobicity and the activities of these two enzymes and found highly significant negative linear correlations. Thus, the potency of circulating bile salts as suppressors of the enzymes regulating bile acid and cholesterol synthesis increases with increasing hydrophobicity, suggesting that the hydrophobic-hydrophilic balance of the bile salt pool may play an important role in the regulation of cholesterol and bile acid synthesis. Glucocorticoids also regulate bile acid synthesis in rat hepatocytes by induction of cholesterol 17α-hydroxylase activity (Princen *et al.*, 1989).

C. Enterohepatic Recirculation of Bile Acids

Only 1 to 10% of the bile acids secreted into bile represents newly synthesized bile acid at any given time (Strange, 1984; Scharschmidt, 1990). Most are derived from the enterohepatic circulation, whereby biliary bile acids delivered to the gut are reabsorbed into the bloodstream, returned to the liver via the portal vein, taken up by hepatocytes, and resecreted into bile. Therefore, efficient uptake of bile acids by hepatocytes is essential to the process of bile formation.

Within the intestine, many bile acids undergo partial dehydroxylation (to secondary bile acids) and may also become deconjugated (Coleman, 1987). During hepatic transport, bile acids derived from the enterohepatic circulation undergo transformation, which may involve the reconjugation of C_{24} bile acids with glycine or taurine, the conjugation of lipophilic C_{23}-nor bile acids with glucuronate, and the rehydroxylation, reduction of oxo groups, and epimerization of iso-(3β-hydroxy) bile acids (Hofmann, 1989).

Bile acids are not passively absorbed from the biliary tract and small intestine due to their hydrophilicity and size. Active absorption takes place in the ileum (Hofmann, 1989). The mechanism of uptake and transcellular transport of bile acids in the gut is different than that of liver.

The low solubility of bile acids requires a means of keeping them in suspension in body fluids. In blood, albumin serves as the major carrier protein for bile acids (Burke *et al.*, 1971; Strange *et al.*, 1981; Aldini *et al.*, 1982; Roda *et al.*, 1982). Bile acids are bound by plasma albumin with high affinity (Stolz *et al.*, 1989). In bile, bile acids are incorporated into mixed micelles, consisting of bile salts, phospholipid, and cholesterol, which form a colloidal suspension.

D. Bile Acid-Dependent versus -Independent Bile Flow

Hepatic uptake of bile acids is the major driving force for bile formation. Bile flow is generally proportional to the total biliary secretion of bile acids. Bile flow that is dependent on bile acid output is designated as bile acid-dependent flow and is defined operationally as the slope of the line relating bile flow to bile acid output (Boyer, 1980; Blitzer and Boyer, 1982; Scharschmidt, 1990). This relationship between bile flow and bile acid output is due to the osmotic effect of the bile acids (Scharschmidt, 1990), which form sodium salts as a result of their hepatic uptake and transcellular transport. Even though the osmotic activity of bile salts is reduced by micelle formation, because the bile salts are accompanied by sodium, the osmolality of a solution of a sodium salt of a conjugated bile acid is about equal to its molarity (Strange, 1984). Therefore, bile acid-dependent bile flow is attributable to the active transport of bile acids, due largely to sodium-coupled hepatic uptake, with water and electrolytes following passively as a result of diffusion and solvent drag (Scharschmidt and VanDyke, 1983). The passive movement of electrolytes and water between blood and bile in relation to bile acid-dependent flow is primarily via the paracellular pathway (Scharschmidt and VanDyke, 1983).

Bile acid-independent flow represents bile flow even in the absence of bile acid output. It is defined as the point of extrapolation of the bile acid-dependent flow back to zero (determined practically by the *y* intercept of the plot of flow versus bile acid output) (Scharschmidt and Van-Dyke, 1983; Scharschmidt, 1990). Although there is no such thing as bile acid-free bile in the *in situ* situation, there are numerous examples of alteration of bile flow without accompanying changes in bile acid output. Certain drugs and hormones increase bile flow without altering bile acid output (Scharschmidt and VanDyke, 1983; Strange, 1984). Bile acid-free perfusion of the isolated perfused rat liver results in a >95% fall in biliary bile acid output but only about a 50% fall in canalicular bile flow (Scharschmidt and VanDyke, 1983).

Bile acid-independent flow is due to the osmotic effect of electrolytes. A substantial fraction of bile formation is due to the active transport of inorganic electrolytes (Scharschmidt, 1990). Bicarbonate secretion plays a major role in bile acid-independent flow (Hardison and Wood, 1978; Barnhart and Combes, 1978; Hardison *et al.*, 1981; VanDyke *et al.*, 1982a; Anwer and Hegner, 1983). Removal of bicarbonate, but not sodium, from the perfusate decreases bile acid-independent flow (Scharschmidt and VanDyke, 1983).

Na^+/K^+-ATPase also plays a role in bile acid-independent, as well as bile acid-dependent flow. For example, thyroid hormone increases bile acid-independent flow via an increase in Na^+/K^+-ATPase activity (Strange, 1984). The two processes of bile formation (bile acid-dependent and -independent) are linked via this sodium transport system, since bile acids can alter the activity of the Na^+/K^+-ATPase (Wannagat *et al.*, 1978; Scharschmidt *et al.*, 1981).

Glutathione is one of the osmotic driving forces in bile acid-independent bile formation in rat liver (Ballatori and Truong, 1989). Bile acid-independent flow increases proportionally to biliary glutathione secretion.

The relative contribution of hepatocytes to bile acid-independent versus -dependent flow may depend on their location within the liver lobule. Zone 1 (periportal) hepatocytes are presented with the highest concentration of bile acids and therefore the greatest transport load. However, there do appear to be functional differences in hepatocytes from different zones. Baumgartner et al. (1987), using antegrade and retrograde perfusion of the liver with taurodeoxycholate, demonstrated that biliary secretion of bile acids was slower in hepatocytes from zone 3 than from zone 1.

Both bile acid-dependent and bile acid-independent flow are reduced in aged rats (Thompson and Williams, 1965; Schmucker *et al.*, 1985; Ferland *et al.*, 1989).

E. Regulation of Bile Secretion

Some bile acids are more potent in stimulating bile flow than others. Nonosmotic factors play a role in determining the choleretic potencies of individual bile acids (Blitzer and Boyer, 1982). Of the common bile acids, the more hydrophobic bile acids stimulate bile flow and cation secretion better than the more hydrophilic ones (Loria *et al.*, 1989). However, micelle formation competence does not account for differences in the choleretic properties of bile acids (Scharschmidt and VanDyke, 1983;

Strange, 1984), since non-micelle formers are not consistently more chole-retic than micelle-forming bile acids (Rutishauser *et al.*, 1980; Sewell *et al.*, 1980; O'Maille, 1980).

Stimulation of bile flow by certain bile acids is attributable at least in part to bile acid-stimulated active electrolyte secretion, especially bicarbonate (Scharschmidt and VanDyke, 1983). For example, ursodeoxycholate and 7-ketolithocholate, which are twice as choleretic as taurocholate at the same concentration, stimulate bicarbonate transport and increase biliary bicarbonate concentration (Dumont *et al.*, 1980; Scharschmidt and Van-Dyke, 1983).

Bile acids may differentially affect Na^+/K^+-ATPase activity, explaining why some bile acids are more choleretic than others at equimolar concentrations (Strange, 1984).

Conjugation with sulfate can also affect the choleretic properties of bile acids. Secretion of sulfated cholic acid is slower and by a different mechanism than nonsulfated cholic acid. Sulfated cholic acid significantly increases bile flow while reducing the secretion of biliary lipids, cholesterol, and protein (Yousef *et al.*, 1989).

Hormones also regulate bile output. Secretin increases secretion by bile ducts and ductules (Scharschmidt, 1990). Bile ductular epithelium actively secretes sodium, potassium, chloride, and bicarbonate against an osmotic gradient (Nahrwold and Shariatzedeh, 1971; Chenderovitch, 1972) in response to secretin. Bile ducts are also capable of absorbing water and electrolytes (Wheeler *et al.*, 1968; Strasberg *et al.*, 1975; Barnhart and Combes, 1978). Somatostatin is anticholeretic, decreases bile flow by ~30%, and also reduces the output of bile acids, cholesterol, phospholipids, Na^+, K^+, and Cl^- (Magnusson *et al.*, 1989). Vasoactive intestinal peptide induces a bicarbonate-rich choleresis (Nyberg *et al.*, 1989), whereby bile volume increases by 65%, bicarbonate output by 250%, and the output of biliary lipids is not affected.

IV. Bile Acid/Salt Secretory Pathways

Due to the enterohepatic circulation of bile acids, 90% or more of bile acids secreted into bile arrive there as a result of transcellular transport by hepatocytes. Most bile acids are rapidly cleared from the splanchnic blood, e.g., a $t^{1/2}$ of 8–12 min in humans (Strange, 1984). Although there is considerable information concerning bile composition and the kinetics of bile salt uptake and secretion, many of the nuances pertaining to the subcellular mechanism(s) whereby hepatocytes effect the transport of bile salts from the blood to the bile remain unresolved.

The consideration of hepatocellular transport of bile salts raises a number of questions concerning the underlying mechanisms. For example, how do hepatocytes take up bile salts, i.e., via specific or nonspecific receptors, carrier proteins, or facilitated diffusion? How does the structure of bile salts influence their uptake? Once inside the hepatocyte, how are the bile salts transported to the bile canaliculus, e.g., via carrier proteins, in vesicles, or by simple or facilitated diffusion? What is the specific pathway(s) utilized, i.e., what organelles or subcellular compartments are involved? What mechanism(s) is responsible for the passage of bile salts across the bile canalicular membrane and into the canalicular lumen?

A. Uptake of Bile Salts at the Hepatocyte Basolateral (Sinusoidal) Membrane

There appear to be several uptake mechanisms in hepatocytes for bile salts, depending on their physicochemical properties. In general, bile salts with hydrophilic properties (conjugated trihydroxy, the predominant type in bile) are taken up via a Na^+-dependent, saturable, carrier-mediated process (VanDyke *et al.*, 1982b; Bellentani *et al.*, 1987), whereas more hydrophobic (lipophilic) species (unconjugated or glycine-conjugated dihydroxy and monohydroxy bile salts) are transported by either (1) nonsaturable passive diffusion or (2) a facilitated carrier-mediated mechanism in common with other organic anions (Table I) (see Section V) (Stolz *et al.*, 1989; Aldini *et al.*, 1989). In contrast to the uptake of hydrophilic bile salts, uptake of hydrophobic bile salts is unaffected by ouabain or removal of extracellular Na^+. Recent evidence from studies using perfused rat liver

TABLE I

Substrates of Hepatocyte Bile Salt Transport Mechanisms

Transport:	Na^+-dependent	Na^+-independent
Mechanisms:	Carrier-mediated	Carrier-mediated
Substrates:	Hydrophilic (conjugated) bile salts	Hydrophobic (unconjugated) bile salts
	Electroneutral lipids	Sulfated bile salts
	Ouabain	Sulfated organic ions
	Progesterone	Sulfobromophthalein
	Cyclic oligipeptides	Bilirubin
	Phalloidin	Glutathione
	antamanide	
	α-Amanitin	
	Somatostatin	
	Drugs	

or isolated rat hepatocytes implicates the importance of (1) bile acid stoichiometry, e.g., the presence or absence of specific —OH groups, (2) cell-surface SH— groups, and (3) side chain length and charge in influencing bile salt uptake kinetics (Bellentani *et al.*, 1987; Kuipers *et al.*, 1988; Aldini *et al.*, 1989; Blumrich and Petzinger, 1990; Hardison *et al.*, 1991).

The transport of hydrophilic bile salts across the hepatocyte basolateral membrane (1) occurs against an electrochemical gradient, (2) is dependent on the cotransport of Na^+, and (3) requires energy (Duffy *et al.*, 1983; Strange, 1984; Scharschmidt, 1990). Na^+ stimulates taurocholate uptake (Simion *et al.*, 1984a). Bile acid uptake is dependent on the extracellular Na^+ concentration and is inhibited by ouabain (a specific inhibitor of the Na^+/K^+-ATPase). Inhibitors of Na^+-coupled ion transport (furosemide and butmetanide) block sodium-dependent taurocholate uptake by isolated rat hepatocytes (Blitzer *et al.*, 1980). Treatment of rats with the ionophore monensin disrupts the Na^+ gradient across hepatocytes, which is essential for the vectorial movement of taurocholate into the cells (Camogliano and Casu, 1989). Bear *et al.* (1987) reported that the Na^+-dependent hepatocellular uptake of taurocholate occurs via an electrogenic process which requires the transport of more than one sodium ion per bile salt molecule. The movement of Na^+ into the cells is energy-dependent and, thus, requires the integrity of the Na^+/K^+-ATPase system.

Although Na^+/K^+-ATPase activity was originally localized primarily to the bile canalicular rather than to the basolateral hepatocyte membrane (Reichen and Paumgartner, 1977), the subcellular fractions used in early studies were of questionable purity. Improved immunochemical methods have identified the basolateral membranes as the major locus of this important enzyme (Scharschmidt, 1990). Perturbation of the membrane's lipid domain apparently alters the efficacy of the constituent Na^+/K^+-ATPase, which reduces bile salt uptake and bile salt-dependent bile flow (Davis *et al.*, 1978; Muller and Petzinger, 1988).

Although the requirement for Na^+ cotransport for efficient uptake of hydrophilic bile salts is well documented, the actual mechanism(s) involved in this process remains unclear. The major mechanism is thought to be carrier-mediated since this meets the following criteria: (1) exhibits saturation kinetics, (2) is competitively inhibited, (3) occurs against an electrochemical gradient coupled with the cotransport of Na^+, and (4) is partially independent of the free bile salt concentration since albumin-bound bile salts are also transported (Strange, 1984; Smith *et al.*, 1987; Forker and Luxon, 1981). Novak et al. (1989) recently demonstrated a similar process for taurocholate uptake in vesicles prepared from the basolateral plasma membranes of human liver.

A plethora of studies suggest that bile salts interact with hepatocyte basolateral membranes via receptors or carriers (Accatino and Simon, 1976; Meier, 1988; Stolz *et al.*, 1989). Zimmerli *et al.* (1989) advocate (1) a multispecific Na^+-dependent basolateral uptake system and (2) a separate organic anion carrier system, both with overlapping substrate specificities. Other investigators have implicated specific bile salt receptors in this uptake process (Anwer *et al.*, 1976, 1977; Simon *et al.*, 1982). Additional evidence for this mechanism has been obtained in studies using compounds which appear to mimic the uptake of bile salts, e.g., the cholecystographic agent iodipamide (Tafler *et al.*, 1986). The receptor- or carrier-mediated process by which bile salts are taken up is distinct from receptor-mediated endocytosis (Petzinger and Frimmer, 1988). Other studies have identified 48- and 54-kDa proteins with significant bile salt-binding capacities in hepatocyte basolateral membranes (Scharschmidt and Lake, 1989). The 48-kDa polypeptide may be associated with Na^+-dependent bile salt uptake, whereas the 54-kDa molecule appears to be involved in the transport of the Na^+-independent fraction. From an evolutionary perspective, a bile salt-binding polypeptide with a molecular weight of 54,000 and an apparent involvement in Na^+-independent bile salt transport has been identified in the liver of the skate (*Raja crinacea*) (Fricker *et al.*, 1987a). Other studies using photoaffinity labeling techniques have identified several basolateral membrane bile salt-binding proteins with molecular weights between 37,000 and 67,000 (Ziegler *et al.*, 1988; 1989). Further evidence of a basolateral membrane bile salt carrier protein has been provided by the elegant studies of Hagenbuch *et al.* (1990), who expressed the Na^+/taurocholate cotransport system of rat hepatocytes in *Xenopus laevis* oocytes. This expression was characterized by taurocholate uptake kinetics similar to those measured in intact liver, isolated hepatocytes, and liver plasma membrane vesicles and was associated with a 1.5–3 kilobase mRNA.

B. Hepatocellular Translocation of Bile Salts

The transient fate of bile salts following hepatocellular uptake remains, perhaps, one of the most controversial aspects of blood to bile transit. The two most prevalent hypotheses suggest either a cytosolic carrier protein or vesicular transport (Boyer, 1980; Blitzer and Boyer, 1982; Coleman, 1987; Meier, 1988). The majority of intrahepatocellular bile salts are localized to the cytosolic compartment (>50%), whereas the remainder are distributed among the various organelles.

Strange and co-workers isolated bile salt-binding proteins from the 100,000 g supernatant of rat liver (Strange *et al.*, 1977a,b). These proteins exhibit binding sites for cholate, glycocholate, chenodeoxycholate, and lithocholate and possess glutathione S-transferase activity. These "carriers" may participate in the facilitated diffusion of bile salts, but a definitive transport function has not been demonstrated. In general, researchers have tentatively identified three major classes of cytosolic bile salt-binding proteins: (1) the glutathione S-transferases (45–50 kDa), (2) Y' binders (identical to 3α-hydroxysteroid dehydrogenase) (33 kDa), and (3) fatty acid binding protein (FABP, 14 kDa) (Stolz et al., 1989). Functionally, the most important cytosolic bile acid binder appears to be 3α-hydroxysteroid dehydrogenase. Indomethacin, which interferes with bile salt binding to 3α-hydroxysteroid dehydrogenase, inhibits overall bile salt transport, while having no direct effect on sinusoidal or canalicular transport of bile salts (Stolz et al., 1989). Glutathione S-transferases comprise 5–10% of hepatocyte cytosolic protein. One of the glutathione S-transferases, ligandin, is thought to be an important binder of other organic anions, e.g., bilirubin, lithocholate, and sulfobromophthalein (Wolkoff et al., 1979; Kaplowitz, 1980). Even if such cytosolic proteins are not intimately involved in transport, they may play a role in reducing the diffusion of bile salts back toward the sinusoidal pole or in preventing the migration of these moieties into other subcellular compartments (Stolz et al., 1989). They also serve to reduce the effective intracellular bile salt concentration, which helps to avoid the toxic effects of bile salts and promotes their uptake by reducing the unfavorable concentration gradient.

The evidence for the vesicular transport of bile salts to the bile canaliculus is less than compelling. An early indication of the involvement of vesicles in the translocation of bile salts was the observation that taurocholate-induced choleresis in rats was accompanied by a >65% increase in the number of 1000 Å diameter vesicles in the pericanalicular cytoplasm, as compared to unstimulated animals (Jones et al., 1979). Furthermore, scanning electron microscopic analyses revealed that the infusion of a non-micelle-forming, choleretic bile acid, dehydrocholate, which does not enhance the secretion of biliary phospholipids, (1) enlarged bile canalicular diameter, (2) increased the number of canalicular microvilli, and (3) increased the number of lateral extensions of the canalicular membrane (Nemchausky et al., 1977; Layden and Boyer, 1978). It was hypothesized that these alterations in bile canalicular morphology may result from the increased incorporation of vesicles into the canalicular membrane during exocytosis and bile salt secretion (Boyer, 1980).

Additional support for the vesicular transport of bile salts is suggested by information concerning the hepatocellular translocation of non-bile acid biliary constituents (see Section V), such as proteins, lipids, and nonphysiological marker molecules (e.g., horseradish peroxidase). For example, autoradiographic and cytochemical data demonstrate that the hepatobiliary secretion of proteins or glycoproteins, such as horseradish peroxidase, insulin, and immunoglobulin A, involves vesicles (Renston *et al.*, 1980a,b; Jezequel *et al.*, 1986; Hayakawa *et al.*, 1990). However, the transit time for bile salts from blood to bile, in the range of 2–4 min (Goldman *et al.*, 1983; Lowe *et al.*, 1984; Coleman, 1987), is much shorter than that of endocytosed proteins, which utilize a vectorial vesicular mechanism for transcellular transport. Therefore, any vesicular transport mechanisms that are involved in bile salt secretion must be of a very different nature from those for protein. Since bile acids clearly do not enter the cell by endocytosis, if vesicular transport exists, it would be a more distal event in the transcellular translocation process (Stolz *et al.*, 1989). The secretion of biliary lipids (cholesterol, phospholipids) occurs largely via unilamellar vesicles and there is evidence that bile salts may be co-transported in the same vesicles (Marzolo *et al.*, 1990). In addition, bile salts may solubilize membrane lipids to form vesicles (Scharschmidt, 1990). The fact that bile salt secretion induces biliary lipid secretion supports this concept, although there is a bile salt-independent fraction of biliary lipids (Barnwell *et al.*, 1987; Rahman and Coleman, 1987; Hofmann, 1990).

The extrapolation of evidence pertaining to the vesicular transport of biliary proteins and lipids to bile salts is speculative at best. For example, Lake *et al.* (1985) reported that, although vesicular transport appears to be an important pathway for the translocation of large molecules, its contribution to bile salt transport seems minimal. The facts that bile flow and taurocholate secretion decline 55 and 45%, respectively, during estradiol-induced cholestasis, yet the biliary secretion of horseradish peroxidase remains unchanged, suggest that the translocation of these different moieties occurs via separate pathways or that vesicular transport may be involved only in the terminal steps of bile salt secretion (Goldsmith *et al.*, 1983).

The cotransport of bile salts and lipids in the same vesicles suggests that mixed micelle formation may occur intracellularly. Reuben *et al.* (1982) isolated mixed bile salt–phospholipid–cholesterol micelle-like material from microsomes, cytosol, and the Golgi apparatus after fractionation of rat livers in the presence of taurocholate. This association of bile salts with microsomes and the Golgi apparatus may reflect a pathway of bile salt transport that involves micelles, and suggests that micelles found in bile may have originated within the hepatocyte. However, the coexistence of lipid-containing vesicles and mixed micelles in bile suggests that vesicle-to-micelle conversion occurs in the canalicular lumen (Cohen *et al.*, 1989). Since the intracellular bile salt concentration (0.2 mM) is substantially below the critical micellar concentration in the canalicular bile (approximately 20 mM), Ulloa *et al.* (1987) suggested that (1) the pericanalicular vesicles represent the transport of biliary lipids independent of bile salts, and (2) mixed micelle formation occurs in the bile canalicular lumen. Additional evidence for this hypothesis is that (1) bile salt secretion can be separated temporally from lipid secretion (Lowe *et al.*, 1984), and (2) colchicine reduces biliary lipid secretion at concentrations having little effect on bile salt secretion (Barnwell *et al.*, 1984).

C. Role of Organelles in Hepatocellular Translocation of Bile Salts

Most evidence for specific organelle participation in the transport of bile salts across the hepatocyte to the bile canaliculus is morphological and has been obtained by perturbing the bile secretory process, such as inducing choleretic or cholestatic conditions. For example, Popper and Schaffner (1970) reviewed the structural alterations in hepatocytes associated with extrahepatic biliary obstruction. Among the changes noted in obstructed liver cells were (1) an enlargement of the Golgi complex, (2) a hypertrophy of the smooth-surfaced endoplasmic reticulum (SER), and (3) an increase in the number of lysosomes. The first quantitative evidence implicating the Golgi complex in bile secretion demonstrated a decline in the amount of Golgi in zone 1 cells following 48 hr of total biliary obstruction (Jones *et al.*, 1976). Since Cooper *et al.* (1974) reported an increase in the Golgi in hepatocytes during enhanced bile secretory activity, the stereological data of Jones *et al.* (1976) suggested that the Golgi complex was involved in the transport of bile to the canaliculus and the structural correlate of this function responded to either enhanced or reduced bile secretion.

Quantitative analyses of rat hepatocyte fine structure following 48 hr of selective biliary obstruction revealed a twofold increase in the amount of Golgi in hypersecretory versus obstructed cells in zone 3 which was interpreted as a structural adaptation to enhanced bile secretion (Jones *et al.*, 1978). These structural changes were confirmed in rats subjected to intraduodenal infusion of sodium taurocholate for 48 hr (Jones *et al.*, 1979). Stereological analysis demonstrated two- and threefold increases in the volume and surface area of the Golgi complex, respectively, in comparison to control animals. Furthermore, the pericanalicular cytoplasm of the taurocholate-stimulated hepatocytes contained significantly more vesicles with diameters in excess of 1000 Å. The origin of these vesicles remains in question, i.e., do they arise from the basolateral plasma membrane, the endoplasmic reticulum, or the Golgi complex? Interestingly, the surface area of Golgi-associated vesicles in the compensatory (hypersecretory) lobes of selective biliary obstructed rat livers is markedly reduced in comparison to that measured in either the obstructed lobes or sham-operated control animals (Jones et al., 1978). Perhaps this reflects enhanced vesicle turnover during bile transport through the Golgi complex in the compensatory lobe(s).

The possible involvement of organelles in the transport of bile salts may be more directly assessed by visualizing the bile salts via electron microscopic autoradiography or cytochemistry. The aqueous solubility of native bile salts results in redistribution of label between different organelles and the cytosol during processing, causing imprecise localization of autoradiographic grains (Blitzer and Boyer, 1982; Strange, 1984). This difficulty can be overcome by using bile salt derivatives that can be cross-linked to protein via fixation. For example, ^{125}I-labeled cholylglycylhistamine, a relatively insoluble neutral derivative of glycocholic acid with secretion kinetics similar to physiologic bile salts (Grandjean et al., 1979; Spenney et al., 1979), was used to demonstrate a lobular concentration gradient for bile salt uptake (Jones et al., 1980).

Subsequent electron microscopic autoradiographic and immunocytochemical studies from several laboratories provided additional evidence for a role of the Golgi apparatus and SER in bile salt secretion (Goldsmith et al., 1983; Suchy et al., 1983; Lamri et al.,1988). The fate of ^{125}I-labeled cholylglycylhistamine was followed after uptake by rat hepatocytes (Goldsmith et al., 1983). Most of the autoradiographic grains, indicative of the bile salt, were localized over the endoplasmic reticulum (40–50%) and the Golgi/lysosome/vesicle compartment (10–20%) within 1–4 min after intraportal injection. On the basis of these quantitative data, Goldsmith et al. (1983) postulated that bile salts localize in the endoplasmic reticulum and Golgi with subsequent vesicular transport to the canaliculus. Autoradiographic studies with another bile salt analog, ^{125}I-labeled cholylglycyltyrosine, demonstrated sequential labeling of (1) the basolateral plasma membrane and SER (30 sec), (2) the Golgi complex (300 sec), and (3) the pericanalicular cytoplasm (300 sec) following intraportal injection (Suchy et al., 1983). Recently, Lamri et al. (1988) used antibodies directed against conjugates of cholic and ursodeoxycholic acids to follow the fate of bile salts by light and electron microscopic immunochemistry. Their data clearly suggest that the Golgi complex and, perhaps, vesicles derived from the SER are involved in the intracellular transport of bile salts to the pericanalicular cytoplasm.

Biochemical studies have also implicated the SER, as well as the Golgi complex, in the bile salt transcellular transport process. Unconjugated bile acids are known to undergo biotransformation in the SER, although similar evidence does not exist for conjugated bile acids (Killenberg, 1978). Glutathione S-transferase activity (a cytosolic bile acid-binding protein) has been localized in membranes of the endoplasmic reticulum (Friedberg *et al.*, 1979). Reuben *et al.* (1982) demonstrated that micelles may be isolated from endoplasmic reticulum and Golgi preparations in the presence of 5 m M taurocholate, suggesting that bile acids move through these organelles as part of the intracellular assembly of biliary mixed micelles (Reuben *et al.*, 1982). Furthermore, studies on the subcellular distribution of bile salts in the liver have identified bile salt-binding sites associated with the Golgi fraction (Simion *et al.*, 1984a,b). Transport by Golgi vesicles and smooth microsomes, in contrast with that of the plasma membrane fraction, was not stimulated by a sodium gradient.

D. Role of Canalicular Membrane in Bile Salt Secretion

As we have already discussed, bile salts cross the hepatocyte basolateral membrane by carrier-mediated mechanisms in concert with Na^+/K^+-ATPase-dependent Na^+ cotransport and move through the cell to the pericanalicular cytoplasm by either carrier-facilitated diffusion or vesicular translocation involving the Golgi complex and perhaps the SER. However, the final step in the formation of canalicular bile, i.e., the movement of bile salts across the bile canalicular membrane, has not been fully resolved. Since uptake of bile salts (clearance from the blood) is more rapid than the maximum rate of steady-state secretion of bile salts into bile, canalicular secretion represents the rate-limiting step in overall transport of bile salts from blood to bile (Boyer, 1980; Strange, 1984; Coleman 1987; Frimmer and Ziegler, 1988).

In spite of their continuity, the bile canalicular membrane is markedly different from the basolateral plasma membrane, e.g., lipid and enzyme composition, other physicochemical properties (Song *et al.*, 1969; Sirica *et al.*, 1975). Furthermore, the bile salt transport systems active at the sinusoidal and canalicular poles exhibit independent ontogenic development (Suchy *et al.*, 1987). Fricker *et al.* (1987b) demonstrated a functional polarity in bile salt transport systems in the rat liver (Meier, 1988; Scharschmidt and Lake, 1989). Using photoaffinity labeling, investigators identified a single bile acid-binding protein of 100 kDa associated with the bile canalicular membrane (Inoue *et al.*, 1984; Meier *et al.*, 1987; Ruetz *et al.*, 1987; Fricker *et al.*, 1987b; Buscher *et al.*, 1988). These data suggested that sinusoidal uptake and canalicular secretion of bile salts are accomplished by different transport systems. Subsequently, Fricker *et al.* (1989) reported that extrahepatic cholestasis was accompanied by (1) increased turnover and (2) redistribution to the basolateral surface of the 100-kDa canalicular bile salt-binding protein. This was postulated to cause increased efflux of bile salts at the sinusoidal surface and, thus, serve as a compensatory protective response of the hepatocyte to prevent potential toxic effects of excessively high intracellular concentrations of bile salts during cholestasis.

Kinetic transport studies in isolated canalicular membrane vesicles have also been used to characterize the bile salt transport system at the canalicular membrane (Meier *et al.*, 1984, 1987; Inoue *et al.*, 1984). The intracellular negative membrane potential of approximately -30 to -40 mV (Graf *et al.*, 1984; Fitz and Scharschmidt, 1987) appears to be a driving force for bile acid transport into the canaliculus (Meier, 1988; Scharschmidt, 1990). Boyer and co-workers suggest that the mechanism underlying the driving force involves changes in the potential difference of the membrane, i.e., a voltage shift, and is analogous to electrodiffusion (Weinman *et al.*, 1989). The electrochemical gradient favors the transport of anions from the hepatocyte interior to the canalicular lumen, and micelle formation in the canalicular lumen (as opposed to the cell interior) also favors bile salt secretion (Blitzer and Boyer, 1982).

V. Secretion of Biliary Constituents Other Than Bile Acids

A. Biliary Lipids

Biliary lipids, as well as bile acids, participate in enterohepatic circulation. The major lipids in bile are cholesterol and phospholipid. Eighty to 95% of biliary phospholipid is phosphatidylcholine, and biliary cholesterol is almost entirely nonesterified (Coleman, 1987). As is the case for bile acids, cholesterol and phospholipid concentrations in bile greatly exceed their solubility in water due to their incorporation into mixed micelles. Cholesterol is both taken up from plasma and synthesized within the hepatocyte. Since cholesterol is the substrate for bile acid synthesis, bile acid synthesis is the major mechanism for degradation and removal of cholesterol.

Biliary lipid secretion is dependent on bile acid secretion (Coleman, 1987). The output of both cholesterol and phospholipid is proportional to bile acid output (Scharschmidt, 1990). At physiological rates of bile acid secretion, the relationship is approximately linear (Hoffman *et al.*, 1975; Crawford and Gollan, 1988). The relationship between biliary lipid secretion and bile acid structure, with regard to water solubility, critical micellar concentration, and hydrophobicity, has been studied. Very hydrophilic (i.e., non-micelle-forming) bile acids produce very little lipid output (Hardison and Apter, 1972; Barnwell *et al.*, 1984). However, Roda *et al.* (1988) report that, although induced bile flow is directly related to the hydrophobicity of the C_{24} bile acids, phospholipid and, to a lesser extent, cholesterol secretion is poorly related to both the hydrophobicity and the critical micellar concentrations of bile acids, suggesting a role for hepatic biotransformation in modulating biliary lipid secretion.

Lipid secretion across the canalicular membrane is vesicle-dependent (Coleman, 1987). Biliary lipid secretion involves bile acid-stimulated microtubule-dependent movement of phospholipid–cholesterol-rich vesicles from the Golgi apparatus to the canaliculus (Hofmann, 1989). Biliary secretion of bile acids can be separated in time from vesicle-mediated lipid secretion in the intact liver (Coleman, 1987). Colchicine, which interferes with vesicle movement, inhibits lipid secretion, while bile acid output is unaffected (Barnwell *et al.*, 1984). Canalicular lipid secretion may involve fusion of precursor biliary lipid vesicles with specific domains of the canalicular membrane (Coleman, 1987).

An increased output of biliary lipid during aging has been reported (Ferland *et al.*, 1989). Alteration in bile acid composition and stimulated hydrophobic bile acid formation do not account for this change. This finding is surprising in light of the fact that other microtubule-dependent hepatocyte transport processes appear to be compromised by aging (Jones *et al.*, 1988).

B. Electrolytes (Nonorganic Anions and Cations)

1. Specific Hepatocyte Transport Mechanisms

The Na^+/K^+ ATPase, localized to the basolateral (sinusoidal) membrane by numerous cytochemical and subfractionation studies (Blitzer and Boyer, 1978; Latham and Kashgarian, 1979; Poupon and Evans, 1979; Boyer *et al.*, 1983; Inoue *et al.*, 1983, Meier *et al.*, 1984; Blitzer and Donovan, 1984), transports three Na^+ out for every two K^+ pumped into the cell. The Na^+/K^+-ATPase thereby creates a steep electrical as well as chemical gradient favoring passive sodium entry (Scharschmidt and Van-Dyke, 1983). If sodium and its coupled solute move in the same direction, as is the case with bile salts, this is referred to as symport or cotransport, whereas if they move in opposite directions, it is referred to as antiport, countertransport, or exchange (West, 1980). Sodium-coupled transport is electroneutral if it does not mediate net transfer of charge, and electrogenic if it does. This transport mechanism, which is coupled to bile acid uptake, is the major driving force in bile formation.

The activity of the Na^+/K^+-ATPase on the hepatocyte basolateral surface can be regulated by a variety of factors. The internal sodium concentration can affect Na^+/K^+-ATPase activity. Influx of bile salts and sodium increases cation pumping, which then decreases bile salt influx. Conversely, a decrease in internal sodium concentration stimulates sodium influx accompanied by bile salt and amino acid influx (Scharschmidt and VanDyke, 1983).

In many kinds of cells, membrane permeabilities to sodium and potassium as well as cation pumping by Na^+/K^+-ATPase are continuously regulated (Schultz, 1981; Grasset et al., 1983) and Na^+/K^+-ATPase activity can increase or decrease over a period of hours or days (Scharschmidt and VanDyke, 1983). Additionally, sodium-coupled transport processes may affect one another. In hepatocytes, an increase in intracellular calcium may mediate an increase in potassium permeability and a decrease in sodium permeability, thereby diminishing sodium influx (Burgess et al., 1981). It is possible that enhanced potassium efflux, in conjunction with accelerated Na^+/K^+-ATPase-mediated cation pumping, may act to preserve intracellular negativity in the face of increased sodium influx and thus maintain the electrochemical sodium gradient in the face of acute increases in sodium-coupled transport (Kristensen, 1980; Grasset et al., 1983).

Hepatocyte Na^+/K^+-ATPase activity has been shown to be modified by bile acids (Wannagat et al., 1978; Ballard and Simon, 1980) and by changes in the lipid content of membranes (Blitzer and Boyer, 1982). Drugs and hormones have also been shown to regulate Na^+/K^+-ATPase activity in isolated liver membrane preparations (Scharschmidt and VanDyke, 1983). Hormones such as norepinephrine, vasopressin, angiotensin II, and phorbol esters stimulate the hepatocellular sodium pump either directly via diacylglycerol formation and protein kinase C activation or indirectly via activation of membrane lipids (Berthon et al., 1985; Lynch et al., 1986).

In addition to the Na^+/K^+-ATPase, there are at least two other transport mechanisms for sodium at the hepatocyte basolateral membrane (Meier, 1988). There is a $Na^+–H^+$ exchange mechanism (Mathison and Raeder, 1983), which is important for the maintenance of intracellular pH (Henderson et al., 1987; Renner et al., 1988) and plays a role in bile acid-dependent bile flow (Lake et al., 1987; Renner et al., 1988). Another mechanism involves $Na^+–HCO_3^-$ symport (cotransport) (Renner et al., 1987; Meier et al., 1988; Gleeson et al., 1988) which, in addition to contributing to the maintenance of intracellular pH (Meier, 1988), also plays a role in bile acid-independent bile flow (Lake et al., 1987).

The basolateral membrane also has a specific transport system for sulfate. Sulfate participates in the broad mechanism of organic anion (nonbile acid) transport (see below). The sulfate-hydroxyl antiport system is linked to the transport of oxalate, succinate, and sulfobromophthalein.

At the canalicular membrane, other electrolyte transport mechanisms operate. There is a $Cl^-–HCO_3^-$ exchange (antiport) mechanism (Meier et *al.*, 1985), which is of great significance for bile acid-independent bile flow (Anwer and Hegner, 1983; Hardison and Wood, 1978; VanDyke *et al.*, 1982a). Mechanisms involved in hepatic bicarbonate secretion are likely to be the same as those responsible for regulation of internal pH (Scharschmidt and VanDyke, 1983). The high bicarbonate content of bile makes it an alkaline fluid compared to plasma. Canalicular membrane ATPase is not stimulated by bicarbonate (Izutzu *et al.*, 1978).

Although chloride appears to be important for organic anion uptake (Meier, 1988) (see below), chloride passively distributes across the plasma membrane and is not involved in bile acid-independent bile flow (Scharschmidt and Stephens, 1980). Chloride–bicarbonate exchange may also occur at the canalicular membrane (Scharschmidt, 1990).

Bile flow is calcium-dependent (Strange, 1984), and the biliary system is almost freely permeable to Ca^{2+} (Hofmann, 1989). There also appear to be two distinct Ca^{2+} pumps, one at the canalicular membrane and the other at the basolateral membrane (Meier, 1988). Bile salts may bind calcium and magnesium during transcellular transport, since the effect of bile salt structure on biliary secretion of calcium and magnesium suggests a secretory link consistent with cation–bile salt binding (Loria *et al.*, 1989). Within bile, Ca^{2+} is bound to bile acid monomers and micelles (Hofmann, 1989).

2. Water and Electrolyte Entry into Bile

The hepatocellular electrolyte transport mechanisms are directly responsible for creating the osmotic gradient for bile formation via the active transport of bile acids. However, it is important to keep in mind that these mechanisms of electrolyte transport at the hepatocellular level, while essential for transport of other biliary constituents, account for only a small fraction of the electrolytes and water actually secreted into bile.

The paracellular route of entry into bile is important in the osmotic regulation of bile formation (Strange, 1984). Much of the movement of water and electrolytes into bile is via the paracellular pathway, and takes place at the level of ductules and ducts in addition to hepatocytes (Scharschmidt, 1990). Water and sodium have been shown to enter bile via the paracellular route (Layden *et al.*, 1978), and the paracellular pathway accounts for most of the electrolyte movement from blood to bile: Na, 95%; Cl, 73%; K, 74% (Graf and Peterlik, 1975). As a result of passive entry of most of the water and electrolytes via the paracellular pathway, the electrolyte composition of bile parallels that of the perfusate, including the tonicity (Scharschmidt and VanDyke, 1983).

Tight junctions regulate paracellular transport. Since the tight junction is negatively charged (Bradley and Herz, 1978), it is less permeable to anions than to uncharged solutes. The negative charge helps prevent organic anions (including bile acids) in the canalicular lumen from diffusing back across the junction into blood (Blitzer and Boyer, 1982). The charge of the tight junction also affects the access of charged fluid-phase markers to the paracellular pathway. For example, the biliary secretion of anionic horseradish peroxidase by both the transcellular and paracellular pathways was less than 50% that of the neutral or cationic molecule, even though uptake from the blood was the same regardless of charge (Hardison *et al.*, 1989).

Transcytosis may account for some water movement from blood to bile (Scharschmidt, 1990). The movement of water through the hepatocyte via nonspecific, vesicle-mediated transcytosis has been evaluated by using fluid-phase markers for vesicular transport (Lake *et al.*, 1985; Lowe *et al.*, 1985). The relative contribution of transcytosis to bulk fluid movement is probably not very significant. Most endocytosed water is returned to the sinusoidal surface (Scharschmidt *et al.*, 1986). The contribution of fluid-phase endocytosis to bile flow has been estimated at 2–4% of bile flow (Yousef *et al.*, 1988). This contribution was not affected by choleresis induced by dehydrocholic acid or cholestasis induced by taurolithocholic acid or bile duct obstruction. In conclusion, fluid-phase endocytosis does not appear to influence bile formation significantly.

C. Organic Anions and Cations

One of the major functions of liver is detoxification (Buscher *et al.*, 1988). Hence, many organic ions are secreted into bile for purposes of elimination from the body.

There are two major transport mechanisms at the hepatocyte basolateral membrane for organic ion transport. One is the same sodium-dependent, carrier-mediated active transport system for which conjugated bile acids are the physiological substrates. The other mechanism is sodium-independent, but still appears to involve carrier-mediated active transport. Both transport mechanisms are utilized by a wide variety of organic ions, indicating their broad range of substrate specificities (Table 1).

The basolateral sodium-dependent bile acid transport mechanism has a high degree of transport flexibility, providing for cotransport of differently charged amphipathic substances by the same transport system (Zimmerli *et al.*, 1987; Meier, 1988). Photoaffinity labeling and kinetic studies indicate that the sinusoidal membrane bile acid-binding proteins mediate the uptake of several different amphipathic anions, uncharged compounds, and even cations (Buscher *et al.*, 1988). Substrates of this transport system include electroneutral lipids (ouabain, progesterone), cyclic oligopeptides (phalloidin, antamanide, α-amanitin, somatostatin), and a wide variety of drugs (Wieland *et al.*, 1984; Petzinger and Frimmer, 1984; Kroncke *et al.*, 1986; Zimmerli *et al.*, 1987; Frimmer and Ziegler, 1988). Phalloidin, a bicyclic heptapeptide, and antamanide, a monocyclic decapeptide, have been shown to interact with the bile acid-binding proteins (54 and 48–49 kDa) at the hepatocyte sinusoidal surface, and compete for binding with bile acids, indicating that the hepatic uptake system for bile acids, phallotoxins, and the cycloamanide antamanide are identical, thus explaining the organotropism of phallotoxins (Wieland *et al.*, 1984).

The functionally separate sodium-independent basolateral organic ion transport system also has a large number of substrates, including sulfated bile acids and other sulfated organic ions (Kuipers *et al.*, 1988), sulfobromophthalein, bilirubin, and glutathione (Meier, 1988). These anions are not substrates for the sodium-dependent transport site of conjugated bile acids (Zimmerli *et al.*, 1987; Frimmer and Ziegler, 1988). Chloride may play an important role in transport of these organic anions (Potter *et al.*, 1987; Wolkoff *et al.*, 1987) via linkage to a chloride antiport (Berk and Stremmel, 1986). Similarly, the basolateral sulfate transport mechanism may involve an anion exchange system that cotransports certain organic ions such as oxalate (Hugentobler *et al.*, 1987).

Endogenous substances, such as bilirubin, diglucuronide, and the glucuronides and sulfates of estrogen, are more concentrated in bile than in plasma (Coleman, 1987). The high biliary : plasma concentration ratio of these organic anions is consistent with carrier-mediated active transport.

Within hepatocytes, sulfobromophthalein and bilirubin are conjugated with reduced glutathione and glucuronic acid and then secreted across the canalicular membrane via a pathway different from the canalicular bile acid transport system (Meier, 1988; Crawford and Gollan, 1988). Conjugated bilirubin and sulfated bile acids use the same canalicular transport pathways (Scharschmidt, 1990). Conjugation with sulfate represents a major detoxification pathway for a variety of endogenous compounds (Scharschmidt and VanDyke, 1983).

Although specific carrier proteins for this group of organic anions (bilirubin, sulfobromophthalein, glutathione) have not been identified (Scharschmidt, 1990), glutathione is known to enter bile via a saturable, electrical potential-driven mechanism (Akerboom *et al.*, 1984; Inoue *et al.*, 1983; Ballatori *et al.*, 1986a,b). Glutathione is a peptide synthesized within hepatocytes that exists in thiol-reduced and disulfide-oxidized forms and plays a major role in enzyme-catalyzed reactions (Meier, 1988). Glutathione levels within hepatocytes are maintained by biosynthesis and sinusoidal efflux into blood (Lauterburg and Mitchell, 1981). Glutathione efflux is dependent on the same carrier-mediated mechanism that is used for unconjugated bilirubin uptake (Ookhtens *et al.*, 1985, 1988; Aw *et al.*, 1986, 1987). However, approximately 20% of glutathione synthesized by rat liver is secreted into bile, and glutathione conjugates are selectively secreted into bile (Meier, 1988). Glutathione disulfide transport is mediated by a carrier present in the canalicular membrane of the hepatocyte (Akerboom *et al.*, 1984).

At the canalicular membrane, secretion of organic anions is via a saturable, carrier-mediated, rate-limiting process (Scharschmidt, 1990). Most drugs are positively charged at physiological pH (organic cations) but very little is known about their biliary secretion (Knodell and Brooks, 1979; Van Der Sluijs *et al.*,1987).

Some organic anions partition into lipid particles, and their degree of vesicular association is correlated with their hydrophobicity. These hydrophobic organic anions are secreted into bile in association with lipid particles (Tazuma *et al.*, 1988).

Many amino acids undergo sodium-dependent uptake and are secreted into bile (Folsch and Wormsley, 1977) but are not required for bile formation. Sodium-dependent transport of L-glutamate occurs at the canalicular membrane, whereas its transport across the sinusoidal membrane is mainly the result of passive diffusion (Ballatori *et al.*, 1986a). Glucose and amino acids, which enter bile as secondary solutes, are reabsorbed efficiently in the biliary ductular system (Hofmann, 1989).

D. Biliary Proteins

Bile contains many intact proteins and polypeptides. Although proteins in bile are a minor component as a percentage of organic solutes, their biological significance is much greater than their low concentration would indicate, because proteins such as hormones, growth factors, and immunoglobulins exert their biological effects at very low concentrations. Proteins in bile originate from either the plasma (such as epidermal growth factor, insulin, and immunoglobulin A) or hepatocytes (including lysosomal enzymes, albumin, secretory component, transferrin, and apoproteins). Albumin, although present in bile at lower concentration than in plasma (Strange, 1984), is none the less the major protein in bile. Newly synthesized albumin and transferrin are secreted only via the sinusoidal pole (Barnwell and Coleman, 1983) and therefore reach the bile by subsequent hepatic uptake from the plasma. The apoproteins of lipoproteins are also produced within hepatocytes, secreted into the blood stream, and later taken up by hepatocytes. Although small amounts of intact apoproteins are found in bile (Sewell *et al.*, 1983), most are catabolized (Jones *et al.*, 1984a) and their breakdown products appear in bile. In contrast, lysosomal enzymes (LaRusso, 1984) and secretory component are secreted directly into bile from the canalicular pole of the hepatocyte, so their route into bile is not via the plasma. Next to albumin, secretory component, the receptor for immunoglobulin A (Fisher *et al.*, 1979), is the major protein in rat bile (Kakis and Yousef, 1978; Mullock *et al.*, 1978; LaRusso, 1984). It is found both free and linked to polymeric immunoglobulin A by disulfide bonds. Although immunoglobulin A is the predominant immunoglobulin in bile, both rat and human bile also contain small amounts of immunoglobulins G and M. In addition to these major biliary proteins, many trace proteins, such as insulin (Renston *et al.*, 1980b) and epidermal growth factor (St. Hilaire *et al.*, 1983), are found as well. There are also many unidentified peptides, probably representing degradation products from a wide variety of proteins that have been taken up from plasma by liver and catabolized.

The majority of biliary proteins, including some of hepatic origin such as albumin, reach the bile by being taken up from the plasma. For those biliary proteins that are not enriched relative to their plasma concentrations (such as albumin), their entry into bile can be accounted for by nonspecific fluid-phase transcytosis or, for small, uncharged molecules, by paracellular leakage across tight junctions. For proteins that are concentrated in bile relative to plasma, such as immunoglobulin A, specific transport systems are utilized by the hepatocyte. These transport systems involve receptor-mediated endocytosis and transcytosis (Fig. 7) (Jones and Burwen, 1989; Burwen and Jones, 1990), whereby specific receptors on the sinusoidal plasma membrane surface of hepatocytes bind plasma-derived proteins, and the protein–receptor complexes are internalized into endocytic vesicles. Movement of these proteins through the hepatocyte is accomplished by microtubule-dependent vectorial vesicular transport (Goldman *et al.*, 1983).

Once taken up by hepatocytes, plasma-derived proteins are either degraded, secreted intact into bile, or utilized by the liver. Although receptors on the hepatocyte sinusoidal plasma membrane determine which proteins are specifically taken up, once endocytosis has occurred, the intracellular destinations of the proteins determine which of the above three outcomes will result.

Our laboratory, among others, has provided much information on the intracellular transport pathways utilized by endocytosed proteins in hepatocytes. The proteins studied include human and rat immunoglobulin A (Jones *et al.*, 1985; Renston *et al.*, 1980a), epidermal growth factor (Burwen *et al.*, 1984; St. Hilaire *et al.*, 1983), insulin (Renston *et al.*, 1980b), asialoglycoproteins (Hubbard and Stukenbrok, 1979), low-density lipoproteins (Chao et al., 1981), and chylomicron and very-low-density lipoprotein remnants (Hornick *et al.*, 1984; Jones *et al.*, 1984a). Several dis-

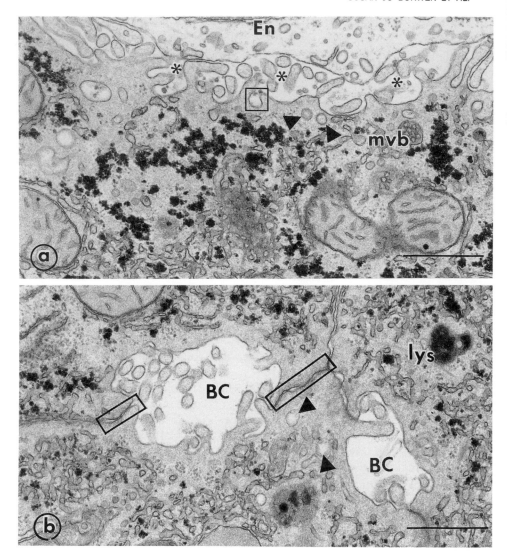

FIG. 7   Electron micrographs showing organelles involved in receptor-mediated endocytosis
and transcytosis of plasma-derived proteins by hepatocytes. (a) An endocytic vesicle forming
at the sinusoidal surface of the hepatocyte is shown within the square. Arrowheads point to
two endosomes. Endosomes that are involved in the lysosomal pathway subsequently form
multivesicular bodies (mvb). Asterisks indicate the space of Disse between an endothelial cell
(En) and the hepatocyte. (b) Multivesicular bodies fuse with primary lysosomes to form
secondary lysosomes (lys), where their contents are degraded by lysosomal enzymes. Endocytic vesicles that bypass the lysosomal compartment can be seen in the pericanalicular
cytoplasm (arrowheads) prior to exocytosis at the canalicular surface. BC, Bile canaliculus;
rectangles, tight junctions. Bar = 1 μm.

tinct pathways have emerged from these investigations (Jones and Burwen, 1989; Burwen and Jones, 1990). Once the proteins are internalized, they do not necessarily share a common fate with their receptors.

Proteins that are primarily degraded by hepatocytes, such as asialoglycoproteins (Hubbard and Stukenbrok, 1979), low-density lipoproteins (Chao et al., 1981), and chylomicron and very-low-density lipoprotein remnants (Hornick et al., 1984; Jones et al., 1984a), are transported by the lysosomal pathway. These proteins are endocytosed in clathrin-coated pits and vesicles, the vesicles fuse to form endosomes, the endosomes mature into multivesicular bodies, and the multivesicular bodies, upon fusion with primary lysosomes, form secondary lysosomes (Fig. 7). Their contents are degraded by lysosomal enzymes, and the degradation products are secreted either back into the bloodstream or into bile.

Proteins that are primarily secreted intact into bile, such as immunoglobulin A (Schiff et al., 1984), are transported by the transcellular pathway. In this transport pathway, endocytic pits and vesicles do not appear to be clathrin-coated. The endocytic vesicles do not fuse, but remain a constant size after formation ($\sim$1000–1500 Å) and travel directly to the bile canalicular membrane (Renston et al., 1980a). They do not interact with other organelle compartments en route, and, consequently, there is no opportunity for lysosomal degradation. These endocytic "shuttle" vesicles empty their contents into the bile canalicular lumen by exocytosis (Fig. 7b).

"Shuttle" vesicles with membrane associated secretory component are continuously formed at the sinusoidal surface and translocated to the bile canalicular surface even in the absence of internalized immunoglobulin A (Mullock et al., 1980). Retrograde perfusion of the common bile duct with immunoglobulin A has shown that vesicle-dependent transcytosis of proteins can take place in the reverse direction, from the canalicular to the sinusoidal pole of hepatocytes (Jones et al., 1984b). These observations indicate that the vesicular transport system for biliary proteins operates continuously and can transport material in either direction.

VI. Summary

One of the liver's principal functions is the formation of bile, which is requisite for digestion of fat and elimination of detoxified drugs and metabolites. Bile is a complex fluid made up of water, electrolytes, bile acids, pigments, proteins, lipids, and a multitude of chemical breakdown products. In this review, we have summarized the source of various biliary components, the route by which they end up in bile, including the underlying subcellular and molecular mechanisms, and their contribution to bile formation. One of the reasons why bile formation is so complex is that there are many mechanisms with overlapping substrate specificities, i.e., many biochemically unrelated biliary constituents share common transport mechanisms. Additionally, biliary constituents may reach bile by more than one pathway. Some biliary components are critical for bile formation; others are of minor significance for bile formation but play a major physiological role.

The major driving force for bile formation is the uptake and transcellular transport of bile salts by hepatocytes. The energy for bile formation comes from the sodium gradient created by the basolateral Na^+/K^+-ATPase, to which bile salt transport is coupled. The secretory pathway for bile salts involves uptake at the basolateral surface of the hepatocyte, vectorial transcellular movement, and transport across the canalicular membrane into the canalicular lumen.

Hydrophilic bile salts are taken up via a sodium-dependent, saturable, carrier-mediated process coupled to the Na^+/K^+-ATPase. This uptake mechanism is also shared by other substrates, such as electroneutral lipids, cyclic oligopeptides, and a wide variety of drugs. Hydrophobic bile acids are taken up by a sodium-independent facilitated carrier-mediated mechanism in common with other organic ions, including sulfated bile acids, sulfobromophthalein, bilirubin, glutathione, and glucuronides, or by nonsaturable passive diffusion. Two major carrier proteins have been identified on the hepatocyte basolateral membrane: a 48-kDa protein that appears to be involved with Na^+-dependent bile salt uptake, and a 54-kDa protein, thought to be associated with Na^+-independent bile salt uptake.

The intracellular transport of bile salts may involve cytosolic carrier proteins, of which several have been identified. Some evidence suggests a vesicular transport mechansim for bile salts. Since bile acids clearly do not enter the cell by endocytosis, formation of transport vesicles must be a more distal event in the transcellular translocation process. Some bile salts appear to be transported within the same unilamellar vesicles that are involved in the secretion of cholesterol and phospholipid. Biliary lipid secretion involves bile acid-stimulated microtubule-dependent movement of phospholipid–cholesterol-rich vesicles from the Golgi apparatus to the canaliculus.

Transport of bile salts across the canalicular membrane into the canalicular lumen is the rate-limiting step in overall bile salt transport from blood to bile. Transport across the canalicular membrane is a saturable, carrier-mediated process and appears to involve a membrane-associated bile salt-binding protein of 100 kDa. Conjugated bilirubin and sulfated bile acids use the same canalicular transport mechanism, which is different from the one utilized by bile salts.

Proteins in bile, although a minor component as a percentage of organic solutes, are of major biological significance. They originate from either the plasma (such as epidermal growth factor, insulin, immunoglobulin A, and some apolipoproteins) or within hepatocytes (including lysosomal enzymes, albumin, secretory component, transferrin, and apoproteins). For proteins that are concentrated in bile relative to plasma, such as immunoglobulin A, specific transcytosis transport systems are utilized by the hepatocyte. These transport systems involve receptor-mediated endocytosis, vesicular intracellular transport, and exocytosis at the canalicular membrane.

Electrolytes are essential for bile formation since they provide electrochemical gradients for coupled transport of specific biliary constituents and osmotic gradients which create water flow. The many specific electrolyte transport mechanisms of hepatocytes are directly responsible for bile formation via coupled active transport of bile salts. However, these mechanisms of electrolyte transport at the hepatocellular level, while essential for transport of other biliary constituents, account for only a small fraction of the electrolytes and water actually secreted into bile. Passive movement via the paracellular pathway accounts for the vast majority of electrolyte and water transport from blood to bile at every level of junctional complex of the liver epithelium.

Despite the great amount of research on bile formation, the molecular basis for the regulation of the many hepatobiliary processes involved in bile secretion remains unclear. The subcellular and molecular mechanisms of bile secretion will continue to be a major focus of research for many years to come.

References

Accatino, L., and Simon, F. R. (1976). *J. Clin. Invest.* **57**, 496–508.

Akerboom, T., Inoue, M., Sies, H., Kinne, R., and Arias, I. M. (1984). *Eur. J. Biochem.* **141**, 211–215.

Aldini, R., Roda, A., Labate, A. M., Cappelleri, G., Roda, E., and Barbara, L. (1982). *J. Lipid Res.* **23**, 1167–1173.

Aldini, R., Roda, A., Simoni, P., Lenzi, P., and Roda, E. (1989). *Hepatology* **10**, 840–845.

Alpini, G., Lenzi, R., Zhai, W.-R., Liu, M. H., Slott, P. A., Paronetto, F., and Tavoloni, N. (1989). *Gastroenterology* **97**, 1248–1260.

Anwer, M. S., and Hegner, D. (1983). *Am. J. Physiol.* **244**, G116–G124.

Anwer, M. S., Kroker, R., and Hegner, D. (1976). *Biochem. Biophys. Res. Commun.* **73**, 63–71.

Anwer, M. S., Kroker, R., Hegner, D., and Petter, A. (1977). *Hoppe-Seyler's Z. Physiol. Chem.* **358**, 543–553.

Aw, T. Y., Ookhtens, M., Ren, C., and Kaplowitz, N. (1986). *Am. J. Physiol.* **250**, G236–G243.

Aw, T. Y., Ookhtens, M., Kuhlenkamp, J. F., and Kaplowitz, N. (1987). *Biochem. Biophys. Res. Commun.* **143,** 377–382.

Ballard, D. C., and Simon, F. R. (1980). *Gastroenterology* **79,** 1002 (abstr).

Ballatori, N., and Truong, A. T. (1989). *Am. J. Physiol.* **256,** G22–G30.

Ballatori, N., Mosely, R. H., and Boyer, J. L. (1986a). *J. Biol. Chem.* **261,** 6216–6221.

Ballatori, N., Jacob, R., and Boyer, J. L. (1986b). *J. Biol. Chem.* **261,** 7860–7865.

Barnhart, J. L., and Combes, B. (1978). *Am. J. Physiol.* **234,** E146–E156.

Barnwell, S. G., and Coleman, R. (1983). *Biochem. J.* **216,** 409–414.

Barnwell, S. G., Lowe, P. J., and Coleman, R. (1984). *Biochem. J.* **220,** 723–731.

Barnwell, S. G., Tuchweber, B., and Yousef, I. M. (1987). *Biochim. Biophys. Acta* **922,** 221–233.

Baumgartner, U., Miyai, K., and Hardison, W. G. M. (1987). *Am. J. Physiol.* **252,** G114–G119.

Bear, C. E., Davison, J. S., and Shaffer, E. A. (1987). *Biochim. Biophys. Acta* **903,** 388–394.

Bellentani, S., Hardison, W. G. M., Marchegiano, P., Zanasi, G., and Manenti, F. (1987). *Am. J. Physiol.* **252,** G339–G344.

Berk, P. D., and Stremmel, W. (1986). *Prog. Liver Dis.* **8,** 125–144.

Berk, P. D., Potter, B. J., and Stremmel, W. (1987). *Hepatology* **7,** 165–176.

Berthon, B., Capiod, T., and Claret, M. (1985). *Br. J. Pharmacol.* **86,** 151–161.

Blitzer, B. L., and Boyer, J. L. (1978). *J. Clin. Invest.* **62,** 1104–1108.

Blitzer, B. L., and Boyer, J. L. (1982). *Gastroenterology* **82,** 346–357.

Blitzer, B. L., and Donovan, C. B. (1984). *J. Biol. Chem.* **259,** 9295–9301.

Blitzer, B. L., Ratoosh, S. L., and Boyer, J. L. (1980). *Clin. Res.* **28,** 273A.

Blumrich, M., and Petzinger, E. (1990). *Biochim. Biophys. Acta* **1029,** 1–12.

Boyer, J. L. (1980). *Physiol. Rev.* **60,** 303–326.

Boyer, J. L., Allen, R. M., and Ng, O. C. (1983). *Hepatology* **3,** 18–28.

Boyer, J. L., Gautam, A., and Graf, J. (1988). *Semin. Liver Dis.* **8,** 308–316.

Bradley, S. E., and Herz, R. (1978). *Am. J. Physiol.* **235,** E570–E576.

Burgess, G. M., Claret, M., and Jenkinson, D. H. (1981). *J. Physiol. (London)* **317,** 67–90.

Burke, C. W., Lewis, B., Panveliwalla, D., and Tabaqchali, S. (1971). *Clin. Chim. Acta* **32,** 207–214.

Burwen, S. J., and Jones, A. L. (1990). *J. Electron Microsc. Tech.* **14,** 140–151.

Burwen, S. J., Barker, M. E., Goldman, I. S., Hradek, G. T., Raper, S. E., and Jones, A. L. (1984). *J. Cell Biol.* **99,** 1259–1265.

Buscher, H.-P., Gerok, W., Kollinger, M., Kurz, G., Muller, M., Nolte, A., and Schneider, S. (1988). *Adv. Enzyme Regul.* **27,** 173–192.

Camogliano, L., and Casu, A. (1989). *Exp. Pathol.* **36,** 37–41.

Chao, Y.-S., Jones, A. L., Hradek, G. T., Windler, E. E. T., and Havel, R. J. (1981). *Proc. Natl. Acad. Sci. U.S.A.* **78,** 597–601.

Chenderovitch, J. (1972). *Am. J. Physiol.* **223,** 695–706.

Cohen, D. E., Angelico, M., and Carey, M. C. (1989). *Am. J. Physiol.* **257,** G1–G8.

Coleman, R. (1987). *Biochem. J.* **244,** 249–261.

Cooper, A. D., Jones, A. L., Koldinger, R. E., and Ockner, R. K. (1974). *Gastroenterology* **66,** 574–585.

Crawford, J. M., and Gollan, J. L. (1988). *Am. J. Physiol.* **255,** G121–G131.

Crawford, J. M., Berken, C. A., and Gollan, J. L. (1988). *J. Lipid Res.* **29,** 144–156.

Daniels, C. K., Smith, K. M., and Schmucker, D. L. (1987). *Proc. Soc. Exp. Biol. Med.* **186,** 246–250.

Davis, R. A., Kern, F., Showalter, R. T., Sutherland, E. R., Sinensky, M., and Simon, F. R. (1978). *Proc. Natl. Acad Sci. U.S.A.* **75,** 4130–4134.

Dubin, M., Maurice, M., Feldmann, G., and Erlinger, S. (1978). *Gastroenterology* **75,** 450–455.

Dubin, M., Maurice, M., Feldmann, G., and Erlinger, S. (1980). *Gastroenterology* **79**, 646–654.

Duffy, M. C., Blitzer, B. L., and Boyer, J. L. (1983). *J. Clin. Invest.* **72**, 1470–1481.

Dumont M., Erlinger, S., and Uchman, S. (1980). *Gastroenterology* **79**, 82–89.

Ferland, G., Tuchweber, B., Perea, A., and Yousef, I. M. (1989). *Lipids* **24**, 842–848.

Fisher, M. M., Nagy, B., Bazin, H., and Underdown, B. J. (1979). *Proc. Natl. Acad. Sci. U.S.A* **76**, 2008–2012.

Fitz, J. G., and Scharschmidt, B. F. (1987). *Am. J. Physiol.* **252**, G56–G64.

Folsch, U. R., and Wormsley, K. G. (1977). *Experientia* **33**, 1055–1056.

Forker, E. L., and Luxon, B. A. (1981). *J. Clin. Invest.* **67**, 1517–1522.

Fricker, G., Hugentobler, G., Meier, P. J., Kurz, G. H., and Boyer, J. L. (1987a). *Am. J. Physiol.* **253**, G816–G822.

Fricker, G., Schneider, S., Gerok, W., and Kurz, G. (1987b). *Biol. Chem. Hoppe-Seyler* **368**, 1143–1150.

Fricker, G., Landmann, L., and Meier, P. J. (1989). *J. Clin. Invest.* **84**, 876–885.

Friedberg, T., Bentley, P., Stasiecki, P., Glatt, H. R., Raphael, D., and Oesch, F. (1979). *J. Biol. Chem.* **154**, 12028–12033.

Frimmer, M., and Ziegler, K. (1988). *Biochim. Biophys. Acta* **947**, 75–99.

Gautam, A., Ng, O., and Boyer, J. L. (1987). *Hepatology* **78**, 216–223.

Gautam, A., Ng, O., Stazzabosco, M., and Boyer, J. L. (1989). *J. Clin. Invest.* **83**, 565–573.

Gleeson, D., Smith, N. D., Scaramuzza, D. M., and Boyer, J. L. (1988). *Gastroenterology* **94**, A542.

Goldman, I. S., Jones, A. L., Hradek, G. T., and Huling, S. (1983). *Gastroenterology* **85**, 130–140.

Goldsmith, M., Huling, S., and Jones, A. L. (1983). *Gastroenterology* **84**, 978–986.

Gorcsky, C. A., Bach, G. G., and Nadeau, B. E. (1973). *J. Clin. Invest.* **52**, 991–1009.

Graf, J., and Peterlik, M. (1975). In "The Hepatobiliary System—Fundamental and Pathological Mechanisms" (W. Taylor, ed.), pp. 43–58. Plenum, New York.

Graf, J., Gautam, A., and Boyer, J. L. (1984). *Proc. Natl. Acad. Sci. U.S.A* **81**, 6516–6520.

Grandjean, E. M., Karlagnis, G., and Noelpp, Y. (1979). In "The Liver: Quantitative Aspects of Structure and Function" (R. Preisig and J. Bircher, eds.), pp. 255–262. Editio Cantor Aulendorf, Berne.

Grasset, E. T., Gunter-Smith, P., and Schultz, S. G. (1983). *J. Membr. Biol.* **71**, 89–94.

Gregory, D. H., Vlahcevic, Z. R., Prugh, M. F., and Sewell, L. (1978). *Gastroenterology* **74**, 93–100.

Grisham, J. W. A. (1962). *Cancer Res.* **22**, 842–849.

Gumucio, J. J. (1983). *Am. J. Physiol.* **244**, G578–G582.

Gumucio, J. J., Balabaud, C., Miller, D. L., Demason, L. J., Appelman, H. D., Stoecker, T. J., and Franzblau, D. R. (1978). *J. Lab. Clin. Med.* **91**, 350–362.

Hagenbuch, B., Lubbert, H., Stieger, B., and Meier, P. J. (1990). *J. Biol. Chem.* **265**, 5357–5360.

Hardison, W. G. M., and Apter, J. T. (1972). *Am. J. Physiol.* **222**, 61–67.

Hardison, W. G. M., and Wood, C. A. (1978). *Am. J. Physiol.* **235**, E158–E164.

Hardison, W. G. M., Hatoff, D. E., Miyai, K., and Weiner, R. G. (1981). *Am. J. Physiol.* **241**, G337–G343.

Hardison, W. G. M., Lowe, P. J., and Shanahan, M. (1989). *Hepatology* **9**, 866–871.

Hardison, W. G. M., Heasdley, V. L., and Shellhamer, D. F. (1991). *Hepatology* **13**, 68–72.

Hayakawa, T., Bruck, R., Ng, O. C., and Boyer, J. L. (1990). *Am. J. Physiol.* **259**, G727–G735.

Henderson, R. M., Graf, J., and Boyer, J. L. (1987). *Am. J. Physiol.* **252**, G109–G113.

Heuman, D. M., Hylemon, P. B., and Vlahcevic, Z. R. (1989). *J. Lipid Res.* **30**, 1161–1171.

Hoffman, N. E., Donald, D. E., and Hofmann, A. F. (1975). *Am. J. Physiol.* **229**, 714–720.

Hofmann, A. F. (1989). *Dig. Dis. Sci.* **34,** 16S–20S.

Hofmann, A. F. (1990). *Hepatology* **12,** 17S–22S.

Hofmann, A. F., and Roda, A. (1984). *J. Lipid Res.* **25,** 1477–1484.

Hornick, C., Jones, A. L., Renaud, G., Hradek, G. T., and Havel, R. J.(1984). *Am. J. Physiol.* **246,** G187–G194.

Hubbard, A. L., and Stukenbrok, H. (1979). *J. Cell Biol.* **83,** 65–81.

Hugentobler, G., Fricker, G., Boyer, J. L., and Meier, P. J. (1987). *Biochem. J.* **247,** 589–595.

Inoue, M., Kinne, R., Tran, T., Biempca, L., and Arias, I. M. (1983). *J. Biol. Chem.* **258,** 5183–5188.

Inoue, M., Kinne, R., Tran, T., and Arias, I. (1984). *J. Clin. Invest.* **73,** 659–663.

Ishii, M., Vroman, B., and La Russo N. F. (1989). *Gastroenterology* **97,** 1236–1247.

Izutzu, K. T., Siegel, I. A., and Smuckler, E. A. (1978). *Experientia* **34,** 731–732.

Jezequel, A. M., Macarri, G., Rinaldesi, M. L., Venturini, C., Lorenzini, I., and Orlandi, F. (1986). *Liver* **6,** 341–349.

Jones, A. L., and Burwen, S. J. (1989). *In* "Handbook of Physiology" (S. G. Schultz, J. G. Forte, and B. B. Rauner, eds.), Sect. 6, Vol. III, pp. 663–675. Am. Physiol. Soc., Bethesda, Maryland.

Jones, A. L., and Spring-Mills, E. (1983). *In* "Cell and Tissue Biology" (L. Weiss, ed.), pp. 685–714. Urban & Schwarzenberg, Baltimore, Maryland.

Jones, A. L., Schmucker, D. L., Mooney, J. S., Adler, R. D., and Ockner, R. K. (1976). *Gastroenterology* **71,** 1050–1060.

Jones, A. L., Schmucker, D. L., Mooney, J. S., Adler, R. D., and Ockner, R. K. (1978). *Anat. Rec.* **192,** 277–288.

Jones, A. L., Schmucker, D. L., Mooney, J. S., Ockner, R. K., and Adler, R. D. (1979). *Lab. Invest.* **40,** 512–517.

Jones, A. L., Hradek, G. T., Renston, R. H., Wong, K. Y., Karlaganis, G., and Paumgartner, G. (1980). *Am. J. Physiol.* **238,** G233–G237.

Jones, A. L., Hradek, G. T., Hornick, C., Renaud, G., Windler, E. E. T., and Havel, R. J. (1984a). *J. Lipid Res.* **25,** 1151–1158.

Jones, A. L., Hradek, G. T., Schmucker, D. L., and Underdown, B. J. (1984b). *Hepatology* **4,** 1173–1183.

Jones, A. L., Hradek, G. T., and Schmucker, D. L. (1985). *Hepatology* **5,** 1172–1178.

Jones, A. L., Daniels, C. K., Burwen, S. J., and Schmucker, D. L. (1988). *In* "Aging in Liver and Gastrointestinal Tract" (L. Bianchi, P. Holt, O.F.W. James, and R.N. Butler, eds.), Falk Symp. No. 47, pp. 181–190. MTP Press Ltd., Lancaster.

Jungermann, K., and Katz, N. (1982). *Hepatology* **2,** 385–395.

Junker, L. H., and Davis, R. A. (1989). *J. Lipid Res.* **30,** 1933–1941.

Kacich, R. L., Renston, R. H., and Jones, A. L. (1983). *Gastroenterology* **85,** 385–394.

Kakis, G., and Yousef, I. (1978). *Can. J. Biochem.* **56,** 287–290.

Kaplowitz, N. (1980). *Am. J. Physiol.* **239,** G439–G444.

Killenberg, P. C. (1978). *J. Lipid Res.* **19,** 24–31.

Knodell, R. G., and Brooks, D. A. (1979). *Gastroenterology* **76,** 1288 (abs).

Kramer, W., Bickel, U., Buscher, H.-P., Gerok, W., and Kurz, G. (1982). *Eur. J. Biochem.* **129,** 13–24.

Kristensen, L. O. (1980). *J. Biol. Chem.* **255,** 5236–5242.

Kroncke, K. D., Fricker, G., Meier, P. J., Gerok, W., Wieland, T., and Kurz, G. (1986). *J. Biol. Chem.* **261,** 12562–12567.

Kuipers, F., Enserink, M., Havinga, R., Van der Steen, A. B. M., Hardonk, M. J., Fevery, J., and Vonk, R. J. (1988). *J. Clin. Invest.* **81,** 1593–1599.

Lake, J. R., Licko, V., Van Dyke, R. W., and Scharschmidt, B. F. (1985). *J. Clin. Invest.* **76,** 676–684.

Lake, J. R., VanDyke, R. W., and Scharschmidt, B. F. (1987). *Am. J. Physiol.* **252,** G163–G169.

Lamri, Y., Roda,, A., Dumont, M., Feldmann, G., and Erlinger, S. (1988). *J. Clin. Invest.* **82**, 1173–1182.

LaRusso, N. F. (1984). *Am. J. Physiol.* **247**, G199–G205.

Latham, P. S., and Kashgarian, M. (1979). *Gastroenterology* **76**, 988–996.

Lauterburg, B. H., and Mitchell, J. R. (1981). *J. Clin. Invest.* **67**, 1415–1424.

Layden, T. J., and Boyer, J. L. (1978). *Lab. Invest.* **39**, 110–119.

Layden, T. J., Elias, E., and Boyer, J. L. (1978). *J. Clin. Invest.* **62**, 1375–1385.

Loria, P., Carulli, N., Medici, G., Tripodi, A., Iori, R., Rovesti, S., Bergomi, M., Rosi, A., and Romani, M. (1989). *Gastroenterology* **96**, 1142–1150.

Lowe, P. J., Barnwell, S. G., and Coleman, R. (1984). *Biochem. J.* **222**, 631–637.

Lowe, P. J., Kan, K. S., Barnwell, S. G., Sharma, S. K., and Coleman, R. (1985). *Biochem. J.* **229**, 529–537.

Lynch, C. J., Wilson, P. B., Blackmore, P. F., and Exton, J. H. (1986). *J. Biol. Chem.* **261**, 14551–14556.

Magnusson, I., Einarsson, K., Angelin, B., Nyberg, B., Bergstrom, K., and Thulin, L. (1989). *Gastroenterology* **96**, 206–212.

Marzolo, M. P., Rigotti, A., and Nervi, F. (1990). *Hepatology* **12**, 134S–141S.

Mathison, O., and Raeder, M. (1983). *Eur. J. Clin. Invest.* **13**, 193–200.

Meier, P. J. (1988). *Semin. Liver Dis.* **8**, 293–307.

Meier, P. J., Meier-Abt, A. S., Barrett, C., and Boyer, J. L. (1984). *J. Biol. Chem.* **259**, 10614–10622.

Meier, P. J., Knickelbein, R., Mosely, R. H., Dobbins, J. W., and Boyer, J. L. (1985). *J. Clin. Invest.* **75**, 1256–1263.

Meier, P. J., Meier-Abt, A. S., and Boyer, J. L. (1987). *Biochem. J.* **242**, 465–469.

Meier, P. J., Lake, J. R., Renner, E. R., Zimmerli, B., and Scharschmidt, B. F. (1988). *Clin. Res.* **36**, 559A.

Möller, O. J., Ostergaard Thomsen, O., and Larsen, J. A. (1983). *Cell Tissue Res.* **228**, 13–20.

Muller, N., and Petzinger, E. (1988). *Biochim. Biophys. Acta* **938**, 334–344.

Mullock, B. M., Dobrota, M., and Hinton, R. (1978). *Biochim. Biophys. Acta* **543**, 497–507.

Mullock, B. M., Jones, R. S., and Hinton, R. H. (1980). *FEBS Lett.* **113**, 201–205.

Nahrwold, D. L., and Shariatzedeh, A. N. (1971). *Surgery (St. Louis)* **70**, 147–153.

Nemchausky, B. A., Layden, T. J., and Boyer, J. L. (1977). *Lab. Invest.* **36**, 259–267.

Novak, D. A., Ryckman, F. C., and Suchy, F. J. (1989). *Hepatology* **10**, 447–453.

Nyberg, B., Einarsson, K., and Sonnenfeld, T. (1989). *Gastroenterology* **96**, 920–924.

Oda, M., Price, V. M., Fisher, M. M., and Phillips, M. J. (1974). *Lab. Invest.* **31**, 314–323.

O'Maille, E. R. L. (1980). *J. Physiol. (London)* **302**, 107–120.

Ookhtens, M., Hobdy, K., Corvasce, M. C., Aw, T. Y., and Kaplowitz, N. (1985). *J. Clin. Invest.* **75**, 258–265.

Ookhtens, M., Lyon, I., Fernandez-Checa, J., and Kaplowitz, N. (1988). *J. Clin. Invest.* **82**, 608–616.

Oshio, C., and Phillips, M. J. (1981). *Science* **212**, 1041–1042.

Pandak, W. M., Heuman, D. M., Hylemon, P. B., and Vlahcevic, Z. R. (1990) *J. Lipid Res.* **31**, 79–90.

Petzinger, E., and Frimmer, M. (1984). *Biochim. Biophys. Acta* **778**, 539–548.

Petzinger, E., and Frimmer, M. (1988). *Biochim. Biophys. Acta* **937**, 135–144.

Petzinger, E., Follmann, W., Acker, H., Hentschel, J., Zierold, K., and Kinne, R. K. (1988). *In Vitro Cell Dev. Biol.* **24**, 491–499.

Phillips, M. J., Oda, M., Mak, E., Fisher, M., and Jeejeebhoy, K. N. (1975). *Gastroenterology* **69**, 48–58.

Popper, H., and Schaffner, F. (1970). *Hum. Pathol.* **1**, 1–24.

Potter, B. J., Blades, B. F., Shepard, M. D., Thung, S. M., and Berk, P. D. (1987). *Biochim. Biophys. Acta* **898**, 159–171.

Poupon, R. E., and Evans, W. H. (1979). *FEBS Lett.* **38**, 134–143.

Princen, H. M. G., Meijer, P., and Hofstee, B. (1989). *Biochem. J.* **262**, 341–348.

Rahman, K., and Coleman, R. (1987). *Biochem J.* **245**, 531–536.

Reichen, J., and Le, M. (1989). *Experientia* **45**, 135–137.

Reichen, J., and Paumgartner, G. (1977). *J. Clin. Invest.* **60**, 429–434.

Reichen, J., Berman, M. D., and Berk, P. D. (1981). *Biochim. Biophys. Acta* **643**, 126–133.

Renner, E. L., Lake, J. R., and Scharschmidt, B. F. (1987). *Clin. Res.* **35**, 412A.

Renner, E. L., Lake, J. R., Cragoe, E. J., VanDyke, R. W., and Scharschmidt, B. F. (1988). *Am. J. Physiol.* **254**, G232–G241.

Renston, R. H., Jones, A. L., Christiansen, W. D., Hradek, G. T., and Underdown, B. J. (1980a). *Science* **208**, 1276–1278.

Renston, R.H., Maloney, D. G., Jones, A. L., Hradek, G. T., Wong, K. Y., and Goldfine, I. D. (1980b). *Gastroenterology* **78**, 1373–1388.

Reuben, A., Allen, R. M., and Boyer, J. L. (1982). *Gastroenterology* **82**, 1241A.

Roda, A., Cappelleri, G., Aldini, R., Roda, E., and Barbara, L. (1982). *J. Lipid Res.* **23**, 490–495.

Roda, A., Grigolo, B., Roda, E., Simoni, P., Pellicciari, R., Natalini, B., Fini, A., and Labate, A. M. (1988). *J. Pharm. Sci.* **77**, 596–605.

Ruetz, S., Fricker, G., Hugentobler, G., Winterhalter, K., Kurz, G, and Meier, P. J. (1987). *J. Biol. Chem.* **262**, 11324–11330.

Rutishauser, S. C. B., Burns, P., and Weil, S. C. (1980). *Comp. Biochem. Physiol.* **66**, 493–498.

St. Hilaire, R. J., Hradek, G. T., and Jones, A. L. (1983). *Proc. Natl. Acad. Sci. U.S.A.* **80**, 3797–3801.

Scharschmidt, B. F. (1990). *In* "The Liver" (D. Zakim and T. Boyer, eds.), pp. 303–340. Saunders, Philadeplphia, Pennsylvania.

Scharschmidt, B. F., and Lake, J. R. (1989). *Dig. Dis. Sci.* **34**, 5S–15S.

Scharschmidt, B. F., and Stephens, J. E. (1980). *Gastroenterology* **79**, 1124.

Scharschmidt, B. F., and VanDyke, R. W. (1983). *Gastroenterology* **85**, 1199–1214.

Scharschmidt, B. F., Keeffe, E. B., Vessey, D. A., Blankenship, N. M., and Ockner, R. K. (1981). *Hepatology* **1**, 137–145.

Scharschmidt, B. F., Lake, J. R., Renner, E. L., Licko, V., and VanDyke, R. W. (1986). *Proc. Natl. Acad. Sci. U.S.A* **83**, 9488–9492.

Schiff, J. M., Fisher, M. M., and Underdown, B. J. (1984). *J. Cell Biol.* **98**, 79–89.

Schmucker, D. L., Gilbert, R., Jones, A. L., Hradek, G. T., and Bazin, H. (1985). *Gastroenterology* **88**, 436–443.

Schultz, S. G. (1981). *Am. J. Physiol.* **241**, F579–F590.

Sewell, R. B., Hoffman, N. E., Smallwood, R. A., and Cockbain, S. (1980). *Am. J. Physiol.* **238**, G10–G17.

Sewell, R. B., Mao, S. J., Kawamoto, T., and LaRusso, N. F. (1983). *J. Lipid Res.* **24**, 391–401.

Simion, F. A., Fleischer, B., and Fleischer, S. (1984a). *J. Biol. Chem.* **259**, 10814–10822.

Simion, F. A., Fleischer, B., and Fleischer, S. (1984b). *Biochemistry* **22**, 6459–6466.

Simon, F.R., Sutherland, E. M., and Gonzalez, M. (1982). *J. Clin. Invest.* **70**, 401–411.

Sirica, A. E., Goldblatt, P. J., and McKelvy, J. F. (1975). *J. Biol. Chem.* **250**, 6464–6468.

Slott, P. A., Liu, M. H., and Tavoloni, N. (1990). *Gastroenterology* **99**, 466–477.

Smith, D. J., Grossbard, M., Gordon, E. R., and Boyer, J. L. (1987). *Am. J. Physiol.* **252**, G479–G484.

Song, C. S., Rubin, W., Rifkind, A. B., and Kappas, A. (1969). *J. Cell Biol.* **41**, 124–132.

Spenney, J. G., Tobin, M. M., Mihas, A. A., Gibson, R. G., Hirschowitz, B. I., Johnson, B. J., and Tauxe, W. N. (1979). *Gastroenterology* **76**, 272–278.

Stolz, A., Takikawa, H., Ookhtens, M., and Kaplowitz, N. (1989). *Annu. Rev. Physiol.* **51**, 161–176.

Strange, R. C. (1981). *Biochem. Soc. Trans.* **9**, 170–174.

Strange, R. C. (1984). *Physiol. Rev.* **64**, 1055–1102.

Strange, R. C., Nimmo, I. A., and Percy-Robb, I. (1977a). *Biochem. J.* **162**, 659–664.

Strange, R. C., Cramb, R., Hayes, J. D., and Percy-Robb, I. (1977b). *Biochem. J.* **165**, 425–429.

Strange, R. C., Hume, R., Eadington, D. W., and Nimmo, I. A. (1981). *Pediatr. Res.* **15**, 1425–1428.

Strasberg, S. M., Ilson, R. G., Siminovitch, K. A., Brenner, D., and Palaheimo, J. E. (1975). *Am. J. Physiol.* **228**, 115–121.

Suchy, F. J., Balistreri, W. F., Hung, J., Miller, P., and Garfield, S. A. (1983). *Am. J. Physiol.* **245**, G681–G689.

Suchy, F. J., Bucuvalas, J. C., and Novak, D. A. (1987). *Semin. Liver Dis.* **7**, 77–84.

Sutton, C. M., and Botham, K. M. (1989). *Biochim. Biophys. Acta* **1001**, 210–217.

Tafler, M., Zeigler, K., and Frimmer, M. (1986). *Biochim. Biophys. Acta* **855**, 157–168.

Tazuma, S., Barnhart, R. L., Reeve, L. E., Tokumo, H., and Holzbach, R. T. (1988). *Am. J. Physiol.* **255**, G745–G751.

Thompson, E. N., and Williams, R. (1965). *Gut* **6**, 266–269.

Ulloa, N., Garrio, J., and Nervi, F. (1987). *Hepatology* **7**, 235–244.

Van Der Sluijs, P., Spanjer, H. H., and Meijer, D. K. F. (1987). *J. Pharmacol. Exp. Ther.* **240**, 668–673.

VanDyke, R. W., Stephens, J. E., and Scharschmidt, B. F. (1982a). *J. Clin. Invest.* **70**, 505–517.

VanDyke, R. W., Stephens, J. E., and Scharschmidt, B. F. (1982b). *Am. J. Physiol.* **243**, G484–G492.

von Dippe, P., Drain, P., and Levy, D. (1983). *J. Biol. Chem.* **258**, 8890–8895.

Wannagat, F.-J., Adler, R. D., and Ockner, R. K. (1978). *J. Clin. Invest.* **61**, 297–307.

Weinman, S. A., Graf, J., and Boyer, J. L. (1989). *Am. J. Physiol.* **256**, G826–G832.

West, I. C. (1980). *Biochim. Biophys. Acta* **604**, 91–126.

Wheeler, H. O., Ross, E. D., and Bradley, S. E. (1968). *Am. J. Physiol.* **214**, 866–874.

Wieland, T., Nassal, M., Kramer, W., Fricker, G., Bickel, U., and Kurz, G. (1984). *Proc. Natl. Acad. Sci. U.S.A* **81**, 5232–5236.

Wolkoff, A. W., Goresky, C. A., Sellin, J., Gatmaitan, Z., and Arias, I. M. (1979). *Am. J. Physiol.* **236**, E638–E648.

Wolkoff, A. W., Samuelson, A. C., Johansen, K. L., Nakata, R., Withers, D. M., and Sisiak, A. (1987). *J. Clin. Invest.* **79**, 1259–1268.

Yousef, I. M., Tuchweber, B., Weber, A., and Roy, C. C. (1988). *Proc. Soc. Exp. Biol. Med.* **189**, 147–151.

Yousef, I. M., Tuchweber, B., Mignault, D., and Weber, A. (1989). *Am. J. Physiol.* **256**, G62–G66.

Zajicek, G., Ariel, I., and Arber, N. (1988). *Liver* **8**, 213–218.

Ziegler, K., Frimmer, M., Kessler, H., and Haupt, A. (1988). *Biochim. Biophys. Acta* **945**, 263–272.

Ziegler, K., Frimmer, M., Muller, S., and Fasold, H. (1989). *Biochim. Biophys. Acta* **980**, 161–168.

Zimmerli, B., Valantinas, J., and Meier, P. J. (1987). *Hepatology* **7**, 1036 (abstr.).

Zimmerli, B., Valantinas, J., and Meier, P. J. (1989). *J. Pharmacol. Exp. Ther.* **250**, 301–308.

INDEX

A

Abscisic acid, 157–148
  isolation, 173, 176
  physiology, 177, 179–186, 189–190
Acetic acid, plant development and, 171, 176
Acid phosphatase, subcommissural organ and, 48, 59
Additive effects, RFLP analysis of plant genomes and, 221
Adhesion, vitamin A and, 27
Adrenal cortex, subcommissural organ and, 106–107
Affinity chromatography, soluble auxin-binding proteins and, 241, 245, 247–249, 261
Affinity column, soluble auxin-binding proteins and, 240–241, 248, 250, 257, 261
Agglutination, cerebellar lectins and, 123, 126, 130, 137–138
Albumin, bile secretion and, 285, 289, 302–303, 307
Aldosterone, subcommissural organ and, 106–107
Alleles, RFLP analysis of plant genomes, 228
  marker-based selection, 205–206, 208, 210–212, 214
  trait localization, 218, 220–222
Allene oxide cyclase in plants, 161, 164
α-Amanitin, soluble auxin-binding proteins and, 249–250, 252, 257
Amines, subcommissural organ and, 57, 104
Amino acids
  bile secretion and, 273, 283, 298, 302
  cerebellar lectins and, 138
  jasmonic acid and related compounds, 159
  subcommissural organ and, 58–59, 62–63, 66, 90
  vitamin A and, 11
τ-Aminobutyric acid, *see* GABA
1-Aminocyclopropane-1-carboxylic acid (ACC) in plants, 180–181

Ammonium sulfate, soluble auxin-binding proteins and, 241–243
Ampulla caudalis, subcommissural organ and, 82–83, 85, 111
Aneuploidy, RFLP analysis of plant genomes and, 207–208
Anions, bile secretion and, 269, 274, 298, 300–302
Antamanide, bile secretion and, 301
Antibodies
  cerebellar lectins and, 123, 125, 130, 148
  β-galactoside-binding lectins, 127–128
  mannose-binding lectins, 131–133, 136–139, 141, 146–147
  subcommissural organ and, 101–102, 108, 110
  ontogeny, 96
  Reissner's fiber, 81
  secretory process, 59–60, 64, 66
  ultrastructure, 49
  vitamin A and, 7
Antigens
  cerebellar lectins and, 123, 127, 131, 136–137, 146
  subcommissural organ and, 101, 108
Apical region, subcommissural organ and, 49–50, 68, 71, 76
Apples, jasmonic acid and related compounds in, 180–181
Arabidopsis thaliana, RFLP analysis of, 205, 209, 214–215
Asialoglycoproteins, bile secretion and, 273, 303, 305
Astrocytes
  cerebellar lectins and, 130, 132, 137, 139, 141
  subcommissural organ and, 56
ATP
  soluble auxin-binding proteins and, 245
  vitamin A and, 22
ATPase, *see also* Na^+/K^+-ATPase bile secretion and, 299

315